Managing the Building Design Process

Managing the Building Design Process

Second Edition

Gavin Tunstall

Architect and Lecturer
Nottingham Trent University, Nottingham, UK

AMSTERDAM • BOSTON • HEIDELBERG • LONDON • NEW YORK • OXFORD
PARIS • SAN DIEGO • SAN FRANCISCO • SINGAPORE • SYDNEY • TOKYO
Butterworth-Heinemann is an imprint of Elsevier

Butterworth-Heinemann is an imprint of Elsevier
Linacre House, Jordan Hill, Oxford OX2 8DP, UK
30 Corporate Drive, Suite 400, Burlington, MA 01803, USA

First edition 2000
Reprinted 2006, 2007

British Library Cataloguing in Publication Data
A catalogue record for this book is available from the British Library

Library of Congress Cataloging-in-Publication Data
A catalog record for this book is available from the Library of Congress

ISBN: 978-0-7506-6791-3

For information on all Butterworth-Heinemann publications
visit our website at books.elsevier.com

Printed and bound in *Great Britain*

07 08 09 10 10 9 8 7 6 5 4 3 2

Dedication

To the memory of my father
Peter William Tunstall, 1916–2004

Contents

Preface

There can be little doubt that towards the latter part of the twentieth century, the creation of many new buildings in the UK had become an excessively confrontational process, encouraging clients, designers and builders to seek to gain advantages from one another rather than to work constructively together. Strict adherence to 'professional' roles and an unwillingness to step over historically defined boundaries discouraged co-operation and collaboration. Blinkered by contracts, time scales and costs, the process often appeared to be cramped in an over-demanding, claims-conscious environment, fixated by narrow aims and responsibilities, seemingly unable or unwilling to reflect a genuine concern with quality or customer care. The Latham and Egan Reports, published in the 1990s described this situation as wasteful and very significantly, that it was contributing to a diminution in the quality of both design and construction. The reports laid the foundations for substantial on-going changes in practice and guidance developed during the past 10 years.

The process of designing and constructing new buildings is a complex activity reflecting the skills, perceptions and expectations of many individuals, who must attempt to respond to technical and philosophical challenges, resolve debates and deal with the inevitable conflicts associated with working together. The associated personnel difficulties and contractual obligations cannot be dismissed lightly, but in an ideal scenario, everyone should be capable of appreciating how and why decisions are taken so that there is a better chance of achieving the best possible results under the prevailing circumstances. Understanding the process of building design in terms of **what should be done** rather than **who should do it** helps to minimise the negative restraints of professional boundaries. This book is based on my experience as an architect, but I use the term *building designer* to describe the process of design and construction of an imaginary new building offering a broad stage-by-stage explanation of the way in which ideas can become reality. Although reference to some technical issues is inevitably based on current UK practice, for the most part my intension is to discuss general principles, which I believe to be universally applicable.

My involvement with the construction industry began at Bath University in 1968, and since qualifying as an architect I have helped to design and supervise the construction of many new buildings. I have been fortunate to have had the opportunity to gain experience with a variety of different building types including public and private sector housing, sports, leisure and museum facilities, as well as a wide range of commercial and industrial developments. I have worked in both large and small organisations for Bath and Nottingham City Councils and in private practices. I have worked on my own as a self-employed, freelance architect, and although

I occasionally occupy the traditional role of team leader or 'director of operations', offering *my clients* a full design and supervision service, more commonly I work as one part of a team, contributing design and construction information to contractors and project managers who are able to offer *their clients* a full design and build service.

As a *building designer*, this represents a significant change in my relationship with both *client* and *builder*. Traditionally, I acted directly on behalf of the client, preparing design proposals for their approval, assembling packages of information, obtaining competitive tenders, arranging for the appointment of a suitable contractor and supervising the construction of their work. More recently, I work for the contractor or project manager, who in many cases has established a relationship with the client before I become involved and who has already negotiated and agreed budget costs and preliminary time scales for the creation of the new building. My involvement is essentially one of offering advice and helping both *client* and *builder* to crystallise their ideas. I produce sufficient information in the form of drawings and outline specifications to define the framework of the building, help to co-ordinate the design work of other specialists and liaise with the statutory authorities in order to gain the necessary permissions and approvals. My direct, personal responsibility for organising the construction process, establishing detailed costing arrangements and site supervision is minimal. As with any form of change in traditionally understood relationships and procedures, there are advantages and disadvantages depending on how the individual is affected by them, but without doubt one of the benefits is that everyone involved must take an active interest in the process itself rather than simply accept 'professional' advice and instructions. In fact to work properly, everyone *needs* to understand the process in order to appreciate what their role *does* or *could* entail if they are to participate to the fullest possible extent.

A few years ago, the flexibility of self-employment enabled me to fulfil a long-standing ambition of involvement in higher education, when I began to contribute to various courses concerned with the built environment at Nottingham Trent University. My students have ranged from those with no previous experience of building to those who have worked in the construction industry for almost as long as I have. The degree programmes have been mainly concerned with the technicalities of construction techniques, contract management, building surveying, architectural technology and residential development, and I have found that most students are interested in design, many aspiring to become building designers or to work in a design and build contracting environment. Even those determined to go into construction management, land buying and marketing have responded to the challenges of design with enthusiasm.

Much of my work with students has been concerned with setting up and managing design projects based on fictitious, but realistic briefs for real sites which students can visit, giving them an opportunity of gaining experience of designing their own buildings and understanding the significance of client, site and authority constraints. My aim has been to encourage students to think about issues so that they might be able to appreciate problems and possibilities. The limited length of subject modules has demanded a relatively simplified approach, introducing concepts which students can elaborate and develop in response to the demands and opportunities which may arise in their specific degree programme and chosen career. Their future in the construction industry may not be as 'hands-on' designers, but if and when they find themselves in positions of power, influence and responsibility they will need an awareness of the design process and the potential role of the designers that they will work with.

My own education at Bath University partially recognised the need for greater co-operation and understanding by arranging for architectural, structural, and mechanical and electrical engineering students to attend combined lectures and co-operate in joint design projects. This was an innovative idea at the time, but I can recall little or no involvement with Planning, Building Control and Fire Officers, quantity surveyors, valuers or other specialist consultants. The buildings we designed were rather like sculptures, sometimes contrived to look attractive but on most occasions ignoring the inconvenience of practical or commercial reality and with even less regard for the lives of future occupants, should our work ever have reached the point of being considered for construction.

Neither did clients and builders feature prominently. Looking back I think to some extent there was an underlying, unstated assumption that these people were the 'opposition', to be tolerated, controlled or beaten, as described by Egan and Latham some 35 years later. And so, when I started to design and supervise the construction of real buildings, problems which I imagined might have been resolved by sensible discussion and negotiation were at the mercy of political policy-making, empire building and contractual enforcement, aggravating rather than oiling the points of friction. I have worked with other architects, quantity surveyors and engineers who have taken the arrogant view that what they say is 'law' and everyone else must do what they are told. This inevitably lead to confrontation which consumed resources and diminished enthusiasm, arguably resulting in the creation of new buildings which were not as successful as they should have been. There were times when it seemed that clients were only interested in minimising costs, builders were only interested in maximising profits and building designers were only interested in design (or really, in being in charge). It sometimes seemed that none of them were particularly interested in the needs of the building's users.

This was not the way that I perceived building design, or a useful philosophy to promote to my students. The need to explain the building design process as I understood it lead to the first edition of this book in 2000, since when I have become a full-time lecturer at Nottingham Trent University. When I embarked on the first edition, I wrote about my own approach to managing the building design process, summarising the 'basics' that I had used myself in commercial practice. I confined my references to things which I thought were relatively timeless, and which might not become too quickly outdated. I did this to help students to understand processes which are fundamental, and to avoid wherever possible responding to 'fashionable' notions of the time. I also thought that the original text would remain current for a lengthy period of time as I had no wish to ever write another word on the subject.

However ... I have reviewed developments in the industry, and added further explanation and discussion points for debate about design, construction and the built environment. I have updated, expanded and filled gaps wherever appropriate, including web page addresses and links, rather surprisingly absent in the first edition. Some of my earlier writing looked forward to electronic communications and the use of the Internet as a source of information, but I had not imagined that the impact would be so dramatic. The recent period of major urban regeneration in the UK has been dramatic too, generating concern, interest and a general change in attitude to design in the built environment. The work of the Office of the Deputy Prime Minister (ODPM), the Commission for Architecture and the Built Environment (CABE) and others is stimulating the industry, self-evidently transforming the design quality of much of its product.

This book is intended for students on any degree programme in building-related studies in the areas of design, construction and management. This is an exciting time to be part of the construction industry.

Acknowledgements

This book was written for my students at Nottingham Trent University. I would like to thank them for their enthusiasm and friendship (tolerance anyway) during the past 12 years. My link with the university began in 1994 when Prof. Terry Lane, Head of the Department of Building and Environmental Health at Nottingham Trent University, asked me if I would like to do some part-time teaching. I am indebted to Terry, and to Andrew Charlett, Prof. Roger Hawkins, Prof. Paul Gallimore and Prof. Roy Morledge for their support in developing this tentative beginning into a permanent full-time position.

Directly and indirectly they have helped me to shape this book, as originally did Prof. Alan Hooper, Dr Ron Blake, Chris Coffey and Mike McCarthy in particular, reviewing content and giving me advice about Planning, building law and their perception of the needs of building students. I would like to add to this by acknowledging the assistance of Dr Adam O'Rourke, who is always looking for ways to do things differently, Phil Hawkins for opening my eyes to electronic communication and web pages and Tony Trevorrow for forcing me to go out for lunch sometimes. Pete Ramsay-Dawber, Paul Collins, Pete Lyons, Alan Fewkes, Tim Fletcher and other colleagues have all offered advice and inspiration. Thanks to Sandra Price at the University library for help with some of the reference sources and to Juanita Gonzales-Metcalf and Dr Rachel Mansfield for being permanently cheerful. I also want to remember my friend Prof. Colin Ferguson who's support and encouragement opened up so many new opportunities.

I would like to mention my friends John Ellis, Chris Hutt, Paul Hyde, Richard Rowe, Pete Smith and Mike Thatcher and thank them for their support, particularly with their challenging and inspirational discussions about architraves, and sometimes, for more useful advice about business practice.

I would like to acknowledge the original support, advice, guidance and confidence given by Eliane Wigzell of Edward Arnold, and especially acknowledge my debt to Eric Johansen of the University of Northumbria who perceptively and crucially transformed the direction of the first edition of this book. I would like to thank Butterworth-Heinemann for publishing the first edition and for supporting the notion of a second edition so soon after the original, and in particular for the help of Sarah Hunt and Lanh Te who have helped with editorial and technical issues.

Finally, I would like to thank my wife, Tricia, who's advice about writing and the structuring of my ideas continues to help me to express my thoughts in a more or less intelligible manner. I would also like to mention my mother Claire, my children, Iain and Eleri (Legs) and her friends, Deborah and Louise for their continued encouragement.

List of Figures

List of Abbreviations

The following abbreviations may appear in this book, or in associated reference sources.

ABE	Association of Building Engineers
ABT	Association of Building Technicians
ACAS	Advisory, Conciliation and Arbitration Service
ACE	Association of Consulting Engineers
APM	Association for Project Management
ARB	Architects Registration Board
BBA	British Board of Agrément
BCIS	Building Cost Information Service
BCO	Building Control Officer
BDA	Brick Development Association
BIFM	British Institute of Facilities Management
BRE	Building Research Establishment
BSRIA	Building Services Research Information Association
BQ	Bills of Quantity
BSI	British Standards Institution
BURA	British Urban Regeneration Association
CABE	Commission for Architecture and the Built Environment
CAD	Computer-Aided Design or Draughting
CAT	Centre for Alternative Technology
CC	Countryside Commission
CDM	Construction (Design and Management) Regulations
CEBE	Centre for Education in the Built Environment
CEN	Comité Européen de Normalisation (European standards)
CIAT	Chartered Institute of Architectural Technologists
CIBSE	Chartered Institution of Building Services Engineers
CIC	Construction Industry Council
CIEH	Chartered Institute of Environmental Health
CIH	Chartered Institute of Housing
CIOB	Chartered Institute of Building
CIRIA	Construction Industry Research and Information Association

CITB	Construction Industry Training Board
COSHH	Care of Substances Hazardous to Health
COW	Clerk of Works
CP	Code of Practice
CPO	Compulsory Purchase Order
CPRE	Council for the Protection of Rural England
D&B	Design and Build
DEFRA	Department for Environment, Food and Rural Affairs
DETR	Department of the Environment, Transport and the Regions
DIY	Do it yourself
DLT	Development Land Tax
DoE	Department of the Environment
DoT	Department of Transport
Dpc	Damp proof course
DTI	Department of Trade and Industry
DTLR	Department for Transport, Local Government and the Regions
EHO	Environmental Health Officer
EIA	Environmental Impact Analysis
EP	English Partnerships
EU	European Union
FMB	Federation of Master Builders
FO	Fire Officer
GIA	General Improvement Area
HBF	House Builders Federation
H&S	Health and Safety
HSE	Health and Safety Executive
IBC	Institute of Building Control
ICE	Institution of Civil Engineers
ICW	Institute of Clerks of Works
IEE	Institution of Electrical Engineers
IHBC	Institution of Historic Building Conservation
IHVE	Chartered Institution of Services Engineers
IMechE	Institution of Mechanical Engineers
ISE	Institution of Structural Engineers
IStructE	Institution of Structural Engineers
IT	Information Technology
JCT	Joint Contracts Tribunal
LA	Local Authority
LI	Landscape Institute
M&E	Mechanical and Electrical
MMC	Modern Methods of Construction
NA	Not Applicable
NBS	National Building Specification
NGO	Non-governmental organisation
NHBC	National House-Building Council
NIMBY	Not in my back yard
NTS	Not to scale

ODPM	Office of the Deputy Prime Minister
OS	Ordnance Survey
PC	Prime Cost
PFI	Private Finance Initiative
plc	Public limited company
PO	Planning Officer
PPG	Planning Policy Guidance
PVCU	Poly vinyl chloride unplasticised (sometimes written as Upvc)
QA	Quality Assurance
QS	Quantity Surveyor
Quango	Quasi-autonomous non-governmental organisation
RDA	Regional Development Agency
RIBA	Royal Institute of British Architects
RICS	Royal Institute of Chartered Surveyors
RTPI	Royal Town Planning Institute
SAAT	Society of Architectural and Associated Technicians
SAP	Standard Assessment Procedure (evaluating energy performance)
SBS	Sick building syndrome
SCI	Steel Construction Institute
SSSI	Site of special scientific interest
SMM(7)	Standard Method of Measurement
T&G	Tongued and grooved
TCPA	Town and Country Planning Association
TPO	Tree Preservation Order
TRADA	Timber Research and Development Association
WWW	World Wide Web (Internet)

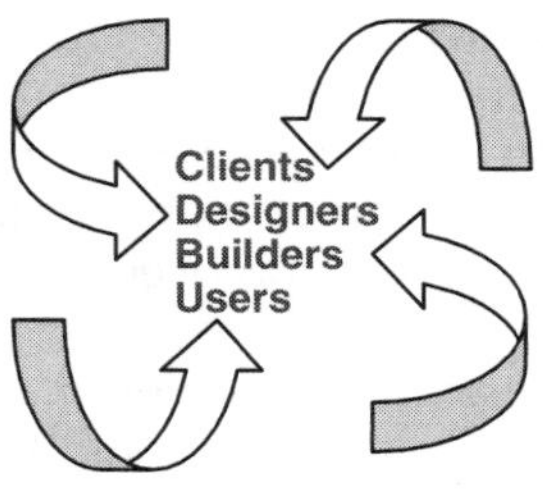

1

About the book

1.1 Introduction

Everyday life for most people is a complex arrangement of individual and collective activities based on needs, desires, demands and choices. Quality of life is influenced by many things, but a common factor in almost all human activity is that it takes place in and around buildings, which possibly affects everyone to as great an extent as anything else that they have to deal with. Every conceivable activity can be made easier or more difficult, inspiring or tedious, enjoyable or depressing depending on the qualities of the buildings that cater for them. There is clearly a connection between people and the built environment that they inhabit; a common, shared experience that has been shown to favourably, or adversely, affect health and well-being.

Academic understanding of the psychology of buildings is a developing subject area, but everyone has their own experience of using buildings; living in houses, being taught in schools and universities, working in shops and offices, enjoying recreation and leisure in cinemas, concert halls and sports centres. Generally, our experience is primarily concerned with the goods, services or entertainments which they contain, which often diverts attention from the design of the buildings themselves. Some buildings are admired landmarks, others are derided 'monstrosities' but the majority fall somewhere in between and rather like the football referee or test match wicket keeper, their performance is taken for granted, only noticed when they make a mistake. This could also be said of those who create the buildings and although there may be perfectly good reasons for their 'mistakes', once they are noticed they become the focus of criticism which can be both disproportionate and misdirected. User dissatisfaction is sometimes expressed by critical comments such as 'whatever were *they* thinking of' or 'why on earth didn't the Planners stop *them* from doing that!'

To the casual onlooker, it might sometimes seem that new buildings arrive mysteriously, rising from the chaos of construction sites, materialising from behind painted hoardings and screened scaffolding. They might also think that buildings are created by builders, who are seen mixing concrete, erecting steel framing and laying bricks. In fact, even the best builders need to know what to do before starting to build, the person paying for the building generally likes to know what the end result will be and there are many statutory authorities to be satisfied as work proceeds. Careful planning, description and explanation is required before construction begins and the work must be supervised, inspected and approved on site to ensure that intentions are properly fulfilled.

These tasks are undertaken by specialists who co-ordinate complex practicalities in the context of an onerous commercial environment. Sometimes the constraints imposed on them lead to mundane results, but their work can be competent, innovative and occasionally extremely imaginative, adding richness to the quality of the built environment, stimulating the lives of future occupants and users. All new buildings result from consideration and control of factors which can be identified and understood together with an awareness of purpose and an ability to pay careful attention to detail.

From the first thoughts about a new building to the time when it can be occupied by its users (Figure 1.1), the design and construction teams work together towards a common purpose. Each person has a particular expertise based on their own aptitude, training and experience. The building designer is part of a team including clients, developers, engineers, quantity surveyors, interior designers, landscapers and the responsible authorities, all of whom to a greater or lesser extent need conceptual ability, an understanding of construction detailing and management skills. They all contribute vital information at different stages of the design process, each dependent on one another. In the same way, the construction process involves builders, subcontractors and suppliers undertaking their work together to turn the designed ideas into a real building.

An appreciation of the way in which new buildings are created is essential for everyone involved in building construction, so that they can see and understand how ideas are generated, how relationships are established and managed, and how their particular involvement contributes to the success of the finished product.

1.2 The purpose of the book

Creating a new building can be seen as a process; one which requires input from a variety of people using a range of skills to arrange or manage factors and resolve contentious issues. This book provides an introduction to this process and to the contribution that can be made by the building designer. This book will summarise the elements of building design and explain the building designer's relationships with other members of the design and construction teams as ideas are developed into proposals which are capable of being built and which will satisfy the needs of the building's users.

The general term *building designer* or simply *designer* is used throughout this book in order to minimise the exclusiveness associated with formal designer titles such as 'architect', 'architectural technologist' and others, which are protected in the UK by professional bodies as a means of recognition of an individual's qualification to undertake professional work. The principal UK building design institutions are the Architects Registration Board (ARB), the Royal Institute of British Architects (RIBA) and the Chartered Institute of Architectural Technologists (CIAT). Membership of a professional institution is a form of quality assurance, based on academic and practical experience. It indicates a certain level of competence to potential employers, and also confers legal authority for some commercial undertakings, particularly the administration of contracts and supervision of building work. Other professional institutions and organisations

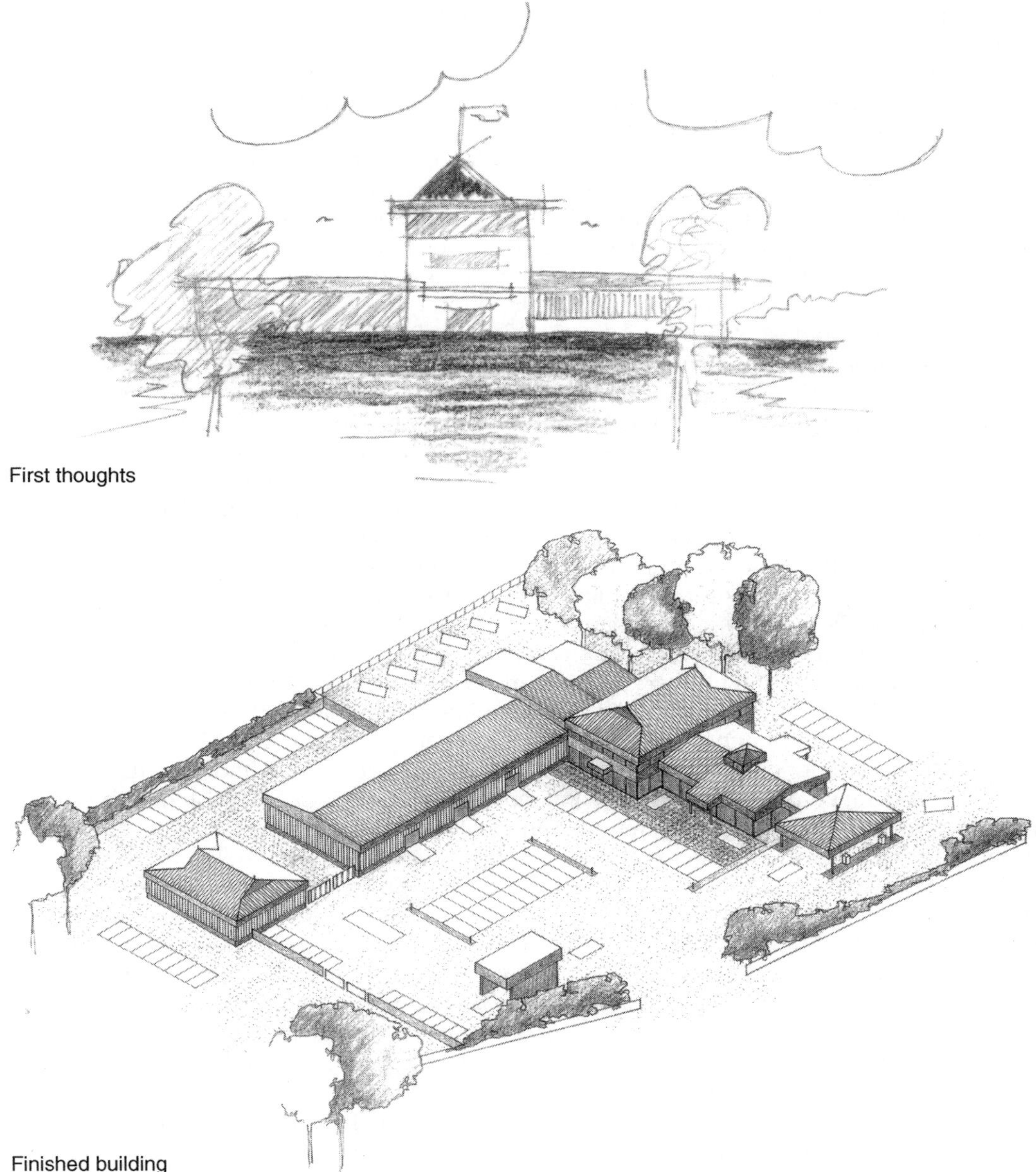

Figure 1.1 From the designer's first ideas about a possible form & style of a building to the completed occupied development.

concerned with buildings and the built environment are listed on pages 345 to 351 at the end of the book.

The content of this book is not intended to be a simple duplication of the mechanics of business practice or of the educational curriculum as they exist *at the present time*, because both subjects are well documented in existing publications, and both the areas are continually subject to revision and development in response to the demands of current commercial forces, political attitudes, formal procedures, standards and legal controls. Specific references appear in places where useful, but should not always be viewed as being exhaustive or definitive, as present-day ideas can quickly become superseded by others. This book is aimed at readers who are principally interested in design or intent on working with, or even managing designers in a professional context who need to understand the design process *as it really is*.

The title of this book is *Managing the Building Design Process* (from an original idea of *Creating New Buildings*). Sometimes buildings or a single building are referred to as architecture or as having architectural merit. There is some debate in academic and professional circles with regard to the criteria which might distinguish '*buildings*' from '*architecture*'. Originally, the architect was the chief builder or master craftsman in an age when the designer and the builder was the same person, and it could reasonably be argued that by definition buildings were architecture. The separation of the two roles in the nineteenth century created a climate of conflict between designers and builders which to some extent still exists today. The elitism of generations of architects has been eroded by changes in business practice and the sheer complexity of many modern buildings, demanding a true 'teamwork' approach to their creation. All buildings can be regarded as either good or bad architecture reflecting the way in which designers and builders have responded to the challenge of creating them. It is not helpful to suggest that the design of buildings can be categorised as somehow important or unimportant as even the most mundane and insignificant of buildings can be handled in a sophisticated way, making a valuable contribution to the built environment.

New buildings are not created in a vacuum. Elements of both design and construction evolve through consideration of a wide variety of influences, including some or all of the subjects listed in Figure 1.2. These topics will be referred to in greater detail later in this book as they become relevant to each stage in the process. They are subjects which can be studied in depth by reference to their own body of specialist literature for those who have the time and interest, but are listed here as a general indication of the spread of knowledge which those who are involved in the creation of new buildings may need to possess or acquire. This is an age of rapid technological, social and political change. The development of new materials and construction techniques, and increasing sophistication in the demands, needs and attitudes of building users continuously changes the emphasis placed on some elements of both design and construction, but it is the significance of each topic and their interaction with one another that influences decisions about the form and function of each and every building.

The design and construction process is changing too, and a client wishing to create or *procure* a new building has various options to consider. The 'traditional' route of procuring a new building based on the client, the building designers and the contractor being separate, independent people is still widely used for small-to-medium-sized projects for 'one-off' clients. However, the formal relationship of the parties to one another involved in this method can lead to circumstances which become adversarial, time consuming and unproductive. The UK construction industry has a history of project performance which is late, over budget and with expensive defects, and in fact the present-day key indicators are little or no better than they were 10 years ago. In 1994, the joint industry/government Latham Report entitled *Constructing the Team*

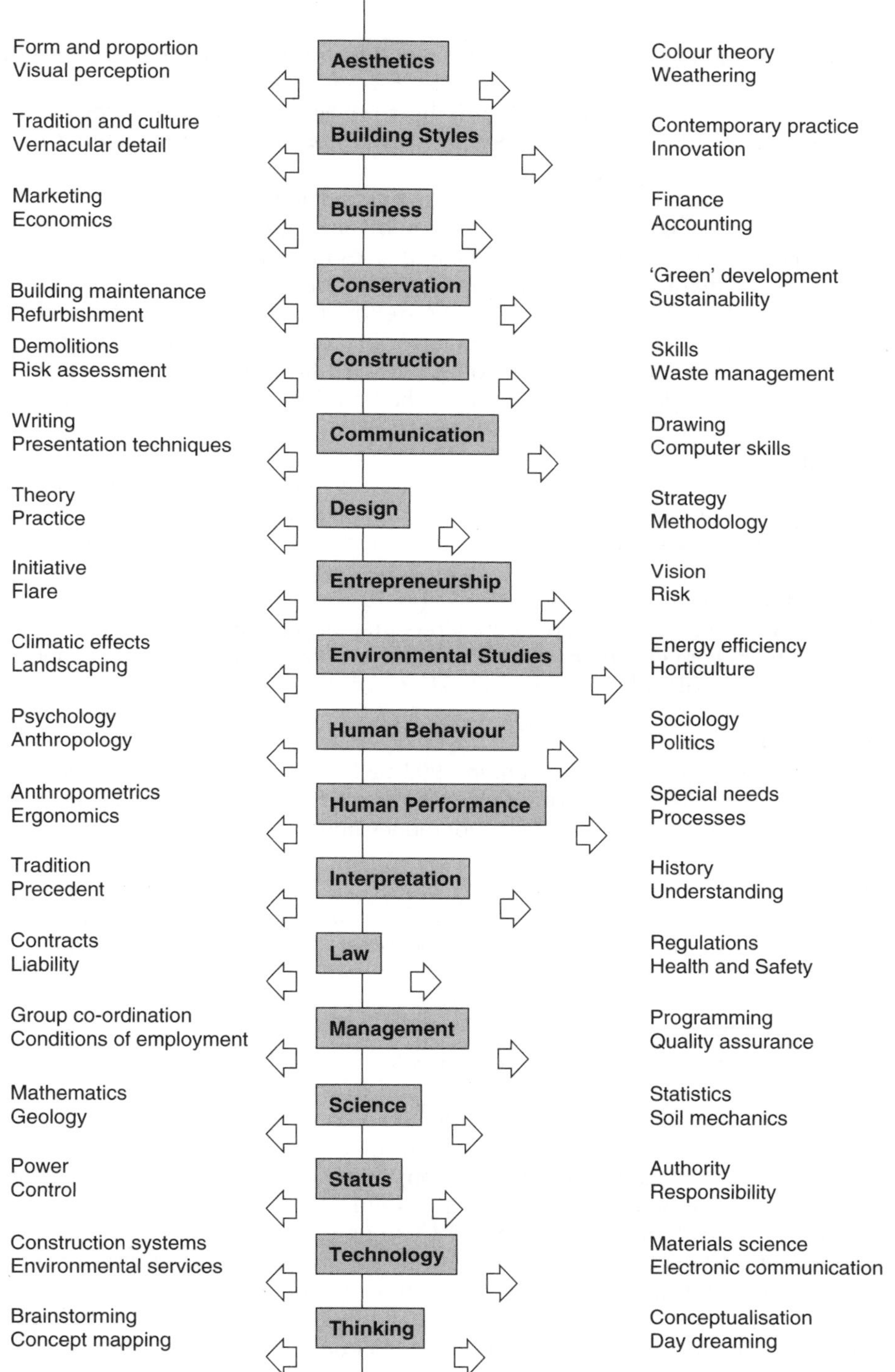

Figure 1.2 Influences on design and construction.

described the construction industry 'ineffective', 'adversarial', 'fragmented', 'incapable of delivering for its customers' and 'lacking respect for its employees'. This report lead to the establishment of the Construction Industry Board, the 1998 Egan Report *Rethinking Construction*, the *Construction Best Practice Programme*, the *Movement for Innovation* and *Constructing Excellence* all aimed at improving performance.

There are other forces too demanding re-examination of traditional notions. The 'fast-tracking' culture and 'short-termism' of commercial interests may be an essential part of change in a modern society, but it may also lead to corners being cut and responsibilities transferred. The idea that buildings might be temporary is a relatively new one, but one which may become increasingly common. Widespread labour and skills shortages are forcing changes in the nature of building design and construction, making off-site prefabrication of standardised buildings a realistic alternative for various types of accommodation. Consequently, procurement of many new-build projects nowadays is through methods other than the traditional route, including the following.

Design and build
An arrangement where a builder, often referred to as a *general contractor*, offers an integrated service to their client from design through the construction process to completion and handover of the finished project. Some 30% of new construction today is created on this basis.

Construction management
Under this form of procurement, independently prepared design proposals are taken forward by an intermediate management company, who organise construction on site on behalf of the client, but who have no directly employed workforce or *in-house labour*.

Private finance initiative
Introduced by the Conservative Government in 1992 by Norman Lamont, the private finance initiative (PFI) system seeks to increase private sector capital funding for public facilities. The idea anticipated efficiency savings, increased value for money and transfer of risk into the private sector.

Partnering
Under this type of procurement arrangement, parties to a contract work towards agreed goals which will benefit them all, demanding trust, openness and co-operation. On the one hand, successfully achieving defined targets such as meeting cost or time limits brings shared rewards, and on the other, failure brings shared penalties.

These are some current procurement options adopted to speed up the process of creating new buildings, to reduce friction between the parties and to ensure that everyone is more closely involved with the development of the project from the outset. For the benefit of readers to whom the process of creating buildings is a new subject, understanding the traditional approach to procurement is a valuable starting point, but consideration will also be given to alternative strategies where appreciation of the distinctions is useful.

Technology has also revolutionised design and communication. It is possible to imagine a time when the demand for a new building can be satisfied simply by pressing buttons on a computer keyboard. It may well be that eventually the client can procure a new building without the involvement of anyone else at all. This could be viable if all buildings were to be more or less the same, and there is some evidence in current residential and commercial development that this is becoming the case. However, for a unique project the premise of this book is that the thought processes leading to or justifying decisions are crucial, and will still be required irrespective of the sophistication of the aids available to those responsible for making them.

1.3 How to use this book?

This book is structured so that the reader can follow the process leading to the creation of a new building, with pertinent issues highlighted at the point at which they would arise in practice. The book is not written as a story of the creation of a single building to be read through in one go, but rather to provide a source of reference for those concerned with design management to help them to better understand the nature of the process. In many sections, lists and bullet points are included to act as prompts or reminders of the need for further thought or action. The format will help both staff and students undertaking academic design projects and professionals creating 'real' buildings for 'real' people. Figure 1.3 is a map of this book illustrating the process of creating new buildings.

This book is divided into three parts: beginning with an introduction to the basics of design, the roles and essential skills of the members of the design and construction teams, and the legal constraints governing the creation of new buildings. This part explains what design might mean and what designers are trying to achieve. The second part explores the ways in which information is collected and analysed, which informs the design process leading to the presentation of realistic proposals. This part shows the way that sketch ideas are developed to the point where they could be built. Finally, the third part deals with the construction period from commencement of work on site to completion of the finished building, ready to be handed over to its future occupants.

The people involved in the process of creating a new building can be categorised as follows:

- *A client or clients*
  One or more people who need a new building and can supply the capital finance required to build it.
- *A building designer or designers*
  Acting as the client's agent, co-ordinating design work and supervising construction or supplying information as part of a design and build package to a contractor or project manager.
- *Consultants*
  Specialists who contribute their own expertise to the design of elements within the new building.
- *Project managers*
  Co-ordinators of design, construction or both, organising the necessary works on behalf of the client.
- *Contractors*
  Builders satisfying the demands of clients, undertaking construction on the basis of an independent design team's specifications and supervision, or who themselves assume direct responsibility for all aspects of design and construction offering a full design and build or project management service directly to their clients.
- *Authorities*
  Independent external bodies empowered to approve aspects of design as ideas are developed, and construction as work on site proceeds.

Considering the number of different buildings used in a day, a week and a year including houses, schools, shops, factories, libraries, sports centres, cinemas, concert halls and churches, it is easy to see the variety which exist in today's built environment, each having its own special requirements.

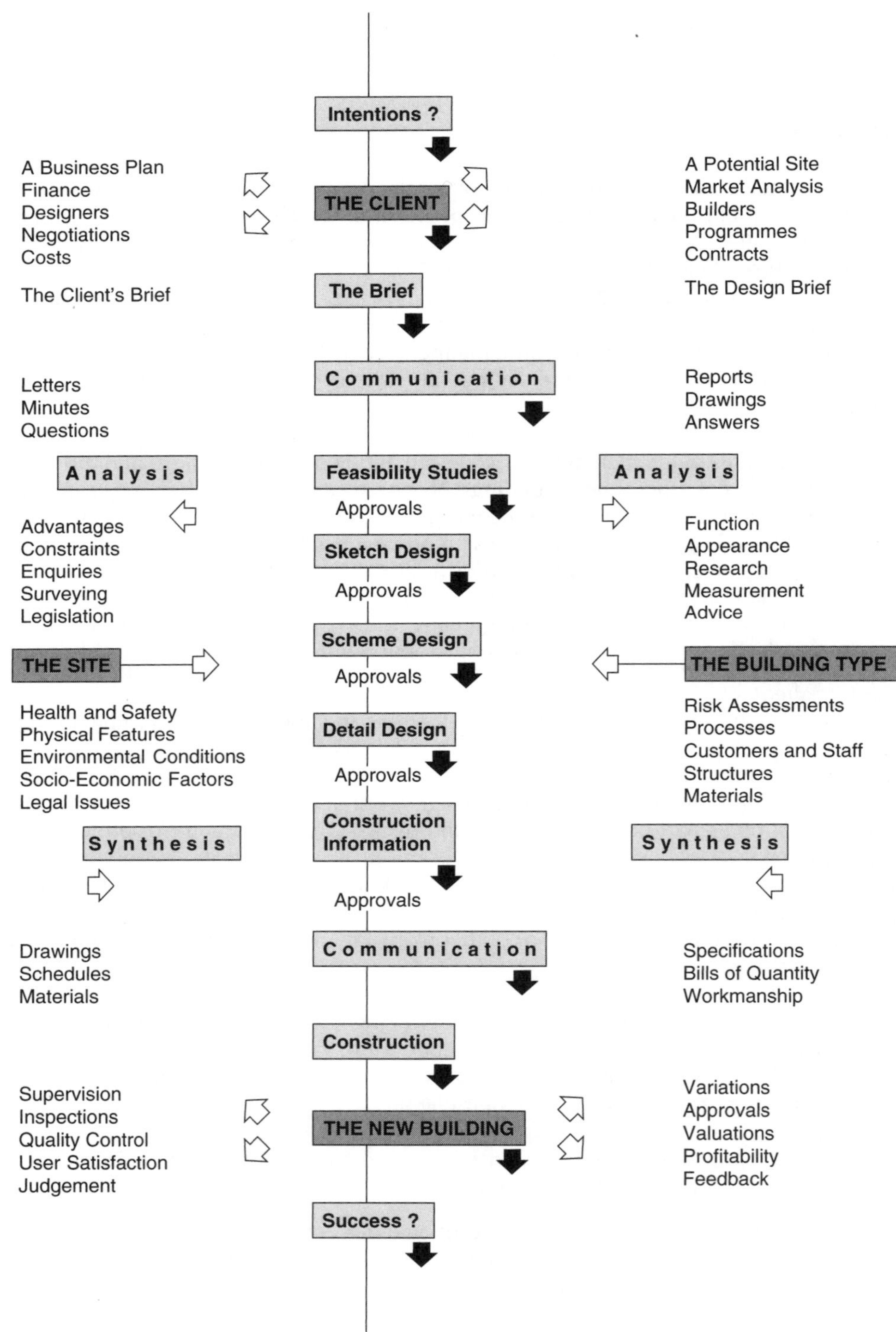

Figure 1.3 A map of this book.

It would be impracticable to cover the nuances of issues pertinent to every type of building in a single book. As a means of explaining the general process, each stage of development will be illustrated by a single case study. In this way, the reader will be able to understand how a new building is created, and to appreciate how the process might be applied to any building type. Additionally, at the end of each section, there will be a brief summary of how the process is illustrated and recorded. Discussion points and references to existing literature will assist the reader to undertake further research into aspects of individual interest.

Throughout this book, words and phrases crop up which are technical descriptions used in the construction industry. Specialist professional activity inevitably has its own language and frequently uses common words and phrases in its own way. In order to make it as easy as possible for the reader, brief explanations of what is meant by these terms will be given when they first arise. Inevitably the reader will want to move forward and jump some sections to suit their interest, and may need to look back for definitions and explanations. Some explanations may be regarded as insufficient, or even unnecessary, but the intention is to explain the process of creating new buildings to those who have not considered the issues before.

The introductory figure of arrows and words found at the beginning of each chapter is included to stimulate thoughts about the interdependence of people, actions or ideas on one another. For example, Clients, Designers, Builders and Users described in Chapter 1 are involved with new buildings, linked together in different ways. In Chapter 6, Design, Time, Cost and Quality have different meanings depending upon the perceptions of those involved. This idea of rotating assessment, attempting to understand significance and value is explained further in Chapter 2 through Analysis, Synthesis, Appraisal and Feedback, leading to a better appreciation of the demands and possibilities of the project.

1.4 The Project File

The illustrated case study is a hypothetical car-dealership, which as a competitive, commercial business provides for sales, maintenance and repair involving a wide range of design issues to satisfy varying customer, staff and community needs. The interlinking and separation of requirements demands a practical solution, and yet there are many examples of such buildings being an attractive addition to the built environment, which even the sternest critics would have to acknowledge as being well designed and worthy of recognition.

In practice, a new car-dealership is commonly developed in conjunction with a major car manufacturer who will dictate aspects of the layout and appearance of the new development in line with their own company style. As with many other commercial building types such as shops, restaurants or pubs, similar buildings appear all over the country, in every town and out-of-town shopping complex, designed to be consistent with the companies' image and market position. The rights or wrongs of this situation are a matter of current concern beyond the scope of this book, but it would not help to discuss general design issues if much of the building's content were predetermined. For the purposes of this book, it will be assumed that the client is independent and not subject to the control of a major manufacturer, so that wherever possible, issues can be examined from first principles. It must be stressed therefore, that the content of this book should not be seen as a handbook for 'real' car-dealership designers.

In practice, the design and construction process generates a considerable volume of information which is periodically exchanged between all the parties involved with the development. Decisions and instructions about future action will be requested or confirmed securing progress, rather like mountaineers hammering in pitons on route to the top. Information must be recorded and retained so that the basis for making decisions is clearly understood and that in the event of a slip, injury from the fall is minimised.

Information is communicated by various means, including discussions, minutes, memoranda, e-mail, letters, reports, drawings, application forms, approval documents, forms of contract and instructions, all traditionally retained in the **Project File**, even though there may be more than one. Large, complex projects can generate several hundred documents, retained in separate files for specific topics such as the client, consultants, Planning negotiations, contracts and site meetings, in such a way that they can be easily retrieved for future reference.

Depending on their purpose at the time, the style and format of documents range from rough notes to formal presentations. For example, details of a conversation or a sketched idea may be roughly recorded, but a formal letter to an external organisation would be typed, and the application for Planning Permission would include carefully prepared drawings. Each document helps illustrate the progress of the project from start to finish, showing what has been achieved and how each stage has been approached and resolved. Typical documents associated with the stage under discussion will be summarised at the end of each chapter as **Project File content**.

Electronic communication and data storage have changed the nature of information transfer for everyone, not least the construction industry, where the speed at which ideas can be developed and confirmed is breathtaking. However, where information has legal implications, it is still common practice to keep a record of hard copies or prints on file to maintain privacy and ensure that it remains available in its original form for future reference.

1.5 Project File content and structure

The traditional Project File contains paper documents secured in a folder. Standard writing paper and most formal documents in use today are A4 in size. Standard drawing sheets are available in multiple sizes from A4 (210 mm × 297 mm), increasing in size to A3 (297 mm × 420 mm), A2 (420 mm × 594 mm) and A1 (594 mm × 841 mm). The choice of paper size for drawings will depend on the content and scale of the drawing, and although occasionally there is a need for drawings to be on sheets larger than A1, they can be unmanageable and difficult to store in a small file. Most drawings would be stored separately in cabinets or electronically.

The usual practice for filing documents is to put them into an A4 folder in reverse order, so that the earliest dated documents appear at the back, and as time proceeds and further documents are added, the most recent are at the front. When the folder is opened as if it were a book, current information is immediately to hand. Documents are secured into the folder so that they remain in date order and so that collections of loose sheets do not inevitably fall on the floor and get mixed up.

Some documents may be more important than others such as the Planning Approval, contract conditions and presentation drawings. As they may be needed for reference as work proceeds, it is useful if they are easily identifiable, either by some form of marking system or by placing them in a plastic wallet. Everyone will have their own preference, depending on how they prefer to administer their projects.

To aid understanding, the Project File content will contain 'quality' rather than 'quantity'. A Project File in real life would contain many mundane standard documents, each of which carry their own importance in practice, but their inclusion in this book would not add to the appreciation of the process. The key documents, forms and drawings will be identified relating to the stage under consideration in each chapter. For example, indicative Project File content may refer to some of the following, listed in alphabetical order:

- Analysis notes
- Briefing information
- Bubble diagrams of thought processes
- Checklists
- Cost implications
- Existing drawings
- Formal applications to external authorities
- Instructions
- Letters
- Lists of sources of information
- Market analysis
- Photographs
- Presentation drawings
- Programming bar charts
- Records of meetings
- Reports
- Sketches and notes
- Specifications
- Summary of first thoughts
- Summary of general aims to be achieved at this stage
- Working drawings
- Useful reference sources

Examples of some of these items relating to the car-dealership will appear in the text at the point at which they are explained. They will not be duplicated at the end of each chapter. A typical business letter, for example, will be illustrated in the section dealing with letter writing, showing the general format of a business letter. Existing publications are available setting out standard letters for all possible circumstances for those who may be interested in using them. The Project File content will identify individuals and organisations to whom letters might be addressed at any particular stage in the process. In the same way, samples of sketches, presentation and simplified working drawings will appear wherever they are relevant and only summarised by title in the Project File content.

1.6 Discussion points

(1) Is there a distinction between architecture and building? Is there a tangible boundary line which requires expert interpretation? Can a building such as a car-dealership be described as architecture?

(2) How does the built environment which people inhabit affect their health and well-being? Why are some buildings psychologically uncomfortable? Is it possible to create buildings which everyone likes?

(3) Is the response to the demand for new buildings better served through speculative development or traditional contracting? How should the development of the built environment be controlled if speculation is encouraged? Is there a relationship between design quality and the state of the general economy?
(4) How should complicated information be communicated? Is there any limit to the time for which records should be retained? Is electronic data processing and handling always advantageous?

1.7 Further reading

Alsop W (2005) *Will Alsop's Supercity Urbis.*
Baden Hellard R (1995) *Project Partnering: Principle and Practice.* London: Telford.
Clifton-Taylor A (1972) *The Pattern of English Building.* London: Faber and Faber.
Egan Report (1998) *Rethinking Construction (Report to the DETR by the Construction Task Force).* London: HMSO.
Elder AJ (1980) *The Rubicon File.* London: Architectural Press.
Emmitt S (2002) *Architectural Technology.* Oxford: Blackwell Science.
Franks J (1998) *Building Procurement Systems: A Clients Guide.* Harlow: Longman.
Glancey J (2000) *The Story of Architecture.* London: Dorling Kindersley.
Gorst T (1995) *The Buildings Around Us.* London: Spon Press.
Groák S (1992) *The Idea of Building: Thought and Action in the Design and Production of buildings.* London: Spon Press.
Harvey R and **Ashworth** A (1997) *The Construction Industry of Great Britain.* 2nd Edn. Oxford: Laxtons.
HRH Prince of Wales. *The Prince of Wales Speeches.* www.princeofwales.gov.uk
Latham M (1994) *Constructing the Team: The Latham Report.* London: HMSO.
Masterman JWE (2002) *Introduction to Building Procurement Systems.* 2nd Edn. London: Spon Press.
Morton R (2002) *Construction UK: Introduction to the Industry.* Oxford: Blackwell.
National Audit Office. *Modernising Construction.* www.nao.org.uk
Nuttgens P (1997) *The Story of Architecture.* London: Phaidon.
Pilling S and **Nicol** D (2000) *Changing Architectural Education: Towards a New Professionalism.* London: Spon Press.
Reiss G (1995) P*roject Management Demystified: Today's Tools and Techniques.* London: Spon Press.
RIBA Bookshops. www.ribabookshops.com/site/home.asp
Smith Capon D (1999a) *Architectural Theory Volume 1: The Vitruvious Fallacy.* Chichester: Wiley.
Smith Capon D (1999b) *Architectural Theory Volume 2: Le Corbusier's Legacy.* Chichester: Wiley.
Turner AE (1997) *Building Procurement.* London: Macmillan.
UK 250.com. *UK Car Manufacturers Websites.* www.uk250.co.uk/CarManufacturers
von Meiss P (1992) *Elements of Architecture: From Form to Place.* London: Van Nostrand Reinhold.

SECTION 1 The Basics

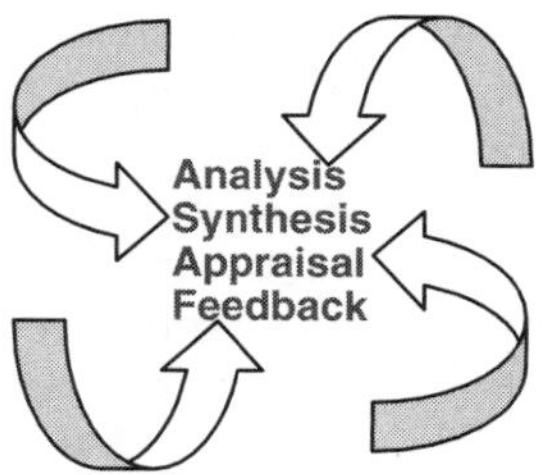

2

Design and the designers

2.1 Introduction

The design and construction of the new car-dealership involves people who fulfil generally understood roles and have professional relationships with one another throughout the duration of the project. Their purposes vary substantially from the creatively proactive to the incidentally supervisory, often taking individualistic views about what is appropriate under the circumstances. The first section of this book reviews the nature and process of design, and outlines the essential skills and potential contributions of the members of the design and construction teams as they communicate ideas and instructions to each other. The section concludes with a summary of the extensive legal constraints governing the design, construction and occupation of new buildings in the UK.

The first chapter begins with consideration of what design might mean, what is involved and what designers are trying to achieve. Understanding the concept of the ingredients or elements involved in product and building design can help to appreciate the way that the design process can move forward towards a workable solution. There are many sources of reference data and good practice guidance to help building designers to realise practical, attractive, safe and economical design solutions, but perceptions of the success of any particular project may depend on the expectations of judges. It is enlightening for the building designer to recognise that clients, users, authorities and the general public may all take different views about the merits of the finished building. The chapter concludes with a review of the nature of professional responsibility and an explanation of the personnel involved in typical design and construction processes.

2.2 The activity of design

Much has been written about design and how designers *do or should* think, but there appears to be no absolute agreement on the nature of creativity or any universally applicable methodology for approaching design. Creativity can take an abstract form with or without obvious meaning, as in the work of the artist who produces extraordinary or unexpected work. It can also describe 'lateral thinking' as a means of solving problems, perhaps leading to innovatory solutions which had not been seen, noticed or recognised previously. Creativity may have nothing to do with innovation at all, but simply a competent, well-organised attempt to produce an obvious, workable solution to a problem which attracts little or no attention. At its most basic level, creativity can just be the act of making anything, which may be good, average or poor.

Teachers and practitioners in different design disciplines all have their own ways and means, and ultimately the commissioned designer must respond to the specific challenges of creating the project in hand on his or her own. Their techniques and styles are based on their own experience of success, and regretfully sometimes of failure. It is important to recognise the impact of success or failure of design on those for whom the design is intended. For some products, it is appropriate that they should be innovative and fresh to momentarily 'catch the eye' of users, to become obsolete intentionally, maybe by mutual agreement as sales, fashion and technology dictate, but for others, a well understood, tried and tested solution will better satisfy long-term needs. Building design falls somewhere between the two; innovation is fine so long as it is relatively practicable (it may be said to 'work'), and something that works is fine so long as it is reasonably attractive. A critically acclaimed building can be a disaster to its regular occupants whilst an ordinary, unnoticed building can be the source of great pleasure. There are *pros* and *cons* to taking a conservative or a radical approach to building design depending on a variety of circumstances, but the success or failure of any building imposes significant long-term consequences on its users, which cannot be dismissed lightly. See sections in Chapters 8 and 9 particularly for more details about how buildings are used.

Before considering specific issues associated with new buildings, it may be helpful to consider what *design* may mean and what the *process of design* may involve. There are similarities in the ingredients of all forms of design and in the way that they can be handled to create a product. Very few products are designed in isolation in the hope that someone will like them. Even the artist working to satisfy him or herself achieving public recognition or professional acclaim, generally needs to sell their work or gain income from exhibitions. The aim of most commercial designers is to satisfy the consumers of their products, who select them because they are useful, attractive and/or economical. In a commercial context it is the consumer who judges the success or failure of the product and their level of expectation is crucial. Competition and market forces mean that product designers must have an even higher level of expectation than their consumers in order, if possible to continue to create better products than their consumers were anticipating.

2.3 The elements of product design

In every area of life from the most basic functional tool to the most elaborate ornament, materials, processes and activities are controlled and organised by design in response to a wide range of human needs, for practical purposes, for information, entertainment and pleasure. In this context, the phrase 'by design' is used to convey a sense of deliberate intent rather than accidental consequence. The word 'design' can be used as both a noun and a verb meaning **a plan** or **an arrangement**; **to plan** or **to arrange**. A *design* as a tangible product, or as an intangible service is a combination of a group of objects and requirements, which have been *designed* to create something new as a result of, or in anticipation of their collective performance and can be described as:

the selection of facts, requirements or perceptions about the properties or behaviour of individual elements which can be combined into a larger whole

The process of *designing* can be seen as reviewing the elements in the context of the way that they have been handled in the past, and consciously searching for an appropriate way of handling them now, perhaps a new way, exploring, comparing and testing alternatives. The idea of elements can be seen on any page in this book. Take a look and consider the following:

- The paper
  Its size, weight, density, colour, texture, absorbency and the way in which it has been cut and bound into the volume.
- The characters
  Their size (point), style (font), **boldness**, *italics,* UPPER and lower case, punctuation, c h a r a c t e r s p a c i n g s, the type and colour of ink used and the continuity of type setting.
- The page
  The headings, paragraphs, graphics, borders, shading, pictures, graphs, charts and bullet points.
- The meaning
  The language, spelling and grammar, the simplicity or complexity of the message.
  Taking all these into account:
  Is it *readable* and is its *appearance* interesting?

In this way it can be seen that elements can be grouped into common packages, some of which are immediately apparent such as materials, textures and colours which can be seen and touched, but others like cost, durability and copyright law are invisible and not so easy to appreciate. However, each element has been carefully chosen to create this page, which together with all the others in the book is a **product**, available to readers; its **users**.

The principle elements of any product are **materials, processes, forms and appearance**, each influenced by numerous factors within three broad areas of concern. *Marketing analysis* and brand understanding help to determine what the product should or could be like in an ideal situation, *technical constraints* reflect controlling standards and realistic investment costs and *commercial viability* assesses the potential for sales and profit. The response of clients, designers, manufacturers and users depends on their perception and evaluation of the importance of any particular factor, which will of course, not necessarily be the same as each other. The product design wheel, illustrated in Figure 2.1, gives some indication of the complexity of interrelationships between people, materials and processes involved in creating a product. The tangible and intangible constraints are relevant to the production of a single brick, a brick wall or a whole building, weighted from imperative through preferable, desirable, optional, immaterial, undesirable and definitely not! Interpretation and perception of these constraints depends on the individual's relationship within the process. For example, consider the likely differences in the desires or expectations of the client and the designer against the demands of technology and manufacture with respect to materials and value for money. As the product design is developed, there will be some agreement, but at many points there will be conflicts, differences of opinion and differences in value judgements, inevitably necessitating compromises. The essential fact, however, is that every element imposes its own possibilities or constraints on the way in which the product has been, or can be designed.

Elements can be described as *factors* or *requirements of performance*, or in many cases can be considered from both points of view. For example:

- the time taken for a particular ink to become touch dry after being placed on the paper is a known **factor**, whereas
- an economical minimum time for any ink to become touch dry to suit the technology of the printing process is a performance or production **requirement**.

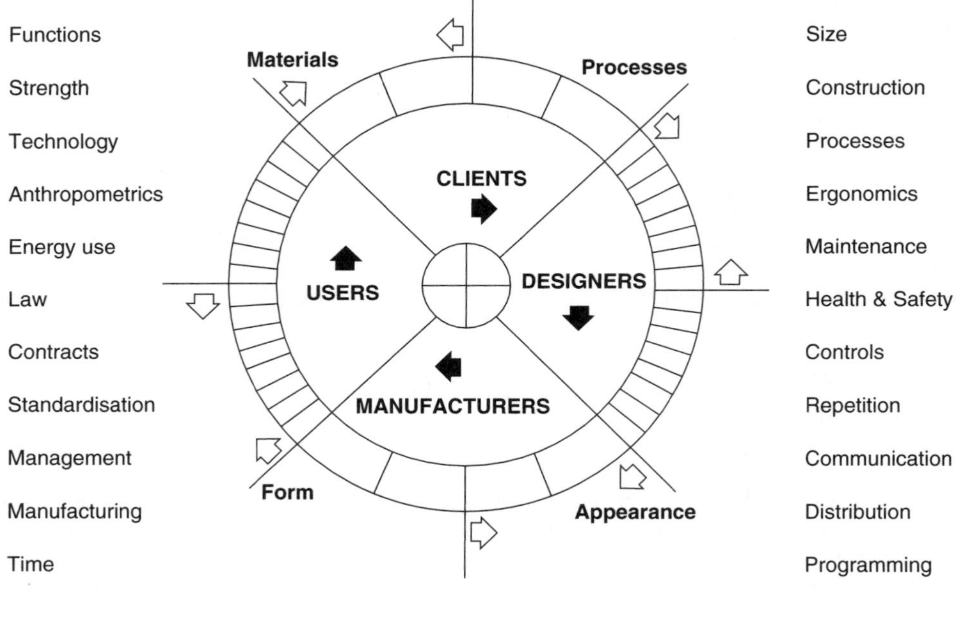

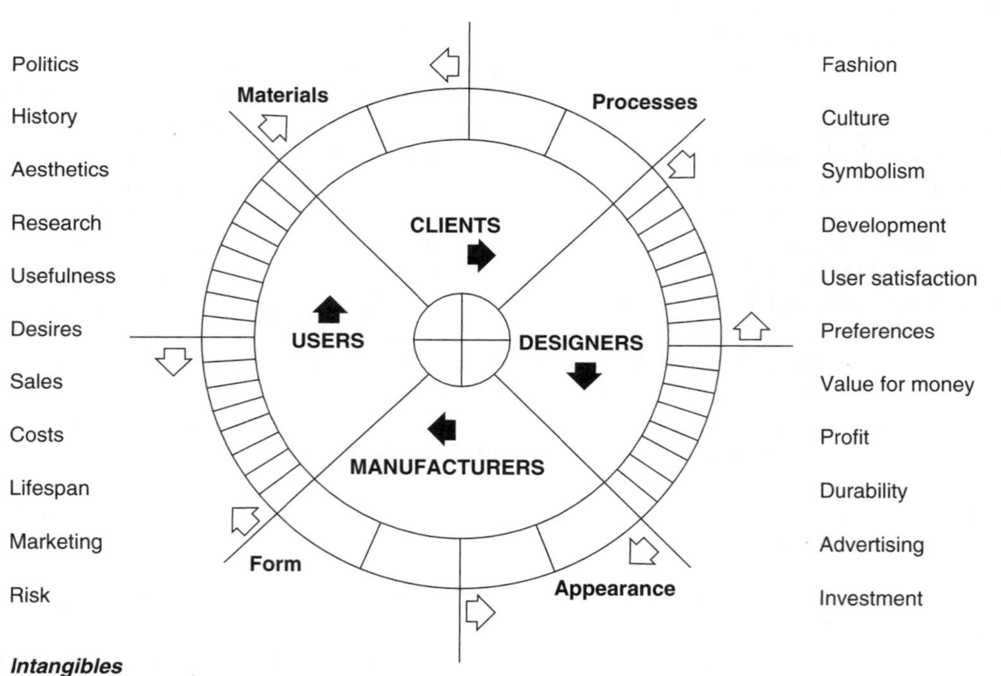

Figure 2.1 The product design wheel.

This is a significant distinction as in the first case the selection of the ink determines the timing of the process, whereas in the second case the process dictates which inks are suitable. Clearly such decisions have a 'knock-on' effect as other elements are considered.

A product can be regarded as the sum of decisions taken about its constituent elements. Look at the pages in this book once again; any piece of paper and any random typeface would do to make a page, but because other elements are missing, or decisions taken about them incorrect, then it may appear to be unsatisfactory, crude or incomplete. Inclusion of more elements can develop and improve it, making it more practical, attractive, readable or even beautiful. The performance and quality of the finished product varies because clients, designers and users take different views about the selection or significance of specific elements. A good designer will create a competent result and may even say something new, exciting or inspiring, depending on what is required or what could be possible, reflecting what both client, designer and user expect of the finished product.

At its simplest level, designing can be seen as collecting and assembling elements as a problem-solving exercise. Think for a moment about the way one might solve or 'do' a jig-saw puzzle. The pieces (or elements) can be randomly laid out with no regard for the relationship between each piece or what the whole picture should look like. The result will be an accidental image of some sort, created as a consequence of the nature of the individual pieces, which have been handled without any appreciation of the way in which they should interact, and with minimal expectation of outcome. It is unlikely that a solution can be achieved by this method that bears any relationship to what the picture should look like.

Clues are needed to solve the puzzle. The finished picture on the box lid is useful, but it is still not easy to locate each piece. However, some pieces can be placed with certainty. For a rectangular picture, if the four corner pieces and all the pieces with straight edges are connected in the right way, the outline of the picture can be established. The remaining pieces could be forced together to complete a picture, but if they don't fit properly, compromises or omissions limit success. Choosing other pieces as the starting point, such as those of the same size or colour will produce a different result, but still it will be incomplete and unsatisfactory.

Ultimately, the puzzle will only be solved when all the pieces are attached to each other in a rational way, arranging them so that location, size, shape, colour and connections are correct and the finished picture is completed. This is essentially a process of applying logic to the problem of organising bits of cut card (tangible elements), but moments of inspiration such as spotting a particular clue that sorts out a difficult area can play an important part.

Although the jig-saw puzzle may be difficult to solve, it has the major advantage that all the pieces are present at the outset, and there is usually a picture on the lid to follow. However, to create a new product, the designer only has some of the pieces; the remainder have to be discovered, or shaped by experience, intelligent guesswork and inspiration. Just like the jig-saw puzzle, some pieces obviously fit together whilst others obviously do not. But unlike the jig-saw puzzle, a clear picture of the finished product is not available. It is only an idea in the client's or the designer's mind, waiting to be conceived in response to many known and still unknown pieces of information. The process of design, the relationship of designer to other people and disciplines, constraints yet to be encountered may lead to the discovery and management of important elements which can generate solutions which were unimaginable at the beginning.

Some of these elements are very important, whilst others are consequential or incidental. Some of them are fundamental, whereas others are relatively trivial. The importance of each one may only become clear as work proceeds and in the light of experience as solutions are tested. It may be necessary to re-evaluate the priority given to an element at the beginning, which eventually merits less attention once the problems are better understood.

There is of course, a significant extra problem for the product designer. Unlike the example of the page referred to already, most products, and certainly all buildings are 3D. This means the designer has to develop an ability to see elements in a 3D context, resolving or balancing **all** the issues associated with design, production and consumption. This is an important concept, described as taking an **holistic** approach, which could be described as the principle aim for the good designer, endeavouring to create a product which fully satisfies its users, and its critics alike.

2.4 The elements of building design

In principle, the elements of buildings are the same as the elements of any product; materials, processes, forms and appearance, selected and arranged to meet the demands and needs of manufacture and use. Although a building could be regarded as one product, it is in fact an assembly of many individual products, some of which are purpose designed to suit special requirements but many of which are obtained from suppliers, chosen from ranges of pre-designed alternatives. Many of the raw materials and components already exist, and it is the way that they are put together that creates the new building, influenced by all the issues previously identified for product design, based on an understanding of the purpose of the building and the needs of its users.

Buildings serve a practical function as enclosed shelters, capable of maintaining a reasonable level of environmental comfort and safety in order to sustain human activity. The selection and use of materials in a local context can be seen throughout the world; African mud-huts, Eskimo igloos and North American Indian tepees, and throughout the UK; Welsh slate roofs, Suffolk thatch and Norfolk flint are all examples of local design solutions based on readily available materials, and understanding the skills developed through lengthy experience.

As well as creating buildings to suit their physical needs, different cultures and successive generations created and continue to develop forms, styles and decorations reflecting additional values and concerns. For example, the Egyptian, Greek and Roman civilisations built elaborate and sophisticated shelters, influenced amongst other things by their perceptions of art, religion and power. Vitruvius, a Roman author in his 'Ten Books on Architecture' offered a definition of the constituents of buildings which is still useful today:

- *Commodity*: function and practicality.
- *Firmness*: construction and durability of materials.
- *Delight*: appearance and attractiveness.

Historically, architectural or building styles survived for long periods of time, gradually evolving over hundreds, or even thousands of years as empires expanded and then contracted. In the UK, design and construction of buildings periodically incorporated the ideas of dominant European civilisations such as the Romans, Vikings and Normans, and later the more peaceful discovery and absorption of the sophistication and beauty of continental architecture, notably at the time of the Italian Renaissance. Many splendid Victorian structures were built by the entrepreneurs of the Industrial Revolution to demonstrate their new wealth and self-importance in society. The use of materials and structures typical of time and place, described as vernacular architecture was familiar to designers, builders and users, unifying the built environment and creating a certainty in the way in which it was understood.

During the twentieth century, however, design and construction changed rapidly as ideas, materials, technology, transport and communications developed and opened up new possibilities. Perceptions of what buildings are for and what they can look like has changed too. Since the 1960s, leisure centres, supermarkets and out-of-town hypermarkets have been invented, pre-cast concrete high-rise housing has come and gone and the design of industrial buildings has been transformed from mere 'sheds' into buildings of 'award winning merit'. We now have mixed use buildings, combining commercial, leisure and residential functions, and redundant buildings being transformed by changing their use to suit contemporary demands. The design of each building type, however, is simply an assembly of elements, and it can be argued that the quality of these buildings has resulted from the ability or failure to recognise the true significance of all the relevant elements at the time that they were designed and constructed. For example, the swift demise of the pre-cast concrete, deck access, urban tower blocks, constructed in a hurry in the 1960s and 70s was perhaps predictable because so many elements were ignored or plain wrong. Radical theories about function associated with inadequate technology and management, created an inevitable time bomb, and many of these developments have already been demolished, some within a mere 20 years of their first being occupied. Learning from mistakes has been a very painful and expensive experience for everyone involved. Interestingly though, the sweeping condemnation of the buildings of this era has recently been reviewed, and a number of examples from the period are now been restored and listed as being of architectural merit.

It is always easy to criticise of course, and with the advantage of hindsight, every building would be perfect. It is said that 'history is a good teacher' and everyone involved in creating new buildings would do well to study past successes and failures so that the same mistakes are not repeated. This can only be achieved by understanding as far as practically possible the true requirements and purposes of buildings so that significant elements are not ignored or forgotten. They can be grouped into the following broad categories.

Basic physical needs
Buildings generally provide a sheltered, warm and secure enclosure or interior environment, appropriate for sustaining anticipated human activity.

Cultural influences
The way that activity is undertaken includes consideration of present-day attitudes, beliefs and political organisation, the meaning of historical precedent, the desire to express symbolic values and the extent of current legal constraint applicable to development.

Means of construction
Creating appropriate environments requires the selection of suitable structures, materials, fixtures and finishes to maintain shelter, warmth and security and to withstand the effects of the local climate and the wear and tear of the regular use.

Appearance
Structures and finishes can be seen and have a visual impact, both on the external and internal environments. The appearance of materials may be naturally finished or decorated, and will almost certainly change over time.

Project-specific matters
The way that any new building can be created depends on the demands of the building type and the constraints of the site on which it is to be located. The requirements and expectations of the client and the costs of construction are factors unique to each and every new building.

All the elements of a new building can be found in these categories, and some will be affected by all of them. It is not possible to decide whether any particular element of the building should be by looking at only one area. For example, consider the front wall of a shop in the high street:

- *What is its physical purpose?*
  Does it maintain internal comfort levels? If it is used to display expensive products, how is security arranged? Does it offer shelter from rain to passers by?
- *How have cultural perceptions influenced its function and appearance?*
  To what extent is the shop front a reflection of present-day trading methods? Does the design say anything about the company's history and beliefs, and is there any symbolic reference to the goods and materials being traded?
- *What legal constraints applied to its design and construction?*
  How was the way the shop works and looks controlled originally? Would current controls mean that it would be different today? Would Planning, Building Control and Health and Safety concerns alter the way the shop could be operated today?
- *How is it constructed and maintained?*
  What materials have been used? How is it fixed to other elements of the structure? Is it holding up the roof? Are there any high-level windows which are difficult to reach and clean?
- *Has it stood up to the weathering effects of the sun, rain and frost?*
  Was the selection and use of materials suitable for the long term? Are elements of the building or its function short term and likely to be renewed in the foreseeable future?
- *Has it been damaged through wear and tear, or as a result of vandalism?*
  Is the shop front capable of being cleaned and/or redecorated?
- *What is its impact on adjacent property?*
  Has it been designed to be sympathetic to other shops in the area, or does it stand out on its own? How does it relate to the public footpath and the road? Is it obscured by parked cars?
- *How much did it cost and was it good value for money?*
  What is the useful lifespan of the building, the materials and the business operation? Is there a relationship between the lifespan of the business and the quality of the building? Is the building energy efficient, the operation sustainable?
- *Does it allow any flexibility for alteration by new occupants in the future?*
  Could a change of use take place without major reconstruction?

There are more questions which could be asked about other issues, factors or elements which may have a bearing on design possibilities for this part of the building, which will inevitably affect other parts of it as well. The product design wheel in Figure 2.1 can be used to suggest some of the questions as more and more implications are explored. Some answers are common to all building types but others must be discovered or determined in relation to the specific project.

2.5 The design process

From the outset of a new project, the designer's preliminary work can be based on very limited information which may be sufficient to justify putting forward ideas for consideration. However, it is very unusual to find that the finished design materialises effortlessly, translating initial ideas into reality without any revision. This quote from Karl Popper 'An Evolutionary Approach' sums up the situation concisely:

'We start, I say, with a problem, a difficulty. It may be practical or theoretical. Whatever it may be when we first encounter the problem we cannot, obviously, know much about it. At best, we have only a vague idea what our problem really consists of. How, then, can we produce an adequate solution? Obviously we cannot. We must first get better acquainted with the problem, but how? My answer is very simple: by producing an inadequate solution, and by criticising it. Only in this way can we come to understand the problem. For to understand a problem means to understand its difficulties; and to understand its difficulties means to understand why it is not easily soluble – why more obvious solutions do not work. We must therefore produce more obvious solutions; and we must criticise them, in order to find out why they do not work. In this way we become acquainted with the problem, and may proceed from bad solutions to better ones – provided always that we have the creative ability to produce new guesses, and more new guesses.'

Designing is a continual process of selecting and organising elements, trying to establish which are the most important and how they might all play their part in the creation of the new product, and inevitably ideas change as possibilities are added or discounted, as proposals are conceived and considered.

The aims of design can be summarised as follows:

- Identify all the relevant elements pertinent to the project.
- Discover or understand how the elements interact with one another.
- Plan or arrange the elements so that they fit together in an appropriate or meaningful way to create a competent product.

It can be argued that trying to meet these aims is a process demanding logic; an intellectual, rational review of the matter, but it is by no means clear that this is how all, or any designers work in practice. Much has been written about brainstorming, mind mapping and even day-dreaming as ways in which unexpected, apparently illogical design solutions appear. However, in many cases, the design development process involves the following actions.

Analysis

Analysis means splitting up the 'whole' into its constituent parts. In the example of the shop front described earlier, it is useful to find out what the essential design criteria is for the major elements of function, appearance, cost, image and so on, which can each be analysed in more detail to determine what they mean, or could mean in relation to creating the shop front.

Synthesis

Synthesis is the re-assembly of the parts into a meaningful 'whole'. The information gained through analysis can be used to suggest a possible design for the shop front.

Appraisal

The proposal for the shop front can be checked to see if it matches the analysis, critically assessed by interested parties such as the client, the Planning Authority and other members of the design team.

Feedback

Critical comments received following appraisal in the form of further information, advice, recommendations, approvals or instructions will either confirm that the proposal is acceptable, or that some elements must be analysed again in more detail. Further examination of the elements leads to a new synthesis, a new design proposal which can be re-appraised and tested once more, leading to more precise feedback so that the design improves, becomes better, more practicable, economical or attractive until at some point it is accepted as being the right solution, or the best solution to proceed with under the prevailing circumstances.

For all but the simplest of design tasks, the process will not be in the progressive linear form of **start, analysis, synthesis, appraisal, feedback, finish**, completing each stage before moving on to the next one. It cannot even be categorically stated which comes first, as ideas and decisions are influenced by each action and there is constant need to go back and test solutions against requirements. For example, preliminary analysis may lead to a practicable solution for the shop front design, which can be presented to the Planning Officer for appraisal. Alternatively, it may be more sensible to obtain the Planning Officer's feedback first so that requirements can be taken into account straight away, and time and effort is not wasted producing a solution which subsequently proves to be unacceptable.

In a sense, it is a circular process, repeatedly rotating through each stage, but rather than returning to the same point in the cycle, if the rotations take the form of a decreasing spiral illustrated in Figure 2.2, then it can be seen that progress leads closer to the centre. It would be meaningless to imagine the process as a target with the bulls-eye as the *perfect design solution*, as for any brief there could be many equally valid alternative proposals, and every designer could produce a different response to the same set of circumstances. However, the centre can be regarded as the *best design solution* that could be achieved under the circumstances. If the spiral took the form of a maze, then reaching the centre would depend on the starting position and wrong decisions at turning points would lead to dead ends. Progress could only be made by retracing earlier steps and starting again. As most building design contains compromise, a good designer tries to get as close to the centre of the target as possible, or has to stop when time runs out. How close the designer can get to the centre depends on ability, skill, conscientiousness, perseverance, inspiration or even sheer luck in taking the correct or best route and being able to recognise where they have already been. It can be a good idea to keep a record of these routes so as not to go down the same dead-end again. Or it might be that revisiting with a fresh piece of information enables the designer to unlock the gate to the next level. Sometimes though, a perfectly sound design proposal can be ignored in the search for a better one, which may not exist. One of the designer's key skills is being able to recognise the target and to know when it has been reached.

To a large extent, the process of designing is a personal activity influenced by the way that the designer's brain works; how they think. Some designer's work predominately in the direction of '*analysis, synthesis, appraisal and feedback*' which is a method of examining problems. They are 'problem solvers' concentrating on working out ways of putting the known elements together to create a product, which is then checked or tested to see if it is a satisfactory solution. For others, the process is predominately reversed in the direction of '*feedback, appraisal, synthesis and analysis*', which is a way of testing solutions. These designers are intuitive, speculating or postulating an idea or possible solution straight away, in advance of the complications of detail, and then checking to see if the answer contains all the necessary elements, or satisfies the brief.

Most designers will use a mixture of both methods as whilst a design solution can come purely from consideration of analysed or perceived problems, it is quite common to find that a proposed design solution redefines the original problem, satisfying needs which were not initially identified or understood. For example, a shop owner may be convinced that the main entrance

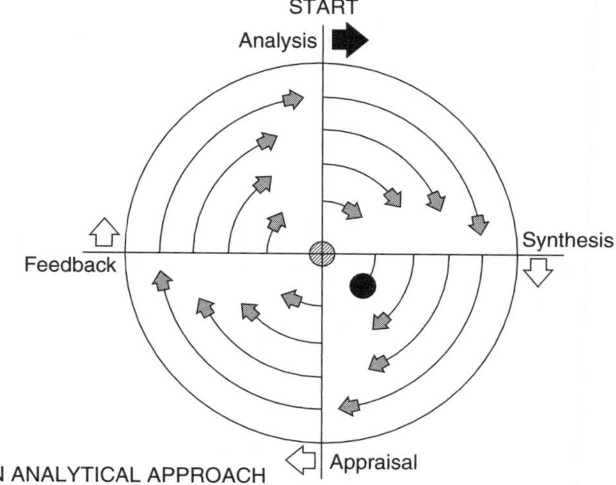

Key:

Repeated analysis, synthesis, appraisal and feedback.

The best solution that the designer could achieve under any circumstances.

The solution that is achieved, representing the best of compromises, the best that the designer is capable of achieving, or the reaching of limits such as cost or time.

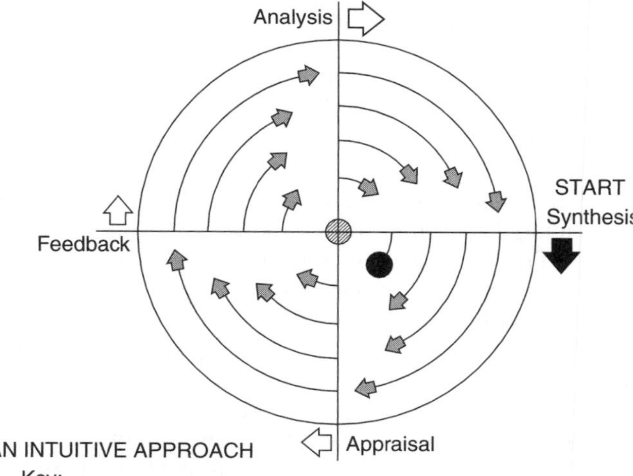

Key:

Repeated analysis, synthesis, appraisal and feedback.

The best solution that the designer could achieve under any circumstances.

The solution that is achieved, representing the best of compromises, the best that the designer is capable of achieving, or the reaching of limits such as cost or time.

Figure 2.2 The design spiral.

door should be in a particular position in the front wall, in relation to the internal arrangement of counters and displays. But the suggestion that the entrance might be located in a different position, could lead to a much better internal arrangement, which the client had not even considered before.

2.6 Design co-ordination

The location of the entrance door to the shop described above, is a relatively straightforward design decision based on analysis of a limited number of pertinent constraints. It is perhaps not difficult to imagine that the complexity associated with larger buildings, or those containing specialised processes such as the car-dealership, necessarily means that design decisions become dependant on understanding many more elements and threads. The co-ordination of multi-disciplinary contributions from many different design specialists, subcontractors and suppliers responsible for systems and individual elements simply to establish a workable design proposal is difficult enough. Bringing everything together to meet time, cost and quality control limitations requires clearly focussed, structured management. The traditional architect or engineer-led process is no longer seen as the only, or even best way of managing such a complicated business, as an independent project manager employed directly by the client may be better equipped to take decisions in the best interests of achieving deadlines, but with the technical expertise to understand the process and inherent difficulties. The project manager's responsibility would be to co-ordinate the delivery of a workable building, planning the process from start to finish, setting a framework for design, timing, costs and associated legal and contractual arrangements. The impact on design quality is debatable. Expedient decisions may limit initial intensions, but equally they may prevent cost escalation. Priority to one area of design might be costly to another, but undoubtedly the demand for rapid progress through both the design and the construction phases for new buildings is creating pressures for project management controls, arguably beyond the capability of the individual building designer. Reference is made elsewhere in this book to current procurement methods designed to offer different ways of creating new buildings.

As well as contractual management, design co-ordination can also be seen as the integration of design criteria, the blending of elements together into the whole. This notion is particularly significant in the development of briefing requirements and the examination of potential solutions from the earliest possible time so that design decisions are based on the best possible information. The holistic approach to building design combines structures, materials, environmental requirements and social expectations into the best possible physical solution under the circumstances. The size, shape and location of everything within the building's shell is in accordance with the 'real' needs as defined at the outset, or as determined as ideas develop. Everything works together; each element is associated with and consequent upon one another. This theme is developed elsewhere in this book as design and construction details emerge. As a simplistic example, imagine that a space in a building is defined by the building designer with no regard to mechanical and electrical services. Windows and doors are located without thought for positioning heaters, power points, lights or other important machinery needed for the intended occupants. Perhaps the building designer introduces a large south-facing, unopenable window because of an attractive view, forgetting the problems of excessive solar gain and glare. Separating responsibility for the design of these uncoordinated elements will inevitably lead to clashes. The uncoordinated bulky heater, unsightly pipework, trailing wires, gloomy corners and airless workspace is likely to result in some very unhappy, perspiring occupants. Design co-ordination is about planning, predicting how everything should fit together without unfortunate clashes; analysis of 'cause and event', before the event.

2.7 Innovation

Design and the design process may be difficult terms to define but they clearly involve 'thinking', and apart from the most mundane, mechanical problems, creative thoughts which sometimes lead to changes in the way that things have been done before. The design and construction of buildings has continually evolved as 'new' materials and processes have enabled the creation of 'new' structures. Even vernacular styles of architecture include useful developments when they become generally accepted. Such advances can be described as innovative, considered at the following different levels:

- Firstly, a gradual or developmental process can introduce improvement, based on previously tried and tested solutions, following set patterns but adding 'extras' or additional quality.
- Secondly, innovation in design can be enterprising, using something in a different way than its original design intension, seeing a fresh use for something existing already.
- Finally, radical innovation proposes something completely new which has not been thought of previously, or introduces leading-edge inventions which have not been previously available.

It can be argued that the very act of designing or creative problem-solving is inherently innovative, but with regard to building design, there are external pressures to encourage designers, and the UK construction industry generally to improve its efficiency and the quality of its services and products. The Egan and Latham reports referred to elsewhere have challenged the industry to re-examine every aspect of its performance for the benefit of its customers and in line with demands for greater sustainability in the building sector. Consequently, innovation is being promoted in the use of materials and systems, in areas of energy conservation, procurement and management, all of which require the building designer to keep up-to-date with advances in technology and developments in manufacturing practice. Incentives and penalties are part of the process of changing attitudes in the construction industry, as for example, with the demand for inclusion of a percentage of Modern Methods of Construction (MMC) for new residential development.

Although not directly relevant to the creation of the new car-dealership, it is worth noting this government initiative in response to demands for faster construction and pressures caused by skills shortages, which is designed to improve the quantity and quality of housing construction through the use of off site prefabrication leading to faster construction, fewer housing defects and reductions in energy use and waste. Such a strategy could become commonplace in the construction industry in the near future. The initiative attempts to address a series of problems facing the construction industry generally. They are subjects of interest and debate including the following:

- The need to increase the speed of the construction process.
- Promotion of off-site prefabricated manufacture to help minimise site defects.
- Matching industry capacity, particularly in line with shortages of skilled labour.
- Health and safety improvements through less worker time spent on site.

- Increased energy efficiency in response to changes in Building Regulations.
- Reduction of site waste and total number of trips to a building site.
- Stabilising or lowering capital and running costs.
- Changing public attitudes and promoting market demand.
- Responding to Planning pressures, notably Planning Policy Guidance Note (PPG3) requirements for greater density.

Few design proposals involve radical innovation. Very few architects actually manage to break the moulds. Paxton's Crystal Palace, the first 'skyscrapers', Le Corbusier's concrete Chapel at Ronchamp in the Alsace Region of Eastern France, the all-glass houses of Mies van der Rowe are examples of whole-building advance. Richard Rogers inside out Beaubourg Centre in Paris, Norman Foster's Lloyds Building in London and the concrete structures designed by Santiago Calatrava in Valencia all re-examine the way in which buildings are 'supposed' to be designed and built. Less obvious innovations come in the form of new materials and technology continually developing in response to market demands. The use of unplasticised polyvinyl chloride (UPVC) and other plastics, optic fibres and high-energy glass become commonplace once they are affordable. Innovative design is essentially an attitude of mind, of enquiry, a 'what if' mentality which searches for improvement or seeks to do something differently from the expected norm. The design team should remain willing to rethink consumer needs, ask questions, challenge the *status quo* and try to develop realistic, but imaginative solutions to design problems. In the best cases, this approach adds value. Occasionally there will be a need to take risks, which is of course the way that ideas and practice move forward from generation to generation.

2.8 Risk

The concept of 'risk' in building design may seem to be an alarming one. The word suggests danger, catastrophe, something to be avoided because it is unpredictable or that has an uncertain outcome. That could be our view of almost anything in life of course; crossing a busy road is 'risky', but we accept the risk because we believe that we can approach the task in such a way that we will be safe; the risk can be eliminated through understanding and management. Crossing the busy road is a finite risk too; once we reach the other side, we are safe until we need to cross another road. Whether or not we are safe from the risk of permanent traffic fumes is another matter.

In the context of buildings, risk appears in various forms throughout the design and construction periods, some of which may continue on into the future, perhaps affecting the lives of generations of subsequent occupants and others involved during the lifetime of the building. Some risks can be eliminated, others controlled within acceptable limits, but others remain indefinitely. Typical risks that the building designer may encounter include some or all of the following:

- *Health and safety*: from the very beginning of the project's life health and safety is a high priority for all concerned. Chapter 4 reviews the impact of the Construction Design and Management Regulations (CDM), which attempts to make design and its consequential implementation safer.
- *Negligence*: failure to fulfil a professional care of duty in respect of undertaking design work. The building designer may miss important survey implications such as the location of an underground high-voltage electricity cable or an area of chemically contaminated ground, both of which could seriously injure an unsuspecting contactor. Setting out discrepancies, boundary inaccuracies and other design errors are likely to result in claims for negligence.

- *Proceeding without authority*: where design work continues without client support or approval. Failure to secure statutory approvals may be extremely damaging.
- *Delays*: failure to submit for approvals at the correct time or proceeding in advance of approvals may cause delays as design work is corrected, or expensive remedial action once construction has started.
- *Lack of co-ordination*: inadequate co-ordination between specialist consultants.
- *Unsupported specifications*: selecting inappropriate or dangerous materials, misusing materials against manufacturers recommendations or misguided innovation.
- *Overrunning budget costs*: failure to meet costs targets at various stages throughout the project.
- *Customer dissatisfaction and/or public dislike*: creating a new building which disappoints, or which actually fails to meet the brief.
- *Long-term risks*: for the subsequent occupiers of any building, there are associated risks in terms of the performance of selected elements. There is a permanent risk of falling down the simplest of staircases, scalding under a hot shower or electrocution from a faulty electrical socket outlet. Risk is also associated with lifespan of structures, materials and installations, which may fail or 'wear out' sooner than expected.
- *Design risks*: risk can also be seen in the positive sense of taking a chance on something, being adventurous and challenging.

Few if any of the risks identified above arise as a result of absolute uncertainty about what might happen. Most are predictable and can be minimised by sensible risk management through identification, assessment, monitoring and control. Creating and understanding a design brief, discussed at length in Chapter 7 is a form of risk management strategy, clarifying expectations and outcomes, evaluating risk and determining how to it might be approached. Clients should be fully advised on what they are getting or could get from the design process, perhaps showing them existing examples as a guide to what they should expect.

2.9 Design guidance

The theory and practice of building design and construction continually develop through research and testing. Existing and new ideas establish principles which can help to inform current performance. For the building designer, there are many sources of reference to help to create the best possible design solutions, ranging from providing inspiration, best practice advice and mandatory requirements, illustrating how things *could, should or must* be done'. Typical areas of design guidance include the following:

- *Published textbooks*: Historical and current practice for construction, details and procedures can be purchased, are held in libraries or in some cases can be accessed electronically. Titles and brief summaries can be viewed on publishers web pages or conveniently at www.amazon.co.uk.
- *Best practice publications*: Professional organisations disseminate information and design exemplars. For example, the Commission for Architecture and the Built Environment (CABE) publish excellent literature on urban design; the Association of Chief Police Officers (ACPO) advise on ways of reducing crime through their 'Secured by Design' material, and various Government publications help with design for disabled access and use.

- *Databases:* Numerous collections of useful material are available in database form including the Barbour Index for construction, materials, manufacturers details, statistics, the Royal Institute of British Architects (RIBA) product selector and the Einstein Surveyor's Channel.
- *Internet*: Web-based material is available in every area of design and construction, and many web pages provide useful links. Most search engines lead to useful material, but don't accept the first answer as definitive, and beware the general enquiry which yields 648,754 hits. It could take several hours to trawl through to very little advantage.
- *Media publications*: Public interest in both design and construction has increased recently following successful television programmes, notably Channel 4's 'Grand Designs'. Other programmes and magazines contain useful guidance about property, particularly for interiors and landscaping.
- *Professional journals:* Various professional magazines present news and promote debate, together with explanations and explorations of theory and practice. As well as technical guidance, many magazines review innovative schemes with reflections on their suitability and success. A list of current journals is given in page 349.
- *Statutory instruments*: The Building Regulations, British Research Establishment (BRE) research publications, British Standards and Codes of Practice (see Section 11.4), National Planning Guidance on principles of development (e.g. see PPG3 described in Section 4.4) outline requirements for new buildings. Local Planning authority design guides dictate expectations for planning applications in their own areas, and for specific development sites.
- *Design research, refereed papers and journals:* Higher-education institutes, government- and industry-sponsored bodies publish information on a wide range of subject areas including design theory, historical and cultural context, current practice, ergonomics, user needs, structures and materials. Research sometimes leads to publication of design 'tools' helping to methodically examine sustainability or life-cycle costing for example.
- *Manufacturers literature*: Manufacturers and suppliers published guidance and recommendations on the use of materials and assembled products, controlling the way that they are expected to be used in order to support guarantees of performance. Failure to comply with limitations can render specifiers and users liable for consequential difficulties.
- *Precedent*: Examination of what has been designed and built already is clearly a good guide to the designer in terms of the potential for repetition and how to minimise the risk of avoidable problems.

In most cases, guidance comes as a result of experience. Advice follows assessment of performance over a period of time, or describes how an element in the design process may perform under certain circumstances. In some cases the information may be objective, factually based and reliable. In others it may be subjective based on someone's value judgement.

2.10 Judging the success of design

The success or 'value' of design clearly depends on the relationship between the product and the judge. Many products are created for general sale to customers for their own personal use. These products can be described as consumable with a limited lifespan and consumers accept that when they are finished with, or worn out they can, and will be replaced. The speed of this process of planned obsolescence varies; some products such as newspapers and magazines, for example, are intended for immediate consumption, replaced swiftly by the next issue and retain little or no value after the initial purchase, but other products like clothes, cars and furniture are each expected to last longer before requiring renewal. If they are carefully looked after or restored, they may well retain a significant second-hand value, but in a commercial context, the principle measure of success for the designers and manufacturers is at the first point of sale, and user demand is an important indicator of the perceived merits of their work.

The performance of mass produced, relatively low-cost products can be assessed objectively by testing, or subjectively through the analysis of market research studies. Refinements can be introduced following research and investment in the construction of prototypes. Advances in technology and fashion or correction of faults can be accommodated in subsequent replacements and indeed, it is widely accepted that many products are designed to wear out rapidly to encourage further sales. Although manufacturers would not admit it, their products are often designed to include elements of in-built obsolescence, encouraging their customers to periodically purchase replacements.

Buildings are not generally like this. They *are* created as a commercial proposition, built to suit the specific requirements of a commissioning client, but as large, relatively expensive products cannot be easily withdrawn for remedial attention as could a car with a faulty handbrake connection or a play in the West End which receives poor reviews. The new building will be judged on completion in a number of different ways including its internal and external appearance and on the comfort of its accommodation, but in other ways its value may not be immediately clear and first impressions may be misleading as problems materialise with time. Initial satisfaction may be diminished through deteriorating finishes or high running costs over the longer term of occupation.

Whilst buildings are designed, constructed, marketed and sold to the first user, they remain in existence for some time affecting the lives of subsequent users and the community around them, part of a larger scene within the street, village, town or city, unlike the newspaper, book and car, which are essentially for individual, private consumption. Buildings are to some extent public property, demanding qualities which will maintain their performance for a long time, or even continue in existence indefinitely. This is a significant factor as the criteria for judgement may not only vary, but be radically revised as time passes. A brief study of the history of building design during the twentieth century will show how design, manufacture and legal control has altered in response to changes in public and professional opinion, sometimes resulting from experience of failure, but more often reflecting changes in design theory, technology, life style and political intervention.

In the face of all these pressures to change, the design team must get everything *right* at the first time of asking, satisfying diverse sources of interest. Most buildings are unique, constructed to deadlines, and are only as good as they can be *at that time*. The first target is a satisfied client, but true success depends on the level of expectation which the building designer, the development team, the local community and the wider society as a whole demands, or is prepared to accept. Here, the building designer will find pressures to confirm or to innovate; to satisfy narrow demands or to explore possibilities.

2.11 Professional responsibility

The term 'professional' can be applied to those who offer goods or services for financial reward, which they attempt to maximise in response to prevailing market forces. The business arrangements associated with commercial design, including general office practice, conditions of employment, efficiency and profitability are essentially private matters to be agreed between employer and employee dictating the costs or fees that can be demanded or negotiated. Being 'professional' in this sense is about 'making money' and is already well documented by other authors.

It is not always compatible with the other definition of 'professional' which is about doing something in a responsible, conscientious manner. This is a concept of both personal and corporate behaviour being controlled or motivated so that customers or clients can be assured that they are being offered, and are receiving the best possible goods or services under the circumstances. Evaluation of the professional quality of building design and the performance of designers is difficult for most clients because they do not know what to expect. They are not buying a product like a bunch of bananas or a television set, which can be seen and chosen from a range of demonstrated alternatives. They are buying a process leading to an end product and must rely on the vision and efforts of their advisers, which may range from brilliant to average to poor. For many clients, procuring a new building is a 'once in a life time' event, of which they have no previous experience and little time in which to undertake research to gain the necessary understanding or appreciation of the processes and the contributions expected of their advisers. They may also be unclear about the part that they should play themselves. The way in which the relationship between client and advisors develops can have a profound influence on the design process and the nature of the outcome itself.

For many products, it can be argued that quality is proportional to cost and generally the client 'gets what they pay for'. This can be true for the bananas if the more expensive ones are bigger, fresher and taste better. It may be true of the more expensive television set, if it has a clearer picture, more elaborate controls and lasts for longer before needing replacement. For many elements found in buildings, it is certainly true. For example, the lifespan of a carpet exposed to heavy wear and tear is directly related to its cost for both manufacture and installation. But competent design is not always directly related to cost. A good designer may quickly see an excellent solution whilst a poor designer may take a long time to discover a mediocre one. An efficient designer incorporates all the relevant elements into the design whilst a disorganised designer misses out critical elements. In this respect, higher design costs will not guarantee the quality of the end product.

When goods like bananas, television sets and carpets are offered for sale, consumers not only have the benefit of legal protection with regard to remedies against faulty goods, but can purchase them with confidence, knowing that they are supplied or manufactured in accordance with applicable controls and standards, guaranteeing that they are of a certain quality, fit for the purpose intended. Measuring and checking can easily verify that the performance of the product is satisfactory. For design itself, defining useful quality assurance references is not so simple. The various institutes concerned with building design publish codes of conduct defining competent performance and acceptable ethical behaviour so that in their view, the *'public may rely on the standard of integrity and professionalism'* with which their registered members can be expected to comply. Some of the issues identified in the codes are specific to membership of each individual professional institute, like maintaining subscriptions and not bringing the 'Institute' into disrepute. They cover business arrangements such as agreement on fee charges and indemnity insurance to protect against the costs of mistakes. These codes are applicable to their qualified practitioners, and can be studied in their publications.

Other issues are matters of good practice, which can be summarised as expectations of behaviour as follows:

- *Operating in a professional manner*
  Undertaking work on the basis of reputation, capability and experience.

Avoiding giving advice on matters which require specialist qualifications.
Offering necessary and appropriate services under the circumstances.
Acting in the best interests of the client and the future users of the building.
To avoid damaging the reputation of others.
Liaising with other advisers appointed by the client.
Having adequate resources to meet commitments.

- *Specifying reliable construction techniques and materials*
  To recommend the use of systems and products which are backed up by a recognised precedent and a source of quality assurance.
- *Meeting cost targets*
  To put forward proposals which match the client's budget and which can be constructed without significant additional cost.
- *Maintaining design and construction programmes*
  To comply with agreed timescales for design and construction work maintaining target dates.
- *Complying with the requirements of interested authorities*
  To formulate proposals in accordance with current legislation, incorporating advice and mandatory conditions imposed by external bodies.
- *Appreciating wider implications on the lives of others*
  To consider the quality of life of subsequent users of the new building.
  To avoid damaging the amenities of adjacent property users and the community around them.
  To endeavour to improve the quality of the local built environment.
- *Keeping up-to-date with current practice, developments and legislation*
  To maintain and improve the knowledge and experience necessary to meet the changing standards and methods.

Implementing and maintaining these issues of good practice will ensure that good, professional working relationships are established with the client and all the other members of the development teams. Professionalism demands a duty of care focusing attention on the purposes and aims of design work so that the process of design and construction moves as smoothly as possible towards the creation of a new building.

2.12 The development teams

Creating a new building involves contributions from many individuals, exchanging ideas and instructions as the project is developed. They can be grouped as clients, consultants, authorities and builders or contractors.

Clients

The client is responsible for commissioning the new building and must approve design decisions as work proceeds. Different types of client include the following:

- *Owners*
  Clients may be individuals, partners or shareholders who's ownership permits them personally or collectively to make decisions.

- *Representatives*
  Directors, managers or in-house specialists responsible for day-to-day activity may have the power to make decisions, but may also be required to seek the approval of the owners.
- *Committees*
  Clients may take the form of a management committee, comprising professional or lay membership, making collective decisions based on professional advice or on their own experience or feelings.
- *Users*
  Prospective purchasers or even tenants may fill the role of the client, making decisions based on their specific preferences. For example, a residential developer may give their prospective purchasers the opportunity to determine aspects of the design and construction of their future house to suit their own particular needs.

Consultants

The consultants supply design and management expertise and can include the following:

- *Measurement surveyors and investigators*
  Land surveyors for the measurement and production of information about the site; sub-soil exploration subcontractors and laboratory testing consultants; ground contamination experts; photographers of underground pipework and aerial views; building surveyors for the preparation of reports on the condition of existing buildings.
- *Quantity surveyors*
  For estimating, cost planning, advising on economy, measuring materials and labours, valuing completed work for payment.
- *Designers*
  Architects, architectural technologists, space planners, interior designers, landscape designers, planting specialists, graphic designers and artists.
- *Engineers*
  For structures, space heating and hot water, gas and solid fuel installations, ventilation, air-conditioning, plumbing, electrics for power and lighting, acoustics, vibration control, specialist machinery and equipment, computers, drainage, highways, lifts, conveyor belts, escalators and moving pavements, fire protection and fire fighting, ground preparation, refuse/waste disposal, internal and external communications, audio systems, water supply and storage for processes and cooling, pumping, security systems, detectors and alarms, mechanical waste disposal, incinerators, macerators, compressed air for processes, spray painting and specialist environments such as clean rooms.

It should be noted that any of these engineers could also be described as designers, who may play a significant role in the creation of the new building. They are distinguished as engineers because they are able to offer specialist services for which the building designer may not be qualified.

- *Valuers*
  Estate agents for advice about land and buildings; specialists for commercial undertakings.
- *Managers*
  Project managers, clerks of work, health and safety planning supervisors and process management consultants.

- *Other specialists*
  Planning consultants; Building Control and commercial development consultants.

Authorities
The authorities advise on design and construction, and have statutory duties to exert control with respect to their individual responsibility. The authorities involved in building development include:

- *Local Authorities*
  County, District, Borough, City and Parish Councils who have interests in Planning, Building Control, rating and licences, grants, domestic drainage, environmental health, waste disposal, street cleaning, highways and transport, highways drainage, street lighting and street naming.
- *The Fire Officer*
  To advise on all aspects of fire safety, to approve design proposals and the subsequent use of the building.
- *The Environment Agency*
  For resolution of issues affecting existing watercourses and potential effects on ground water.
- *Health and Safety Executive*
  To advise on site safety, inspect and approve site practice, prosecute in the event of injury or death as a result of non-compliance. To help to improve health and safety generally.
- *Police*
  To advise on security issues for the design and management of premises and areas.
- *Magistrates*
  To approve licensing arrangements.
- *Service Providers*
  For the supply and use of electricity, gas, water and telecommunications.
- *Home Office and Other Governmental Agencies*
  In connection with specific types of development such as law courts, prisons and defence installations.
- *Heritage Organisations*
  For advice and approvals about development in sensitive locations or for buildings of national or historic significance.
- *Insurance Companies*
  For advice about protection of buildings and their contents, particularly from theft and fire damage.

Contractors
A contractor is a person who undertakes or contracts to provide the materials, equipment (plant) and labours required to construct the new building. The contractors personnel may include:

- *Managers*
  Directors, contract or project managers, site agents or general foremen and trade foremen.
- *Finance and administration*
  Estimators, programme planners, buyers and valuation surveyors.
- *Site engineers*
  For setting out and dimensional co-ordination.

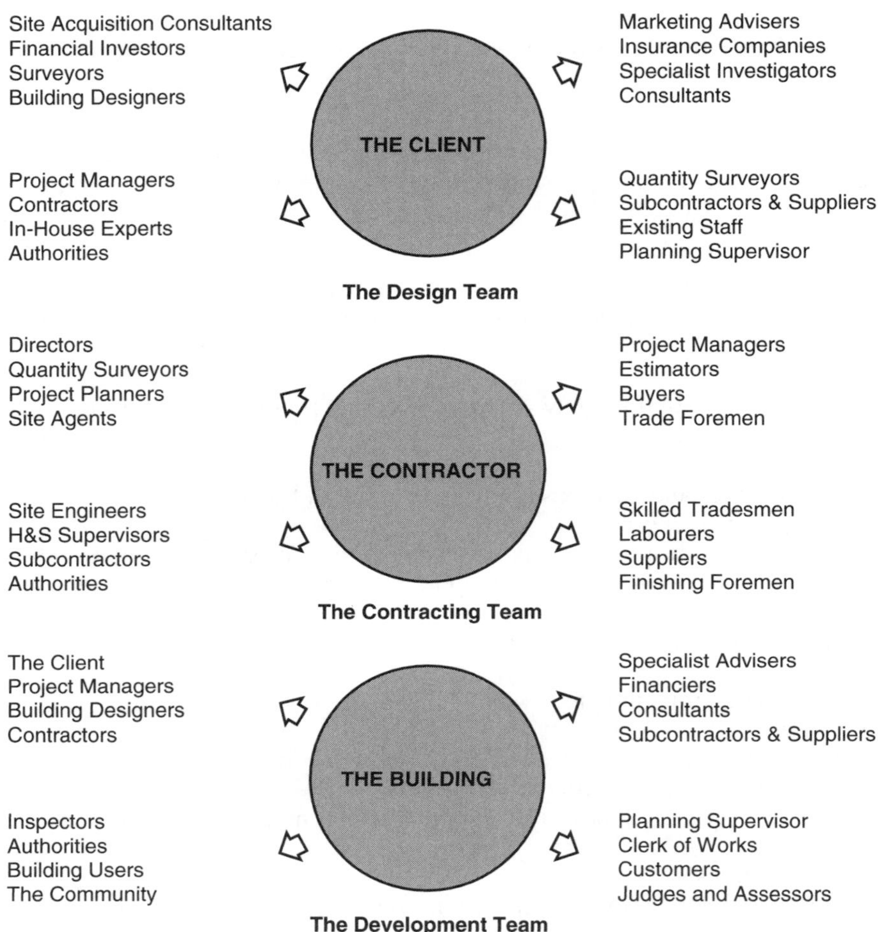

Figure 2.3 The development teams.

- *Operatives*
  Trades people (bricklayers, carpenters, joiners, plumbers, etc.), labourers.
- *Subcontractors*
  Subcontractors are independent companies employed for the duration of their work on the specific project for specialist trades such as roofing, steel frames and cladding.

As the processes of design and construction are considered in later chapters, reference is made to the development team, the design team and the contracting team. These descriptions as shown diagrammatically in Figure 2.3 are used to indicate collective responsibility for certain areas of work depending on the client's chosen method of procurement. The *development team* is used to refer to everyone involved with the project from start to finish. The *design team* is used to describe those responsible for the design and preparation of construction information and the *contracting team* undertake construction on site. For design and build and project management methods, these two teams are combined into one, whereas in the traditional method they are separate.

2.13 An outline of the design and construction process

As we described in the previous chapter, the procurement of new buildings takes place in a variety of alternative ways, including:

- *Speculatively*
  By a developer on an entrepreneurial basis for unknown clients or occupants who have no specific involvement in the design or construction processes.
- *Design and build*
  By a developer who provides all necessary design and construction expertise to a specific client based on negotiation or competitive tender.
- *Project management*
  By a manager who does not have any direct, personal involvement in the building works, but co-ordinates the design and construction teams on behalf of a specific client.
- *Consultant management*
  By a contractor for a specific client based on information provided by consultants, who may also supervise construction.
- *Private finance initiative*
  By a contractor or a consortium of contractors who finance the development for a specific client who subsequently rents or leases back the building when it is completed. This method may also include a facilities management contract to maintain the building in use for an agreed period of time.
- *Partnering*
  By several separate companies, or client and developer together, combining their expertise and financial resources in a joint venture.

The building designer's involvement varies with each type of development depending on defined levels of responsibility required by the client. Figure 2.4 illustrates the main differences between the *design and build* and the *traditional* procurement processes. The traditional stages in the creation of a new building can be defined as **pre-contract** prior to commencement of work on site, describing the building in theoretical terms and **post-contract** once construction has commenced.

The pre-contract stage includes the following:

- *Inception*
  Meeting the client, receiving the client's brief, starting to collect survey information, initial design ideas and programming the design period.
- *Feasibility*
  Formulating the design brief, including contributions from all the consultants, considering basic options.
- *Outline proposals*
  Establishing a concept in principle from the design brief requirements, obtaining outline advice from interested authorities.

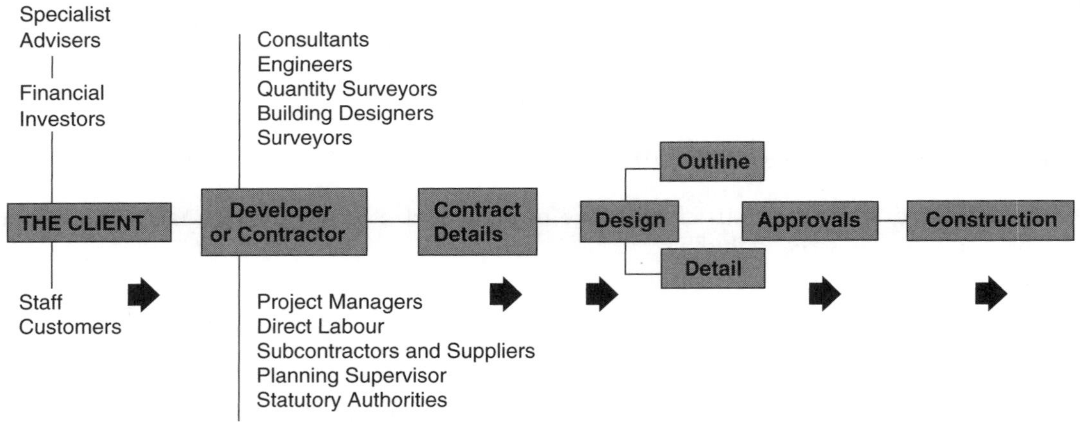

(a) The Design and Build Procurement Process

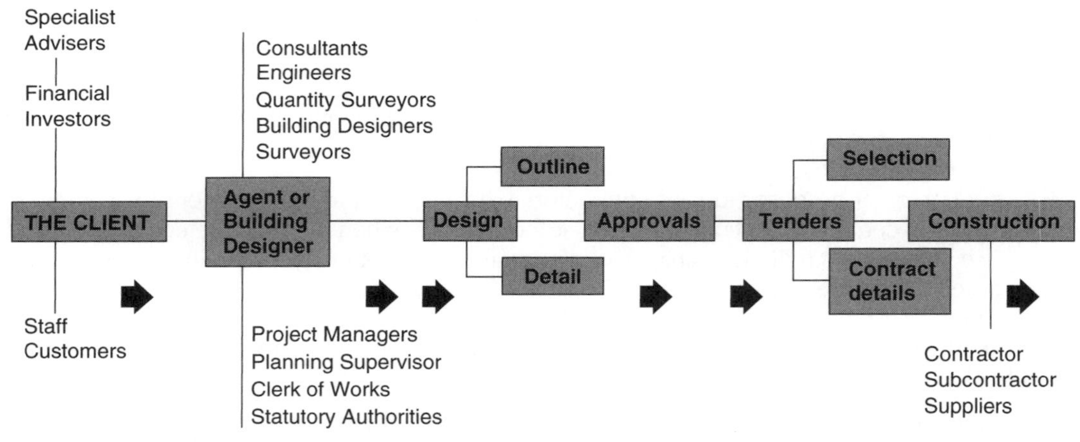

(b) The Traditional procurement process

Figure 2.4 Procurement processes.

- *Scheme design*
  Developing an agreed idea into a coherent working proposition, obtaining approvals from interested authorities.
- *Detail design*
  Fully developing the idea, incorporating specialist design work for structures, electrical and heating installations, etc.
- *Construction information*
  Detailed working drawings and specifications defining all the elements in the new building.
- *Measurement*
  Preparation of bills of quantity with numerical measurement of all the materials and labours required to construct the new building.

- *Tendering arrangements*
  Obtaining competitive prices from a number of selected contractors.
- *Pre-contract planning*
  Analysis of tenders and exchange of contracts between the client and the successful contractor to construct the building as designed; confirmation of construction methodology and construction programming.

The post-contract stage includes the following:

- *Construction on site*
  Supervision, inspections, approvals and valuations.
- *Completion*
  Handover to client and user occupation, correction of defects, completion of contract requirements and settlement of the final account.
- *Feedback*
  Lessons for the next project and for the future.

2.14 Discussion points

(1) Is design about 'sparks' of ideas or patient methodology? Why should building designers challenge the notion that 'the customer knows best'? Why do different designers think of different solutions?
(2) Will increasing labour shortages and use of prefabrication lead to standardisation of new buildings? Will standardisation eliminate the need for 'one-off' building design? Will factory production lead to 'perfect' buildings which never wear out?
(3) What criteria could or should be used for judging the quality of the built environment? Who is entitled to judge the quality of a completed building? How should building designers and developers respond to public opinion?
(4) What is the value of innovation in terms of user benefit? What are the main barriers to innovation in the construction industry? Does design guidance stifle good design and limit expectations?

2.15 Further reading

ACPO: Association of Chief Police Officers. *Secured by Design.* www.securedbydesign.com
Allen E (1995) *How Buildings Work: The Natural Order of Architecture.* London: Oxford University Press.
Allinson K (1997) *Getting There by Design: An Architects Guide to Project and Design Management.* Oxford: Architectural Press.
Baxter M (1999) *Product Design.* Cheltenham: Stanley Thornes.
BBC History of Architecture. www.bbc.co.uk/history
Brett P (1997) *An Illustrated Dictionary of Building: A Reference Guide for Practitioners and Students.* Oxford: Butterworth-Heinemann.
Brunskill R (2000) *Vernacular Architecture: An Illustrated Handbook.* London: Faber and Faber.
Brunskill R (2004) *Traditional Buildings of Britain.* London: Cassell & Crawley.

Building Research Establishment: Modern Methods of Construction. (www.bre.co.uk)
Burden E (2005) *Illustrated Dictionary of Building Design and Construction.* London: McGraw Hill.
Buzan T *Mind Mapping.* www.mind-map.com/EN/people/tony_buzan
Ching F (1997) *A Visual Dictionary of Architecture.* London: Wiley and Sons.
Clarkson J (2000) *Born to be Riled.* London: BBC.
Construction Industry Council *Design Quality Indicator (DQI).* www.dqi.org.uk/DQI
Curl JS (1993) *Encyclopaedia of Architectural Terms.* London: Donhead.
Curl JS (2000) *A Dictionary of Architecture.* Oxford: OU Press.
Ghirardo D (1996) *Architecture After Modernism.* London: Thames and Hudson.
Gordon D, **Stubbs** S and **McDonald** B (1991) *How Architecture Works.* London: Van Nostrand Reinhold.
Gray C and **Hughes** W (2001) *Building Design Management.* Oxford: Butterworth-Heinemann.
Green R (2001) *The Architect's Guide to Running a Job.* Oxford: Architectural Press.
Health and Safety Executive *Designers Still have a Long Way to Go.* www.hse.gov.uk/press/2004/e04052.htm
Jenks M and **Dempsey** N (eds) (2005) *Future Forms and Design for Sustainable Cities.* Oxford: Architectural Press.
Jodido P (2001) *New Forms: Architecture in the 1990s.* London: Taschen.
Jodido P (2005) *Architecture Now.* London: Taschen.
Lawson B (1997) *How Designers Think.* 3rd Edn. Oxford: Architectural Press.
Lawson B (2004) *What Designers Know.* Oxford: Architectural Press.
McCloud K (1999) *Grand Designs: Series 1.* London: Channel 4 Books. www.channel4.com/4homes/ontv/grand-designs
Newman O (1972) *Defensible Space: People and Design in the Violent City.* London: Architectural Press.
RIBA: All about architecture. www.architecture.com
Robbins E (1997) *Why Architects Draw.* London: MIT Press Secured by Design.
Rogers R (1991) *Architecture: A Modern View.* London: Thames and Hudson.
Smithies KW (1981) *Principles of Design in Architecture.* London: Van Nostrand Reinhold.
Summerson J (1996) *The Classical Language of Architecture.* London: Thames and Hudson.

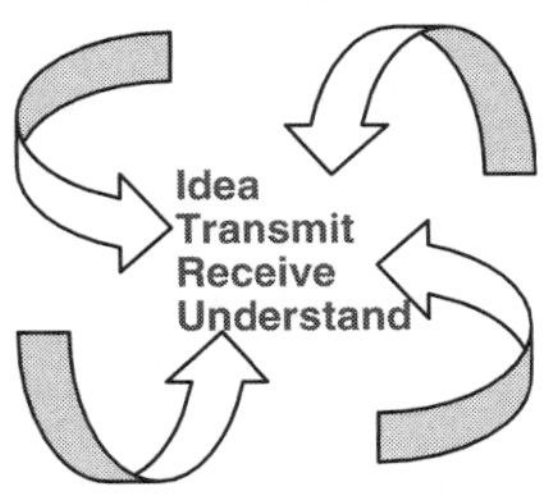

3

Communication

3.1 Introduction

Professionalism described in the previous chapter relies on forming successful working relationships, a two-way process of communication between people, building up confidence in each others ability to understand what is happening. Building design, and later construction are not abstract processes undertaken in individual isolation. They are both reliant on a collective approach to developing and managing information. At various times throughout the lifespan of the project, ideas are discussed informally or presented formally for consideration. Documents are issued to interested parties recording outcome for guidance or future reference. Some are routine, may be of limited significance, but others are important, marking key points along the way as the new building materialises, firstly on paper and then on the ground.

Communication skills are at the heart of all professional behaviour, essential for clarity and accuracy of thought and action. For the building designer, good communication can be persuasive too, encouraging support for proposals as they are presented for approval. The extent and quality of personal organisation often instils confidence, conferring authority in any given situation. Good writing, attractive drawings and articulate presentations can make a big difference to the way in which ideas are received and response is given. This chapter covers the principles of traditional and electronic letter and report writing, and outlines the way that construction drawings are used to explain what the new building will look like and how it will be constructed. There is guidance on preparing for and giving verbal presentations to an audience, and suggestions on how to organise and manage meetings, reviewing how information is disseminated and retrieved.

3.2 Exchanging information

Building design and construction are activities which involve presentation and manipulation of information and ideas as the processes move forward, usually referred to as 'progress'.

Individuals and groups frequently need to exchange information, ask each other questions, provide answers, make suggestions or proposals, persuade each other to adopt courses of action, confirm requirements and issue instructions. They can do this by:

- ***Talking***
  Speaking to each other face to face, informally between individuals or formally in group meetings, using telephone or video linkup for conferencing.
- ***Writing***
  Confirming information using letters, faxes, e-mail, reports, meeting notes, minutes, specifications and through intranet and extranet arrangements using web pages.
- ***Using images***
  Explaining ideas using 2D and 3D drawings, static models and computer-generated 3D-animated models, traditional and digital photography, and PowerPoint slide presentations.

In whichever way individuals in the development team exchange information with one another, the outcome is likely to have an affect on the others who were not directly involved in the exchange. Consequently many of the discussions and agreements made between individuals must be made available to others, or even to everyone. From the beginning of the project, it is useful to have an understood and agreed strategy for:

- collecting relevant information and confirming its accuracy;
- presenting ideas to one another in order to explain what may be possible and to secure necessary approvals;
- establishing a reliable system for the management and distribution of information, so that agreed ideas are understood as a common basis for proceeding.

To be successful, communication must be understood by recipients. Simply passing information from one point to another may not be sufficient if the form, style and content is not appropriate in the circumstances to convey the necessary meaning. Poor communication may be of little value, counterproductive, misleading, confusing, negligent or even dangerous. The form, content and the relative importance of communication will depend on its purpose at the time. For example, informal discussions between the consultants exploring possibilities permits flexibility as design work develops, but confirmation of specific data such as required floor loadings must be formally agreed and circulated as the basis for further detailed design work in order to avoid wasted effort in the event of subsequent misunderstanding. Rough sketches of ideas can be useful for informal discussion, but the professional quality of presentation drawings can influence the way in which ideas are judged. A good idea may be lost on a poor drawing. Equally of course, a poor idea can be disguised by a good drawing.

Verbal instructions to the bricklayers on site to change the mix of the mortar may be neglected or implementation could lead to a construction failure. Responsibility for either eventuality cannot be determined on the strength of 'who said what to whom' at some indeterminate time in the past. Instructions should always be confirmed in writing and ideally, illustrated on drawings. Instructions demanding attention and action must be supported by documentation confirming agreement of what was actually said, decided or required. Recording and archiving this kind of communication is an important part of the process of creating a new building which may have legal significance later on in the event of disputes.

Communication is also about personal presentation, style and authority. Effective speech, writing and drawing is needed to lead and to listen, to persuade and convince, and to advise and instruct. Communication skills influence the way that decisions are taken and how the work of others is co-ordinated, monitored and approved. Good communication creates confidence which is an essential ingredient in any team activity.

3.3 Written communication

As well as creating a record of exchanges, communicating ideas in writing formalises a view of issues which others can study at their leisure. The writer's aim should be to create a record of information as it arises, confirming questions, answers and statements, establishing reference points about the course of action being taken or which could or should be taken as work proceeds. Written communication as a follow up to conversation, gives recipients the chance to consider the issues much more carefully and the opportunity to respond after due reflection. Subject matter can range from a single issue to a description of aspects, or even the whole of the project. Confirming decisions in this way is important as it:

- reminds the writer and the recipient about things which have been said, requested or decided;
- guides the work of others so that the basis for their contribution is clear;
- confirms the need for others to respond if progress is to be maintained as planned;
- transfers responsibility for decisions onto those who have taken them;
- formalises concerns or reservations which may have future implications.

The two-way nature of communication referred to previously means that although the writer may have a clear idea of what they wish to say, the reader's understanding of written communication is likely to depend on:

- them having the time to read the document fully;
- them having the inclination to study the document in order to extract significant meaning;
- their degree of familiarity with the language used, including technical jargon;
- them possessing the knowledge and experience to interpret or fill in missing gaps.

For example, in describing the intricacies of alternative heating systems, the heating engineer will be able to understand all the technical details in conjunction with construction, air changes, 'U' values and applicable legislation, whereas the client may only be interested in the cost. The client's building maintenance manager will need to know how the system works, but the managing director's only interest is that it *does* work. Therefore, the content of written communication to each person must be adjusted so that the information which they require is clear and not obscured by incomprehensible or essentially irrelevant detail.

Consider too that in conversation, expressions of desire, humour or regret can be adjusted or balanced in response to the attitude of the listener, but there is no immediate opportunity to do this when the comments are made in writing. The recipient can only read the words and interpret, or misinterpret them at face value. When writing any document, the following points will help the writer to express themselves and assist the recipient's understanding.

- *Construct the content of the document in a logical order*
  Before starting to write the document, make a rough list of the issues or topics of interest, writing down all the points that might be included without worrying about the structure. Once the possible content is comprehensive, highlight the important issues and group associated points together. Concentrate on explanation of the most important issues first and if they lead on to other issues or are linked to them, organise them consecutively adding supporting detail as necessary.
- *Keep points in paragraphs*
  Deal with one point separately and comprehensively before starting to describe or discuss the next one. Including completely unrelated points in the same paragraph can be very confusing for readers, and can lead to an impression that the writer does not know what they are talking about and therefore, the content is not worth reading.
- *Be pertinent, brief, simple and clear*
  Be certain about the reasons for writing the document and focus on the content ensuring that it addresses the subject(s) or answers the question(s) as concisely as possible. Create an outline structure of the whole document before starting to write but précis the content down to its basic form. Check the whole document, each paragraph and each sentence, and eliminate unnecessary duplication, elaboration and over-complication. Read each statement carefully and make sure that the intended meaning is obvious.
- *Think carefully about the readers*
  Imagine being the reader and think how they might receive the message. What they are looking for or might do with the information, may help to decide on style and content, on what the report should or may be should not contain.
- *Avoid the use of technical terms or jargon*
  Select words and language which the reader can be expected to understand. The use of slang descriptions like 'brickie', 'chippy' or 'sparks' may be meaningless to a lay person. Always state terms in full the first time they appear before using acronyms. The use of BCO for example is confusing, but easily avoided by writing Building Control Officer (BCO) the first time, then BCO thereafter.
- *Maintain a consistent style in describing the writer throughout the document*
  For example:
  I am going to carry out a site survey (the writer), or
  we are going to carry out a site survey (on behalf of the writer's company), or
  a site survey (it) **is** going to be carried out (no definition of the who is the surveyor).
  The use of 'it', the third person singular implies impartial objectivity but does not indicate who is responsible for taking action.
- *Adopt an appropriate tone*
  Choose a style of writing to suit the recipient. Decide on active/passive, formal/informal, unconcerned/sympathetic. Be accurate and assertive but not aggressive or rude.
- *Be specific*
  Words like appear, seem, perhaps and maybe can be confusing and unhelpful to the reader. If there are doubts about issues, explain what they are without unduly raising false hopes. Decide what you want to say and be clear about your key messages. If you do not understand what they are, it is probable that your readers will also be confused. Make sure that you do sufficient research to fill missing gaps in information.
- *Distinguish between subjective and objective observations*
  Clearly separate indisputable facts from personal perceptions, views or observations, and ensure that factual statements can be supported with evidence.

- *Avoid sweeping generalisations and clichés*
  Grouping all the possibilities of outcome into simple statements can be deceptive and even wrong. For example, statements such as 'it will be alright to construct the new roof in the summer because it never rains in July', or describing a particular builder as 'the best thing since sliced bread' are both meaningless and incorrect. Also avoid conclusions based on limited or skewed evidence (biased data).
- *Write out in rough first*
  Establish key messages using single words or short phrases. Ignore grammar, spelling and eloquent sentence construction, focussing initially on content and structure. This helps to clarify thoughts so that the final writing should be much easier to complete.
- *Repeat key messages*
  'Tell them what you are about to tell them, tell them and then remind them of what you've told them' is a familiar edict for verbal or written communication, as most people tend to remember more of the first and last things they are told.
- *Check the document carefully when it is completed*
  Before sending a document to anyone, check accuracy, style and meaning, and make sure that there are no spelling or grammatical errors in the text, or that any words have (*been*) missed out. Careful proof reading is essential as to many readers, the absence of *been* will be a source of irritation diverting their attention from the content. It can be useful to let someone else check the document for obvious errors before sending it off.
- *Ensure that free handwriting is legible*
  Do not assume that because you can read your own handwriting that everyone else will be able to do so. It can be very frustrating wasting time trying to interpret scribbles and can often lead to dangerous misunderstandings.
- *Make documents look interesting*
  Help readers to access material by using highlighted paragraph headings, bullet points, numbered lists, small chunks of text, good graphics, diagrams and pictures, which can all brighten up unrelieved blocks of words.

3.4 Informal and formal writing

The way in which ideas and information are transmitted between people continues to change. Since the first edition of this book the use of electronic mailing as a way of making rapid contact and exchange has become more or less universal, and the use of mobile phone texting is developing. Communication by both of these methods is really equivalent to scribbling a quick note, and as such can be seen as 'informal' in the sense of taking care about the quality of writing. Poor grammar, spelling errors and abbreviations are relatively unimportant to the understanding and value to the recipients. Mobile phone texting at present, is essentially for personal use, but e-mail has become a significant working tool for most businesses. As an immediate link to wherever recipients are located, there are clearly valuable management benefits to commercial organisations, although storage and retrieval of information transmitted in these ways is not always reliable. The development of extranets and intranets is discussed in Section 3.7.

The traditional use of letter and report writing can be seen as 'formal' ways of communicating which do have the merit of creating a permanent document for subsequent reference. Letters are normally used to communicate short statements, usually best limited to a single page of text so that the message can be read and understood by the recipient as quickly as possible. This is an important consideration in a commercial context, particularly at a senior management level when time is likely to be at a premium. A lengthy, badly organised letter may not receive the attention it actually deserves. If further explanation of the subject is necessary, it can be presented separately as an appendix or in the form of a report with a brief outline of its content and relevant recommendations in a covering letter.

A typical business letter is as follows:

- *You or your company name and address*
- *The date of writing*
- *The name and address of the company to whom you are writing*
- *The name of the person to whom the content of the letter is targeted*
- *Opening greeting*
  Dear Sir, Sirs, Mr, Madam, Ms or Miss
  or Dear Dr Blake (using a title)
  or Dear Ron

The choice of greeting depends on the degree of familiarity that exists between writer and recipient. A friend might be addressed informally as Ron, but a working colleague more professionally as Dr Blake. Some organisations take the view that correspondence is from or to their most senior manager, and should not be addressed to individual employees. For example, a letter to the Planning Authority should be formally addressed to the Chief Planning Officer, who need not be named, but for the attention of Mr Jones, referring to dealings with him as the officer directly concerned with the project.

- *The purpose of the letter:*
  Re: New Car-dealership at Old Mills Lane
- *The content:*
  I am writing to you to in connection with observations made by your Mr Jones about some issues of concern in connection with the Planning Application. The most important matter is the location of the paint spray shop in relation to the neighbouring property.
  I appreciate his concern and am looking at alternatives. If you are in agreement, I can alter the layout at the rear of the site so that the access does not face the boundary. I understand that this would satisfy his objection. I have suggested this to Mr Jones, who is happy to agree to this change as a minor amendment to the original proposals.
  The other comments raised are matters of detail which I can confirm have already been satisfied and are included on the revised drawings. I enclose a separate summary of all the details referred to in Mr Jones letter.
- *Closure*
  Yours Faithfully (if the opening is a formal title)
  Yours Sincerely (if the opening is a name)
- *Signature*
  Gavin Tunstall
- *Name, in case signature is illegible:*
  Gavin Tunstall
  (including title or position in the organisation, if relevant)

In some instances the content of the letter may be asking for a relatively short response, as for example confirming the date of an arranged appointment. In such cases, the reply can be handwritten on a photocopy of the letter and returned to the sender. This saves the time and effort of constructing a new letter and keeps the relevant points together for easy reference. In this respect, e-mail exchanges create a running record of correspondence about issues which can be very useful if retained.

If the message is longer than a few simple points suitable for a letter, it may be necessary to produce a report enabling the author to explain complicated issues to a reader(s) in a form which gives them time to fully understand them. Unlike a letter, a report is not designed to elicit a rapid response. A report enables a writer to describe, summarise or discuss subject matter at length, and to include supporting analysis, charts, tables, graphs, etc. In business terms, it has the purpose of explaining or justifying the reasons for taking a certain course of action, offering conclusions or recommendations. As a prompt to taking decisions or the need for further action, it is this element of information which the recipient may need first, and it can be useful to present the outcome at the beginning. The detailed arguments can then be studied at length as and when desired.

A typical report is as follows:

- *You or your company name and address*
- *Title of report: Development at Old Mills Lane*
- *Date*
- *Summary of purpose*
  It is useful to describe briefly the purpose of the report or why it has been produced.
- *Contents*
  Indicate chapter headings, sections, paragraphs, photographs, drawings, charts, etc.
- *Executive summary of conclusions*
  A précis or outline of opinions, recommendations, proposals or outcomes.
- *Content*
  The body of information should be carefully constructed in the form of a logical progression or argument through the issues which are being raised. Points of information should be clear and identifiable, structured in numbered or headed paragraphs. Long, unbroken sentences and paragraphs can be very difficult to read and intended meaning easily lost.
- *Appendices*
  Include supporting documentation if not included with the text.
- *Conclusions*
  Summarise the outcomes if not given at the beginning.
- *Bibliography*
  A list of reference sources and acknowledgements where existing publications have been used to support argument.
- *Glossary of terms*
  An explanation of technical words, phrases and abbreviations, if necessary, to help the reader to understand terminology. For a highly technical document, this could appear at the beginning of the report.
- *Index*
  A detailed list locating key words and phrases, usually only for major works such as reference books.

As well as the qualities of the writing itself explained at the beginning of this chapter, the quality of the presentation of the report can have a bearing on its value. The general layout and appearance of the material in the report can influence the way that it is received by its readers. They may be impressed by its style, accuracy and neatness, by the organisation and numbering of its paragraphs, by the attractiveness of its drawings, photographs, charts and graphs. They may equally be put off by its scruffiness, chaotic organisation and tatty illustrations.

A report should be bound and presented with a titled front cover. Pages should be numbered and supporting illustrations cross-referenced to the text to which they belong. The use of colour should be considered not only for clarification but because it creates additional interest, making the reader more receptive to the ideas in the presentation. The writer should always remember that any issued document represents the individual or company who has prepared it, and the recipient will form an opinion about their ability from its content and appearance.

3.5 Drawings

It would be difficult and complicated to attempt to fully describe the design proposals in words. Elements of construction and selection of materials can be defined with written specifications (see Section 11.4), but explanation of layout, appearance and the assembly of elements is more easily understood when it is shown visually. In the construction industry, visual ideas are communicated to other members of the development team on 2D and 3D drawings, prepared mechanically by hand (using a drawing board with tee square and set square or parallel motion) or electronically with computers (computer-aided drawing or design, CAD). Helpful annotations can be added to drawings, but without careful control, they can easily become confusing.

The majority of drawings used in the construction industry are *technical drawings*, drawn to scale using the metric system of dimensioning, to show how parts of buildings fit together, enabling accurate measurement of dimensions and quantities. Technical drawings are described as being 'mechanical' and are relatively easy to construct. For the building designer, it is useful to have some free-hand drawing ability, but there is no need to be deterred by the lack of 'artistic ability' as mechanical drawing skills can be learned, and an individual style can be developed through experience.

Drawings associated with design and construction communicate in two different ways:

(1) *As a general indication to show what a building will look like, or what being in spaces might feel like: experiential presentation drawings*
These drawings are intended to give an idea of what could be visible from a fixed viewpoint. They need not be totally accurate, showing only style or a feeling of reality and can include visual clues such as people, cars and trees to give a sense of place and scale. These drawings are used to explore design options without the complications of the detail needed for construction, and to explain proposals to people who either do not understand technical drawings, or who are not interested in constructional detail at this stage.

(2) *As a detailed explanation of how the building's parts are arranged and constructed: scaled working drawings*
These are the majority of the necessary drawings showing elements of construction. They are line drawings indicating how individual parts fit together accurately so that their size and number can be measured. They illustrate information in vertical and horizontal layers and can also be used to indicate elements which are not visible. It is very important to remember that with these drawings, every line has a meaning.

For the building designer, there are five levels of drawings, prepared and used at different stages:

Survey drawings: as existing
Design work for some projects may be done in the abstract, but generally design ideas are sketched against the constraints of a particular site; a plot of land or an existing building. The first drawings needed are measured-site plans, or floor plans, external elevations and vertical sections of existing buildings. Figure 3.1 is an example of typical survey information in drawing form.

Sketches or feasibility studies: proposals
The sketch of the front of the building illustrated in Figure 3.2 is rough and inaccurate and lacks dimensional definition, but is useful for investigating possibilities. Floor plans and sections may be developed to the same level to test the general feasibility of the idea.

Outline design drawings
The outline design drawing in Figure 3.3 progresses the idea for the front of the building in sufficient detail to indicate the main elements in principle. This is an accurate, scaled drawing of the selected idea developing the theme of the design although the other sides of the building may not all be illustrated at this stage. Floor plans and sections may be advanced in the same way.

Scheme design drawings
Figure 3.4 shows the appearance of the front of the building accurately, so that all the members of the design team have an agreed basis for developing subsequent detail. All the other sides of the building would be shown to the same level of detail at this stage along with fully defined floor plans and vertical sections.

Working drawings
Once the scheme has been accepted by the client and the relevant statutory authorities (see Chapter 4), a working drawing, as shown in Figure 3.5, is completed with exact dimensions and construction detailing, showing how all the parts of the building are to be put together. All the other sides of the building would be shown to the same level of detail at this stage. Site plans, floor plans, sections and individual construction details would all be developed to this level.

3.5.1 2D and 3D drawings

A full description of the techniques involved in preparing visual impressions and technical drawings is beyond the scope of this book. There are many ways in which the elements of the building can be illustrated using both manual and CAD. Traditionally, drawings were prepared using drawing boards with set squares and tee squares to enable horizontal and vertical lines to be drawn on paper using pencils and fine-nibbed ink pens. Draughting techniques were/are very much a matter of individual style. CAD has largely superceded manual drawing for many reasons, which are outlined later in the chapter. Whichever method is used at any particular time, the essential point to remember is that the purpose of any drawing is to communicate information at that stage in the design and construction process so that readers or viewers can understand it.

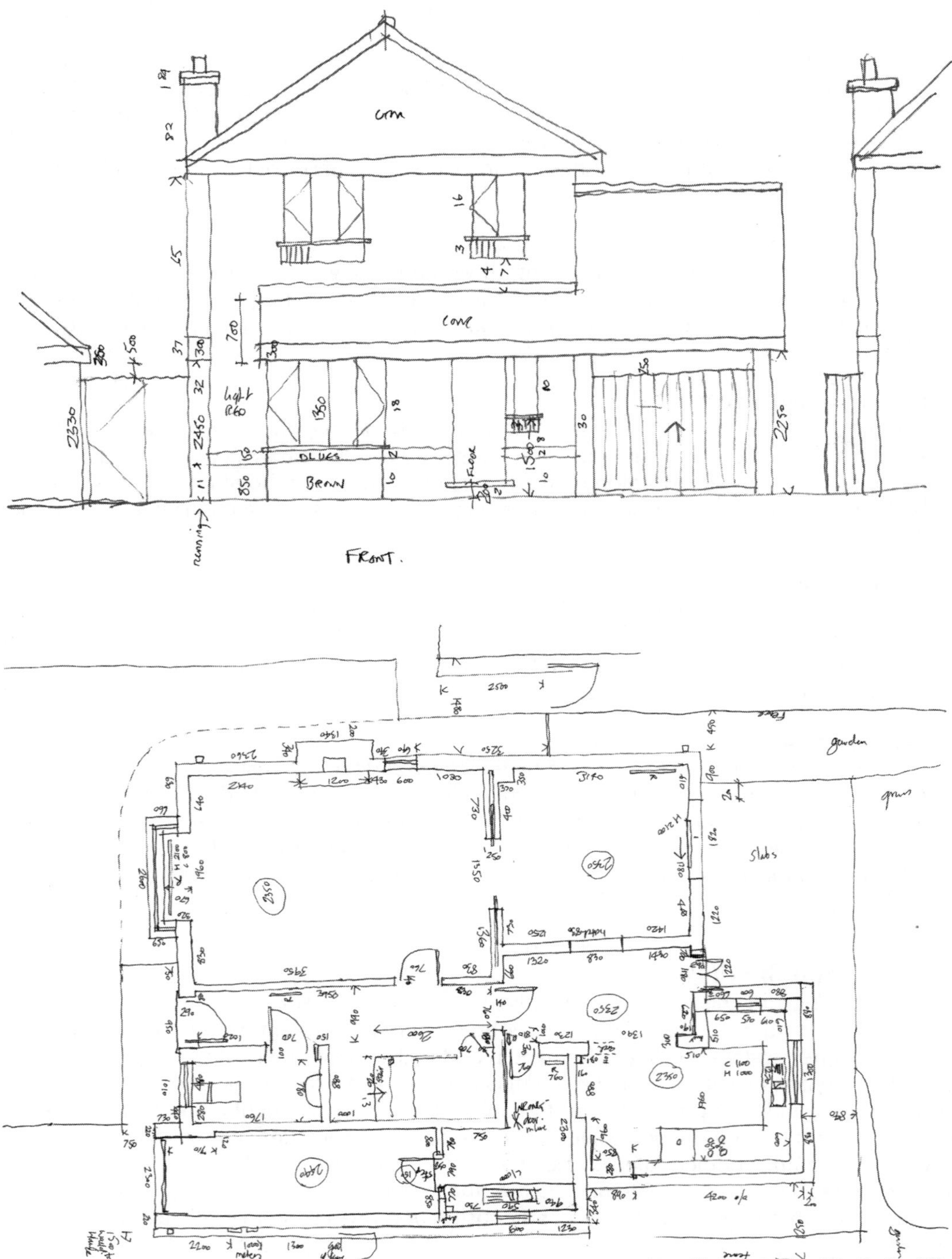

Figure 3.1 Survey drawings.

Figure 3.2 Sketch drawings.

Figure 3.3 Outline design drawings.

On most occasions, construction drawings are 2D illustrating information in layers. Selected elements are defined by lines and the drawing constructed so that all of the element or every part of the building is viewed at 90 degrees to the eye. Simplified drawings showing only those elements of the building that are visible help everyone to understand intensions. Typical 2D drawings take the following forms:

- *Plans: horizontal layers looking down or up*
  Plans are most commonly drawn to show site layouts and building floors, identifying and locating elements. Additional layers can be used to illustrate drainage layouts, services and pipework positions, landscaping and planting detail. Figure 3.6 shows a typical floor internal plan showing walls, door and window openings, ceilings and roofs, floor finishes, fixtures and fittings. Variations in level, such as the slopes on a pitched roof are 'ironed out' as if they were flat. Extra information can be added about hidden elements on layers underneath which are

Figure 3.4 Scheme design drawings.

not actually visible, such as the foundations underneath the ground floor walls, or the floor joists under the first floor decking.

- *Elevations: vertical views looking at the sides*
  Elevations are generally drawn to show the external appearance of the building, including wall surfaces, doors, windows and roof, illustrating what these elements would look like. Figure 3.6 shows a small section of the front of a building. For most buildings, there are four principle external elevations looking at the building from front, back and both sides. Internal walls can be drawn in elevation too where illustrations of design information is needed.
- *Sections: cuts through the building showing the insides*
  The cuts can be imagined either horizontally or vertically, depending on what information they are intended to show. They can illustrate elements which would normally be hidden like the construction inside a cavity wall or the way that a pipe is fixed in a void. A horizontal and

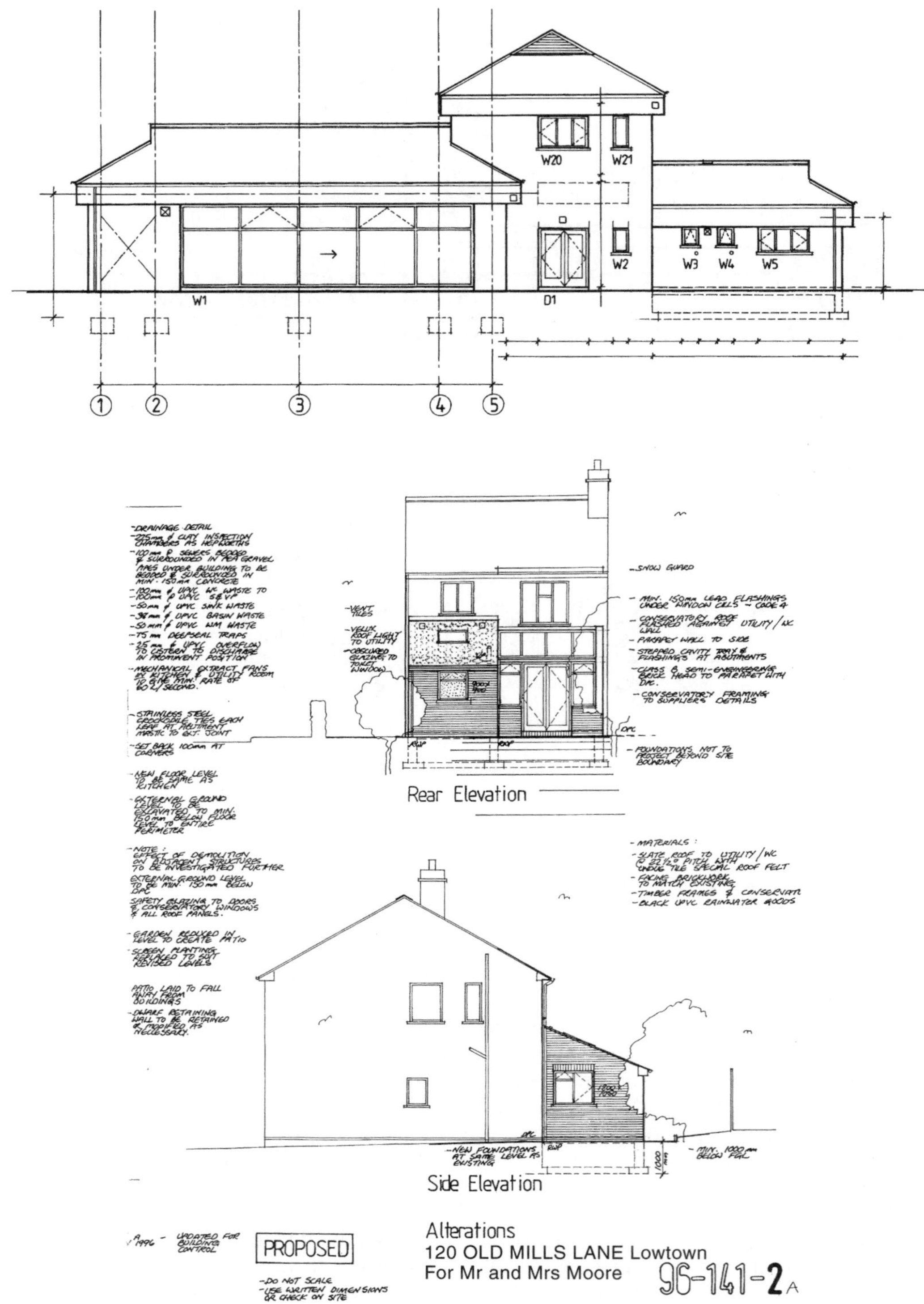

Figure 3.5 Working drawings.

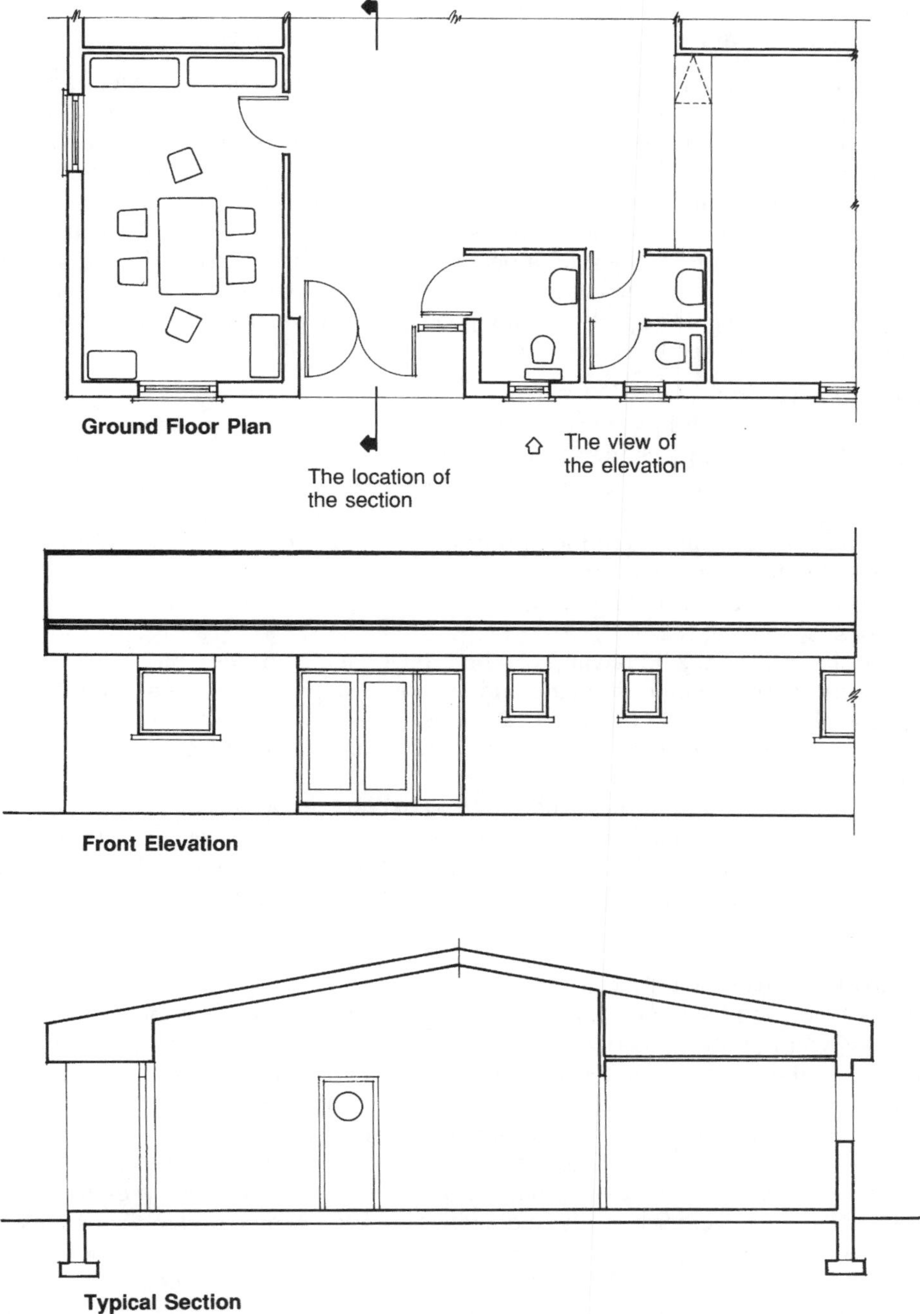

Figure 3.6 Plans, elevations and sections.

vertical section will show the construction of floors, walls and roof. Figure 3.6 includes a cross-section of the whole of a building from one side to the other.

The drawing of a roof plan and four elevations defines and explains the extent and nature of the external surfaces of a rectangular building. A floor plan, an elevation and a section of a building can be used to indicate all the surfaces, the internal volume and some of the elements of construction. Figure 3.6 shows a corner of a building only as a part explanation. A drawing of the whole building would need to be much larger than the space here permits.

Sometimes design intentions can be described more usefully on 3D drawings enabling easier appreciation of overall form and visual appearance without having to study and imagine how all the individual plans and elevations fit together. Such drawings give a more realistic impression of the building or its parts, advantageous to those unfamiliar with traditional technical drawings. There are three basic options:

(1) *Perspective drawings*
The natural view of building shown in Figure 3.7 is affected by perspective, and although representations can be mechanically drawn, they are laborious to construct and cannot be used for measurement due to the diminishing size of distant elements and surfaces. Constructing good perspective drawings requires considerable skill and without this, the results can be so appalling that the effort will be counterproductive. A poor drawing will divert attention from what may actually be a good design. However, a good perspective can be a powerful way of illustrating design intensions.

(2) *Projections*
Mechanically produced 2D drawings can be used to create images which have a 3D appearance. The most common are isometric and axonometric projections illustrated in Figure 3.8, which use the plans, elevations and sections referred to already as the basis for their construction. They are relatively easy to draw and retain the advantage of being to scale.

(3) *Enhanced 2D views*
2D plans, elevations or sections can be used as drawn, but by adding visual clues, a feeling of 3D can be created. Figure 3.9 shows an enhanced front elevation. Projections and enhancement both have the value of economy, as the same conceptual design drawings can be developed as a framework for construction by the addition of further detail information in due course.

3.5.2 Presentation drawings

The methods employed by individuals in the design team as they work their way through the design process will not always be exactly the same, but their aim is to create and describe the new building. As the design team consider and discuss ideas, they will produce many sketches, scribbles and doodles, and exchange them as a way of rationalising their thoughts in the search for solutions. From time to time, all the rough work must be co-ordinated and translated into coherent drawings, summarising the design development at any particular stage, illustrating the whole building, or typical parts of it so that the proposals can be formally presented to anyone with an interest in the development. These drawings are usually prepared in the first instance for the client, and subsequently for the Planning Authority and other statutory bodies who are required to approve the design. Presentation drawings are sometimes described as 'finished' drawings in the sense that they are complete, showing specific intentions rather than the many options which may have been considered on the rough sketches. Rough sketches are essential in the first place because it is not possible to design and produce finished drawings at the same

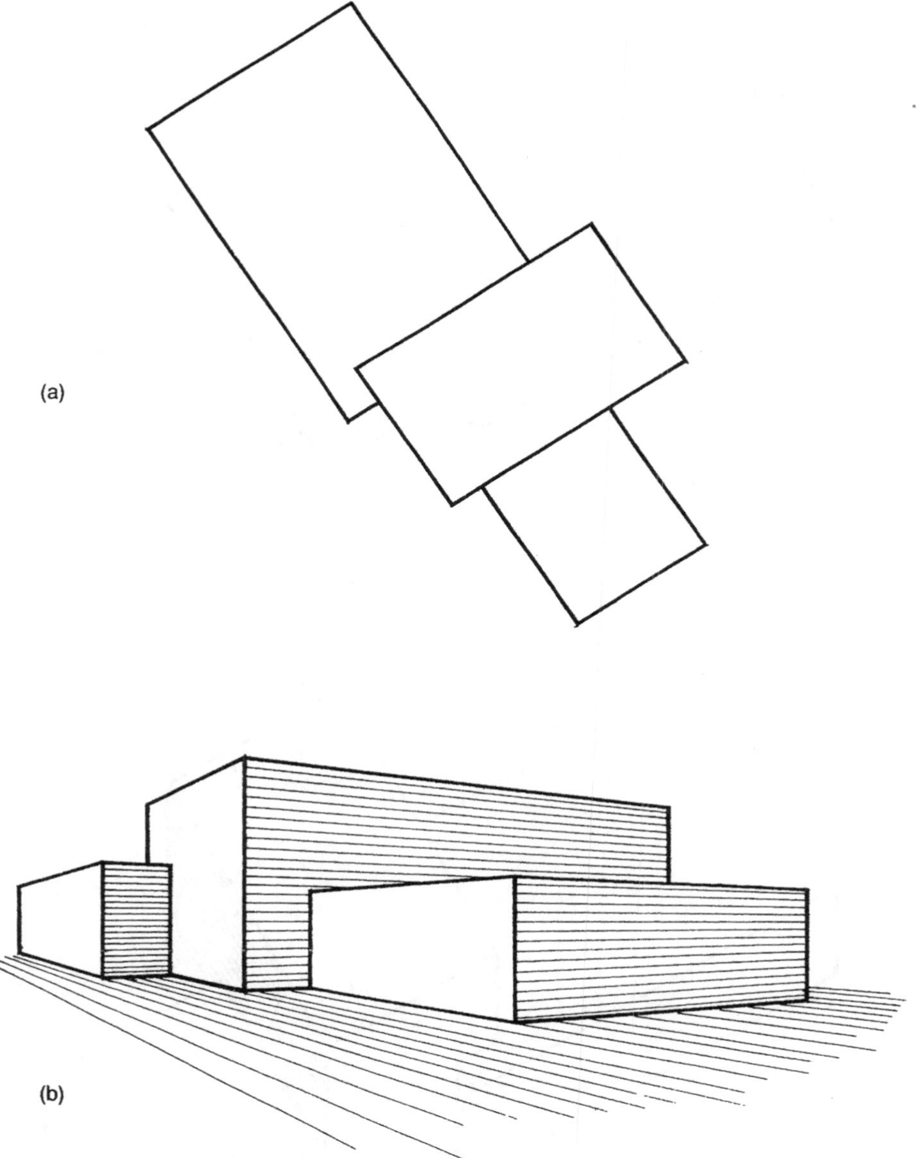

Figure 3.7 A Perspective drawing. (a) The floor plan aligned with the angle of view. (b) The building as it would actually be seen.

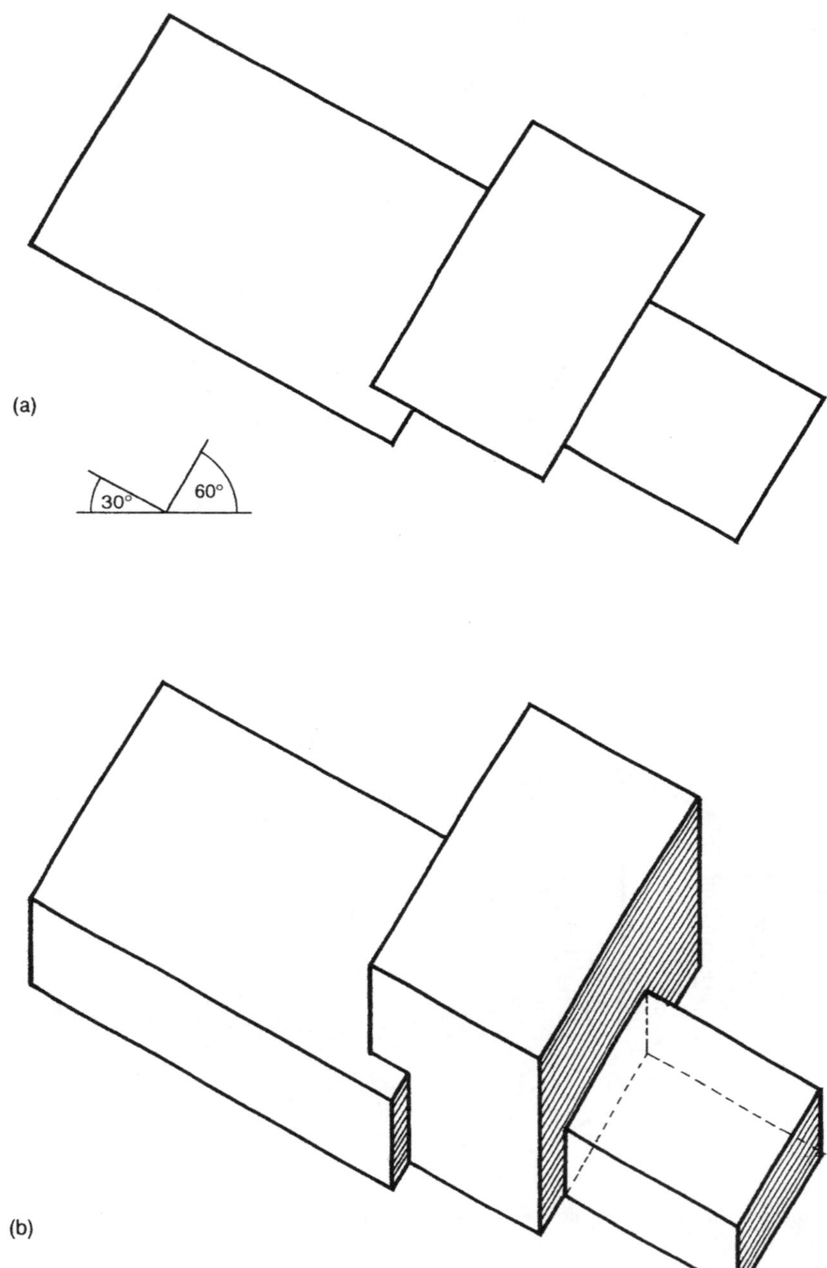

Figure 3.8 An axonometric projection. (a) The floor plan. (b) The 'hidden' construction lines are removed to create a 3D appearance.

Figure 3.9 An enhanced elevation.

time, but rough sketches will be incomplete, lacking sufficient clarity to provide the necessary level of explanation to others who are either not familiar with, or who do not need to know about the background thinking. As with written communication, readers can only extract meaning from the document as it appears in front of them. Letters, reports and presentation drawings will be scrutinised at leisure and must be complete if the intended meaning is to be clearly understood.

Presentation drawings serve two important purposes:

(1) *To co-ordinate a collection of design ideas into a coherent statement*
These drawings bring together a resolution of design input at a particular moment in time, helping members of the design development team to comprehend proposals so that they are satisfied of the viability of carrying on to the next stage.

(2) *To persuade or convince someone to give permission to carry on to the next stage*
Presentation drawings are needed to explain intentions to anyone with an interest in the proposals, to sell the merits of an idea so that they approve progression into the next stage. They may be persuaded of the design teams ability to understand the nature of the design problem and their capability of bringing a solution to fruition.

Preparing information in a presentation format is a skilled activity, and irrespective of the client's procurement arrangements it is likely that these drawings are key points in the whole process. These are the framework drawings which are essential to lead the work of the design team. For example, the client can only be sure that the building is as they would like it to be when all the spaces are defined, when all the walls, doors and windows are shown, and when access and external site planning can be seen and understood. The structural engineer cannot design foundations and the steel framework, or the services engineers position power, heating and lighting equipment until the layout and volume of the building is clarified. The statutory authorities cannot determine the appropriateness of proposals until they are properly illustrated. The information, requirements and constraints contributed by everyone in a variety of forms must eventually lead to finished, presentable drawings or some other visual device to communicate intentions.

To many clients, the quality and appearance of presentation material, particularly at the design stage is very important. Proposals presented in a way that they can understand give them confidence in the design proposal and engender enthusiasm and delight at the prospect of their new building actually being realised. Of even greater importance is that understanding gives them a much better opportunity to appreciate exactly what the design intentions are so that they can decide whether or not it is what they want. If the presentation drawings are too technical in content, or too much orientated towards construction detail, the design concepts can be obscured. The traditional plans, elevations and sections used by designers are fine for those with experience of *visualising* the building in this way but for many clients who are unfamiliar with technical drawings, initial presentation of ideas in this way will not be very helpful to them. Consequently there is a risk that although they agree to the proposals at any given point, they will want changes to be made later on when they actually realise the implications. Changes requested by the client can be incorporated as the proposals are developed on paper, but once construction has started on site changes become much more problematic. Of course, members of the design team may well be able to say that construction is exactly in accordance with design documentation, and in strict contractual terms this is an essential safeguard, but in the spirit of trying to give the client what they want (or what is most suitable for their actual needs) the onus must be on the designers to make sure that the client is fully aware of design intentions from the earliest possible time.

3.5.3 Presentation techniques

As well as the use of the basic types of drawings described above, there are a number of other techniques that can be used to help to explain ideas, particularly to those with little or no experience of understanding technical drawings, or who need to be made aware of ideas quickly.

Enhanced drawings

Technical drawings, created to communicate detail about construction, can be made easier to read if they are simplified or enhanced in a variety of ways. Titles, notes and dimensions should be added to minimise confusion, clearly distinguishing between the building and other lines on the drawing. The inclusion of visual clues, as illustrated in Figure 3.9, which would be expected in a photograph, such as people, cars, trees, planting and clouds, add touches of reality, and a 2D elevation can be brought to life by the adding shadows, as if the building were being viewed on a bright sunny day.

Collages and montages

Collages are compositions made by using cut and paste methods to assemble a number of existing images. Figure 3.10 shows pictures of equipment, cars and people assembled in a 2D composition to give an impression of reality in 3D. This is not a commonly used technique to illustrate the building itself but is often used by interior designers for instance, to present decorative colour schemes with proposals for furnishings and the fabrics for curtains and carpets. Pictures of sanitary goods, light fittings and ironmongery could be presented for consideration rather than bland, meaningless technical specifications. A montage adds 3D elements to the composition, introducing texture and depth, modelling some aspect of the design in an even more realistic manner.

Superimposed photography

This can be a very useful way to show what the new building will look like in its immediate environment. Sketches of the proposed building can be superimposed onto photographs of the

Figure 3.10 Ideas taken from existing car-dealerships in Nottingham to give a flavour of the building type.

site from various vantage points so that as far as is practicable the appearance of the building is shown as it really will be. Figure 3.11 is a typical example. The same photographs can be used by the designer or an artist to create perspective views of the new building in relation to the familiar surroundings which are easily understood. This can be a useful design tool, helping the designer to better understand possibilities and constraints. Aerial photographs are excellent for this purpose as the whole site can be seen at once. For a large development, or redevelopment of parts of an existing site, aerial photographs can be very useful at meetings so that all those present can clearly appreciate the areas of the site under discussion.

Static modelling

For complex buildings and site layouts it may be worth making or commissioning a card or timber model, constructed to a small scale. Not only can a model, such as the one illustrated in Figure 3.12, enable the client and the design team to better understand the layout, form and appearance of the building, but it can be photographed from different positions and angles to test and explain particular views. Realistic interior views can be created using a 'modelscope' tube, which is inserted into the model and attached to a camera. These photographs can also be used as the basis for producing drawings if needed, particularly if linked with a computer.

(a)

(b)

Figure 3.11 An example of superimposing a design idea onto a photograph to illustrate extensions by Lu Haoyang. (a) Existing elevation. (b) Artist impression of design idea.

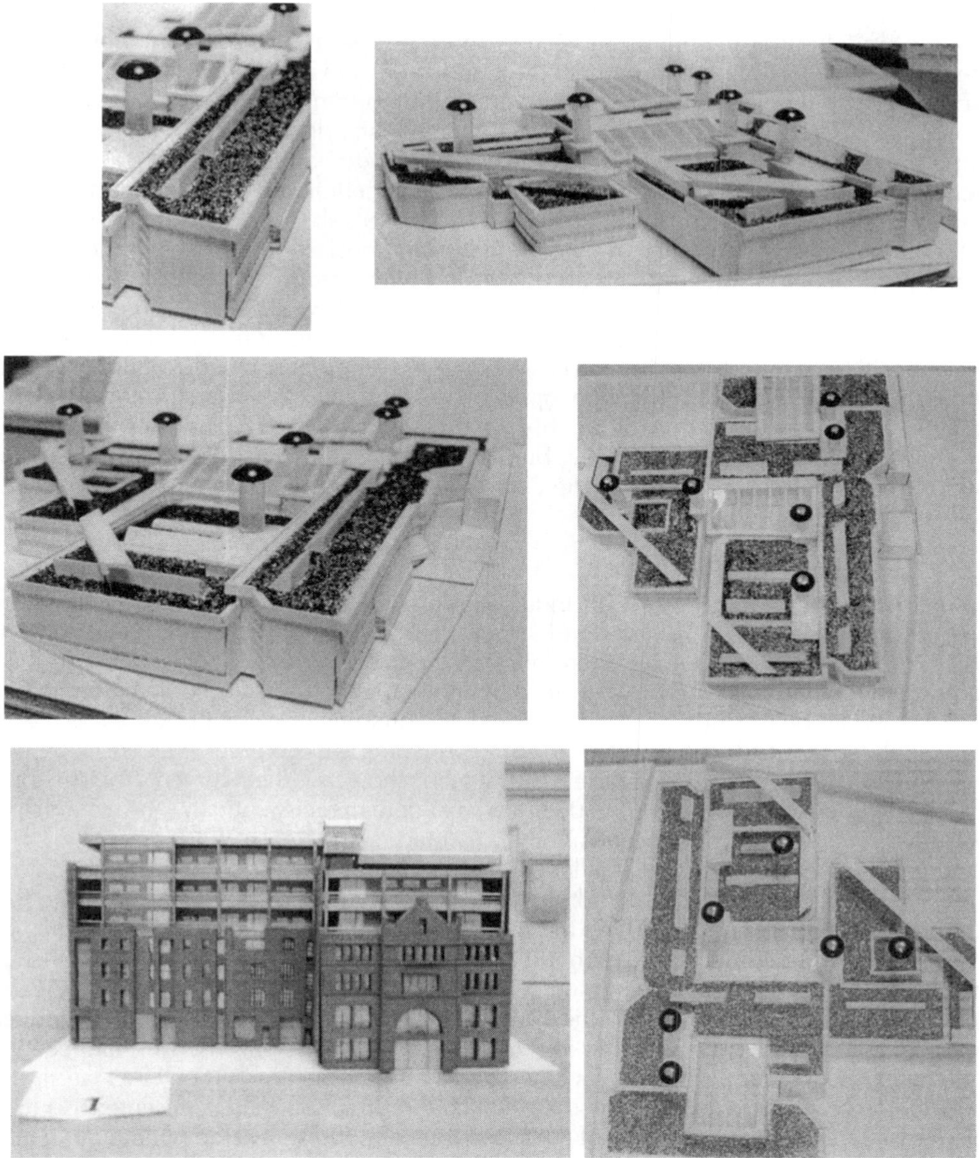

Figure 3.12 A simple model which can be viewed from various angles to get ideas about the design. Models by Chris Netton.

Computer-generated animated modelling

Computer modelling is becoming more and more accessible to designers and clients. 3D images of the new building can be generated to allow viewers to walk around and through the spaces as if they already existed. Any part of the building can be examined in this way, gaining a feel for the spaces and also improving understanding of how parts of the building are interlinked, or how service arrangements can be fitted into the structure. Computer modelling is important as a design tool as well as an excellent aid to presentation. The use of computers in this way helps the client to discuss changes to the design much more quickly than traditional drawings because they can understand design proposals so much better. The type and complexity of images that can be created with computers would be almost impossible to achieve by hand. For example, using a computer it would be possible within quite a short period of time to present the client with perspective views of the building taken from 10 different positions around the site, impressively and comprehensively explaining proposals. Until recently, this was not an option open to the designer as production of presentation material to this level could not reasonably have been offered. Perhaps over and above any other development in the way that new buildings are created, it is this facility that can lead to the fullest possible involvement of the client and all the other members of the design team in the design process.

3.5.4 Dimensions and scale

Measurement in the UK construction industry is metric: kilometres, metres, centimetres and millimetres for linear length, hectares and square metres for area and cubic metres or litres for volume. For most drawings, elements are normally represented to their 'real' size, so that they are proportionally correct when related to one another. This is described as being to scale. A scale of 1:1 would be the full size, the real size of the building or element. A scale of 1:2 would show everything reduced to half real size, but everything still correctly related in proportion. Scales of 1:5, 1:10, 1:20, 1:50 and 1:100 would show all the elements of the building progressively reduced in size down to 100 times smaller than reality, but always retaining relative proportions when compared with one another. As the scale increases from 1:1000 to 1:10,000 the elements become smaller and smaller as the area being viewed becomes larger and larger. So for example, at 1:1, a drawing could illustrate a door handle; a drawing at 1:50, the room in which the door is located; at 1:100, all the rooms in the building; at 1:500, the building on the site; at 1:1000, the site in its neighbourhood and at 1:10,000, the neighbourhood in the city.

These scales are recognised for communicating various levels of detail, traditionally based on manual technical drawing (in fact in the UK, the imperial system of measurement in feet and inches was only superceded by metric in the 1970s and may be still used in connection with 'old' drawings). CAD permits drawing at any scale at all, (see Section 3.6 on CAD later). The key point is that a scale should be selected so that the information being presented is clear and comprehensible. For example, a plan of the site layout at 1:500 shows the building in relation to the remainder of the site, but it would be difficult at this scale to show elements of construction such as cavity walls and column encasements (which would be too small to show at all). A construction section of a building's eaves could be fully detailed at a scale of 1:10, but it would be pointless to draw the whole building to this scale as most of it would be empty space.

As a general guide, the level of detail can be expanded by using the following hierarchy.

Location drawings

These drawings will be at the smallest scale showing the whole building in its context with plans, elevations and sections locating all the principle elements. For example, a scale of 1:1250 is commonly used to show the relationship of the building to its neighbours. A scale of 1:500 will show the building within the site and a scale of 1:100 or 1:50 will show the position of the walls and doors on plan, the position of the windows on elevation and the space between floors in section. These drawings can be used to refer to elements illustrated in greater detail on other drawings.

Component drawings

Components are separate parts identified on the location drawings which can be illustrated as complete units. Elements like windows, doors, staircases, kitchen units and door handles can be fully described independently of each other at an appropriate scale for their size and complexity, varying from 1:100, 1:10, 1:5, 1:2 or even 1:1 showing the information which is needed so that they can be valued, manufactured or ordered.

Assembly drawings

Assembly drawings are used to show how all the components are to be placed or secured together, to create finished elements of construction like roof eaves, window jambs and door thresholds. The scale of these drawings depends on the complexity of the arrangements being illustrated, but anything less than 1:50 is likely to be too small. Many manufacturers now make drawings available to designers through the Internet, which can be downloaded and used directly to show details of their product range, and their recommendations for use in typical circumstances.

Illustrations of each of these types of drawing are shown in Figures 11.1–11.4 in Chapter 11 when construction information is prepared.

3.5.5 Managing drawings

As ideas are developed, information is regularly exchanged between the members of the design team. Possibilities and constraints come to light, adding to or revising what had previously been agreed. The earlier drawings serve an important purpose at the time that they are prepared, but they can quickly become out of date. In this respect, it should be appreciated that 'drawings' are a flexible tool. For much of the design period, and even sometimes during the process of construction, they are a means to an end rather than the end itself, helping to establish the framework of ideas so that they can be created in reality. The importance to everyone involved is that agreed or approved amendments are incorporated onto existing drawings, and circulated to all the members of the design team so that they are all working together towards the same end.

Changes to drawings are called *revisions*, updating information to reflect proposals at that point in time. Clearly it is important that revised drawings are issued quickly to avoid confusion and wasted effort. Implementing revisions on drawings is a relatively simple task for those which are prepared electronically on computers, but traditional methods of writing and drawing on hard copy are much more difficult to update. Graphics and text can quickly be deleted, added or repositioned electronically, but there is a limit to the number of times that original paper drawings can be altered before they are ruined. The author of any drawing must take care to identify and protect the 'master' copy as additional information is added or alterations are made.

Sometimes drawings require so much alteration that they are replaced by completely new ones. The originals become redundant and should be marked '**SUPERCEDED**', and stored for reference only so that they are not used or issued in error. This problem can be managed by maintaining an up to date list or register which is regularly issued to all members of the design team. It is good practice to use the drawings register as an issue slip, so that the recipient can check that all the drawings in current use are up to date. They can then also decide if other drawings, which they have not received might be of some value to them.

Some alterations are so fundamental and important that drawings must be revised and issued immediately. For example, changes in floor size or ceiling heights must be circulated quickly as they affect all the consultants' work. Other revisions may be minor, such as enlarging a window or repositioning a door which do not merit a reissue of the drawing in their own right. If drawings were circulated on this basis it would not only be wasteful, but would overwhelm recipients with unnecessary paperwork. It is better to mark up minor changes on master hard copy prints as they arise. At some point, the relevant drawings can then be revised and reissued. The length of delay before reissue will depend on the needs of the drawing at the time.

It is important to indicate the status of drawings when moving from sketches or preliminary design drawings to ones which are to be used for construction. Drawings should be clearly marked '**NOT FOR CONSTRUCTION**', or '**FOR CONSTRUCTION PURPOSES**' to distinguish between ideas which are still under consideration and those which have been fully agreed for implementation.

3.5.6 Numbering drawings

As design work proceeds and the number of drawings begins to multiply, they need to be carefully and systematically identified so that they can be conveniently stored and easily retrieved for reference. There are alternative ways that this can be done, but consider the following system:

- **21**
  The number given to a new drawing.
- **100-21**
  The project number **100** added so that it does not become confused with drawings belonging to other projects.
- **D-100-21**
  The prefix **D** is added to show it is a drawing produced by the building designer, as opposed to the structural engineer or mechanical services consultant.
- **D-L-100-21**
  The prefix **L** is added to show it is a location drawing. The prefixes C or A would be added for component or assembly drawings.
- **2006-D-L-100-21**
  The prefix **2006** can be added to show the current year.
- **2006-D-L-100-21   A**
  The suffix **A** is added to show that the drawing has been altered or revised, containing information which was not on an earlier copy. '**A**' is the first set of alterations, changed to '**B**', '**C**', '**D**', etc. as subsequent changes are made. 2 March 2006 might also be added to the revision letter to confirm the date when the revision was first added.

A title block included adjacent to the number may also include details of the scale and sheet size of the drawing.

- **Floor Plan;   1:50   A1**

and should also include:

- **The name of the organisation who have produced the drawing.**
- **The Client's name and the project title.**
- **The author's name, and sometimes confirmation that the drawing has been checked and approved for issue.**
- **The date that the drawing was prepared.**

This format helps to retrieve drawings from a filing system, keep them in order, and to locate missing drawings being used by other members of the design team. It also helps recipients to know that they are in possession of the complete, current drawing. It can be particularly important later on when copies of drawings are being used on site, where access to specific, accurate, up to date information is absolutely essential.

All the drawings prepared for the project should be carefully recorded on a register with their numbers and current revision letters. Although the first drawing can be number 1, it is not essential to maintain a continuous numbering structure. It is better to keep drawings of a type together rather than mix them numerically. So for example, drawings:

- **1–20**: sketch drawings, even though there may only be 12 of them
- **20–40**: plans
- **40–60**: elevations
- **100+**: building designer's working drawings
- **200+**: structural engineers drawings
- **300+**: services engineers drawings
- and so on.

This system is particularly useful because it allows the design team to insert later drawings of the same type into the register, so that it is much easier to locate specific drawings when they are needed. Another method of filing and storing drawings is to keep all the drawings of the same size together. This is easier said than done because the size of a drawing depends on its purpose, but mixing A1 and A4 drawings together in a plan chest will make it difficult to find the small ones.

3.5.7 Printing drawings

The design team's work generally involves creating master drawings, prepared mechanically on translucent tracing paper, acetate or film, or electronically on a computer. The information on these drawings builds up over time, developing ideas incorporating advice and detail from others. Drawings prepared by means other than CAD are 'art work', valuable, sometimes irreplaceable documents which must be protected, looked after carefully, maintained in a reasonable condition. Even drawings prepared using CAD must be carefully stored and labelled so that they remain fully accessible. Whichever system is used to create the drawings, once completed they must be reproduced so that multiple copies can be circulated to the members of the design team and later, issued for construction on site.

The mechanics of reproduction depend on the systems available, but the two currently used economical options are photocopying and dyeline printing. Photocopying is a photographic process generally used for opaque papers where the copied image is a reflection of the drawing or text. It is economical for small-sized prints but relatively expensive for large drawings. It can however, reproduce drawings very quickly and can be used to produce coloured copies. Dyeline printing is used for translucent papers where the image is created by shining a bright light through the drawing onto a chemically treated paper. The light burns off the chemical where there are no lines or areas of shading, which when exposed to an ammonia vapour, develops the chemical left behind into a copy of the drawing. It is the most common way of producing copies of large drawings, but is quite slow, labour intensive and can leave a sharp odour in the air for some time. It will not produce coloured copies, but there is a choice of paper giving a blue or black appearance to the lines and text. The two systems have the following characteristics and limitations.

Readability

Care must be taken to see that lines or shading are dark enough to be picked up by either system. Pencil drawings do not always reproduce well. The usual practice is to use dark black ink, but it is as well to test selected style before going too far to see that the printed result is actually readable.

Distortion

Printed drawings should not be scaled as both processes can distort the length of lines. Where dimensions are important, they should always be written. A graphic scale can be included on the drawing for reference to check the extent of distortion.

Colour

If master drawings contain colour, it will appear on black and white prints as a tone, ranging from light grey to black. The balance of the drawing in colour is not necessarily the same when printed. For example, blue lines in photocopies disappear altogether and light yellow lines in dyeline copies come out black. It is useful to have a sample of colour effects for both systems before adding colour to master drawings. Controlled use of colour can achieve excellent effects in printed tones, producing a variety of shades of grey.

Lifespan

The choice of printing paper depends on the purpose of the print. The image will eventually fade if exposed to prolonged artificial light or strong sunlight. The inks will run and smudge if prints get wet. Paper copies should always be used for issue and file record purposes. Master drawings should never be folded as the fold lines may show up on prints. The title panel on large drawings is usually located in the bottom right-hand corner so that it is visible when the drawing is folded.

3.6 Computer-aided draughting (CAD)

The development of computers and electronic communications has changed the way that drawings can be produced. Economical and sophisticated equipment, programmes and communication systems are now in general use throughout the industry. Complex drawings can be prepared as shown in Figure 3.13 would be difficult or even impossible to create manually and access to data bases through the Internet has opened up a virtually limitless library of detailed information which can be easily downloaded and edited.

For those who are unfamiliar with CAD, the advantages include the following:

- *Handling large drawings*
  The size of manually produced drawings is limited by the size of the drawing board and the ability of the draughts person to reach the extremities. There is no limit to the size of CAD drawings, although the size of the monitor limits how much of the drawing can be seen at any one time and there is a limit to the size of papers that can be handled by printers.
- *Changing the scale of drawings*
  Manual drawings can only be produced at a single scale. Larger- or smaller-scale drawings would have to be drawn separately. CAD drawings can be converted to any scale or unit of measurement without having to redraw the original.
- *Designing to a grid*
  CAD drawings can be constructed to a grid based on absolutely accurate dimensions. Prototype design solutions can be tested for 'real-life' performance, and in some cases, CAD information can feed directly into the manufacturing process.
- *Improves design team co-ordination*
  Information can be immediately transferred between members of the design team in a matching style and format, so that drawings can be developed more accurately and precisely with

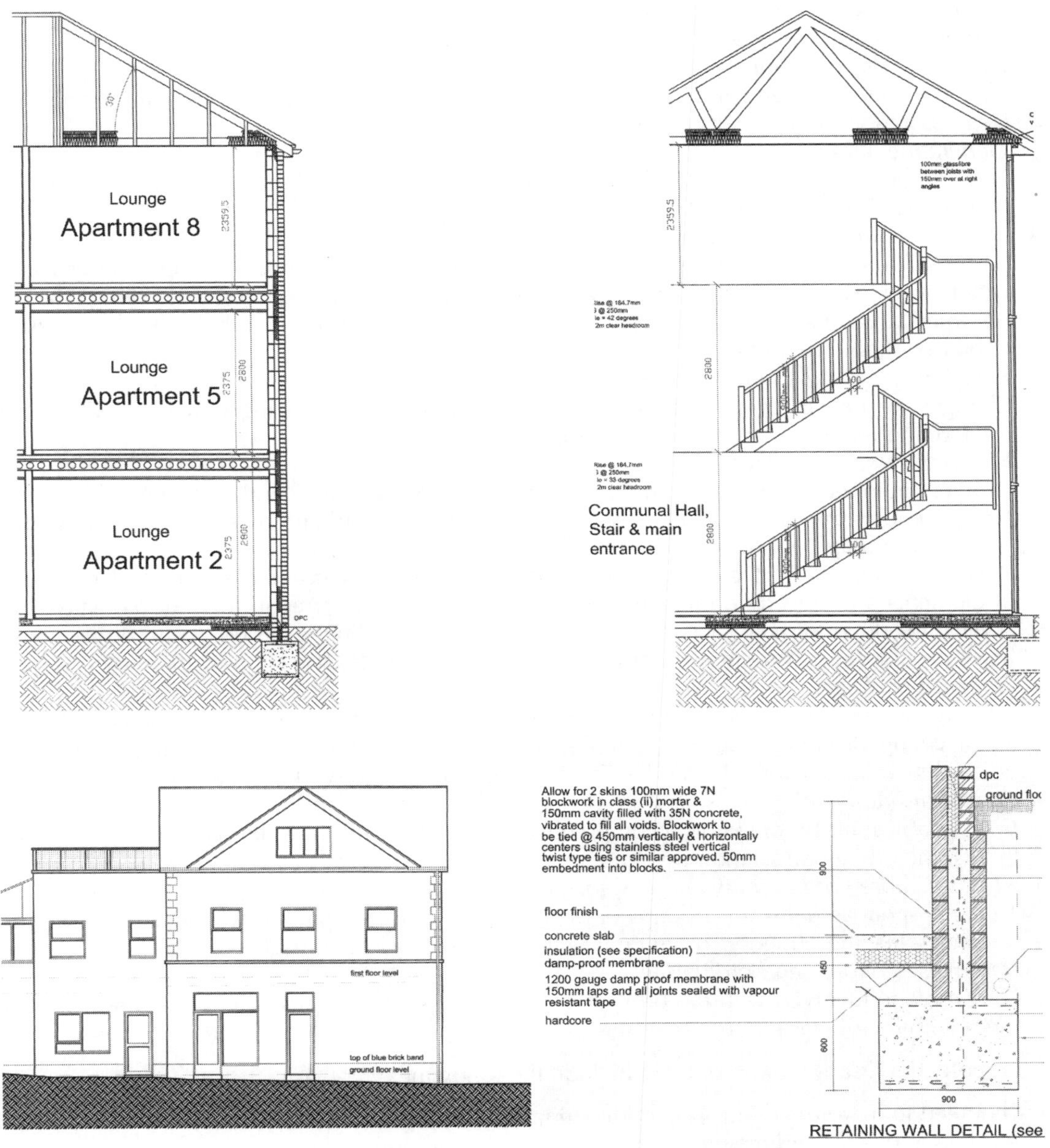

Figure 3.13 Examples of computer generated drawings (CAD). Illustrations by Tim Fletcher.

much less risk of error or omission of important elements. Drawings can be published on the Internet or a local intranet, aiding rapid communication.

- *Direct use of survey information*
  Electronically obtained survey data can be obtained in a form that can be used immediately for preparation of design drawings.
- *Generating alternative views*
  The computer can transform basic information to generate alternative views of proposals, particularly 3D images based on 2D plans and elevations.
- *Direct measurement of quantities*
  It is possible to quantify elements of construction as design work proceeds related to specification data bases. This may be a useful way of testing the economy of design ideas. Automatically generated schedules, quantities and cost estimates may help the design team to better rationalise ideas.
- *Revisions generate a new drawing by default*
  Changes to the design can be accommodated by adjusting existing drawings, which are printed and issued as clean, fresh drawings.
- *Economic storage and retrieval*
  Limitless numbers of drawings can be stored and retrieved conveniently without the need for large, cumbersome filing cabinets. Access to drawings is rapid and organisation simplified.
- *Takes the 'drudge' out of repetitive drawing work*
  Repetitive elements such as columns or window reveals can be copied from a standard library of elements and quickly duplicated rather than be laboriously reconstructed manually. The copies are dimensionally accurate eliminating the possibility of accumulated errors. CAD helps designers to easily move, copy, enlarge, reduce, merge, stretch, transform, etc.
- *Standard packages*
  CAD supports packages of standard graphics, text and visual aids, which can be easily managed, changed, edited, etc. It is simple to change lines, fonts, symbols, shading, figures, colour, etc. Manufacturers CAD details can be imported into working drawings.
- *Standardisation of drawing style*
  CAD enables all the drawings to be prepared in the same format, standardising drawn elements, including consistent titling and annotation.
- *Releases more time to think and design*
  The time spent actually constructing drawings can be reduced, particularly once revisions are necessary leaving designers more time to think about the design itself.
- *Improves economy and profitability*
  For everyone involved with the project, swift transfer of information saves time and reduces the possibility of wasteful errors or misunderstandings.

The potential disadvantages of CAD include the following:

- The need to learn new skills in a commercial environment.
- Significant capital and maintenance investment in equipment.
- Continuous operation of keyboards and peering at monitors can be unhealthy.
- Use in commercial offices can be divisive, separating designers and CAD operators.
- Exploratory ideas become 'fixed' too quickly before options have been fully considered; conceptual design and absolute accuracy are not necessarily compatible.
- Encourages the extravagant perception that proposals can be revised indefinitely, irrespective of programmes and previously agreed positions.

3.7 Intranets and extranets

Electronic communication is now a familiar part of commercial life. The use of e-mail and the Internet is common practice. There is a risk that writers of e-mail and texts may wrongly presume their readers attention, and that they demand an instant response or reaction. Like any form of written communication, there is a need for filing and retrieval systems to prompt action, but e-mail does create its own useful record, accumulating responses in chronological order telling the whole history of an exchange.

Many organisations are now using the Internet as a way of sharing information and computing resources within their organisation, and externally for the benefit of those with an interest in the organisation. An **intranet** is a private network accessible to staff or employees working in groups or on common projects. It can take the form of web pages which are only accessible to those within the organisation. When part of an intranet is designed to be accessible to anyone outside the organisation, it becomes an **extranet**, a private network using Internet technology in such a way as to control the secure flow of information. A business such as a car-dealership might use an extranet to link suppliers and other business partners rather than the web page which advertises to its customers.

3.8 Organising and managing meetings

Throughout the design and construction periods, meetings take place to present and discuss ideas, identify and resolve problems, establish target dates and monitor progress. The main objectives of meetings are to:

- ensure that key personnel are fully informed;
- obtain expert advice and opinion;
- explore implications, enabling and encouraging everyone present to express their point of view about the issues under consideration;
- receive and consider feedback from any relevant source;
- attempt to resolve problems with a reasonable degree of collective agreement, recognising and recording disagreement or dissent;
- take decisions about the course of future action.

Meetings will be held with the client, consultants, authorities and builders to achieve some or all of these objectives. In order that meetings are efficient and meaningful, they should be organised so that:

- those present are aware beforehand of the issues which need to be discussed;
- the outcome of discussion is recorded and distributed in a useful manner.

Constructively organised and structured meetings are more likely to achieve desired results than those which are *ad hoc* and chaotic. Topics to be discussed can be identified in the form of an *agenda*, issued in advance of the meeting so that participants can consider their position and undertake research beforehand leading to more profitable discussion. The agenda is a comprehensive list of matters that must be resolved at the meeting and helps everyone to concentrate on clearly defined issues, one at a time and in a logical order.

Details of the content or conclusion of discussions at the meeting are summarised in the form of *minutes*, confirming what was said or decided. When issued to individuals, the minutes are a clear record of matters which directly affect their work, and for senior management the minutes enable them to extract key information about the progress of the development, essential for future planning and allocation of resources.

Meetings which are of general public interest are formally structured and may be recorded verbatim, including every word uttered. For example, parliamentary debates in the House of Commons are recorded in Hansard and will even refer to the 'hear, hears!'. This would not be appropriate for the relatively small, less formal, private meetings associated with the construction process, where it is normal to record brief summaries of the topics, highlighting agreed decisions. However, if decisions are contentious, those with strong objections may wish to have their views recorded for future reference in the event of subsequent problems. The value of any meeting between any or all of the members of the development team can depend on two significant factors. Firstly, the discussion is directed towards meaningful conclusions and secondly that conclusions are correctly recorded.

3.8.1 Chairing meetings

The function of any person who chairs a meeting is essentially to maintain order so that individuals can make their contribution in turn and as far as is possible, to focus discussion on specific points avoiding confusion, misunderstanding and time wasting. The role of the chair is sometimes that of an unbiased referee. For example, the Speaker in the House of Commons, who does not contribute to the debate but acts as a defender of the constitution, periodically demanding 'Order!, Order!' as unruly Members clamour for attention. Alternatively, the chair can act as a leader, prompting discussion and directing outcome where appropriate, permitting flexibility and allowing for the fullest possible discussion.

Any member of the development team can act as Chair, but in practice it is often undertaken by the person who has convened the meeting, or the person responsible for co-ordinating the work of others or issuing formal instructions. Although this person may have a clear idea of what they expect the outcome of discussions to be, there is still an obligation to act fairly to all parties, and they should not be too dictatorial. For some meetings, the chair could also take notes recording the outcome, but there is a risk that in concentrating on one subject, the significance of another is missed. It is better that the chair should be in a position to concentrate on the discussion taking place. When chairing meetings, a number of points should be considered.

- *Be prepared*
  Everyone present should have an agenda so that they are aware of the topics to be discussed. A clear idea of the purpose of the meeting and the consequences of possible alternative outcomes helps to concentrate minds. Take the meeting seriously, do not arrive late, leave early or spend time thinking about something else.
- *Consider setting a time limit for the meeting*
  Agree the period of time available to conduct the business of the meeting so that members appreciate that they must not spend too much time on some points at the expense of others.
- *Remain objective and focused on the agenda*
  Try to ensure that each matter is dealt with in turn as identified in the agenda. Encourage discussion which is relevant, if possible guiding it towards useful conclusions. Once points have been made, try to discourage members from making them again and again, going round and round aimlessly. Avoid wandering off and digressing into areas which are not directly relevant. The chair can usefully summarise arguments, and if necessary call for a vote to determine decisions.

- *Once decisions have been agreed, do not re-open discussions on the subject*
  Direct the members of the meeting towards positive conclusions and move forward. Back tracking can cause confusion, although the chair must exercise discretion in the light of later observations if they are sufficiently important. In some cases, decisions may be left to the chair's discretion, subject to information becoming available after the meeting has finished.
- *Ensure that decisions taken are sensible*
  Do not allow the members of the meeting to make decisions which are clearly wrong or dependent on further research. Be aware that unverifiable statements may influence others. It may be much better to defer a decision until the next meeting when more information is available or when members have considered matters more carefully. Equally, if decisions are essential, ensure that the necessary information is available.
- *See that action for future work is correctly apportioned*
  Make sure that it is clear *who* is to do *what* following any decisions taken at the meeting. Do not assume that individual members will automatically respond unless the matter for which they are to be responsible has been specifically highlighted. Make sure that decisions are converted into action by agreed dates.
- *Give members an opportunity to raise other matters not on the agenda*
  When all the items on the agenda have been discussed, it is usual to finish the meeting by enquiring if there is *any other business*. Members should feel free to raise concerns or queries about other matters which are not on the agenda, but may be of interest.

3.8.2 Keeping records

Someone present at the meeting should be responsible for taking notes of the outcome of discussion so that after the meeting they can be distributed to all those present at the meeting, and to any other interested parties. The minutes should accurately record the views expressed at the meeting, but in some cases may be used to simplify debate into concise statements. At the next meeting, all parties are invited to accept the minutes as a true record, and request amendments if they are not satisfied. When recording notes, some points to consider include:

- take everyone's name and position at the beginning;
- listen carefully and record conclusions accurately;
- take down information under headings;
- collect all relevant points together.

During the meeting, discussion may flow from one point to another and back again. The minute should summarise facts about each issue as comprehensively as possible irrespective of the fact that discussion was fragmented:

- Develop a shorthand method for key words to aid memory for subsequent writing-up.
- Allocate comment to individuals where appropriate.
- Write up the minutes as soon as possible after the meeting as content of conversation can easily be forgotten particularly when it is recorded in scribbled notes.
- Circulate the minute as soon as possible after the meeting to remind the members of the meeting to attend to issues which are their responsibility in time for the next meeting.

The example given below shows a typical format of a set of minutes. The number of items in each section depends on the extent of discussion, but for the sake of clarity the observations in this example are limited and simplified.

Project title

New car-dealership at Old Mills Lane for (Client)

Meeting title

Minutes of site meeting No 4, Thursday 2nd March 2006, held on site

Present:

Robert Green	**(RG)**	**Client**
David White	**(DW)**	**Structural Engineer**
Gavin Tunstall	**(GT)**	**PGT Designers**

Apologies for absence:

Received from Ron Blake (RB) Company Director who is looking for new opportunities in Uganda.

The previous minutes are read and agreed. It is not necessary to literally read them out loud, but there is an opportunity here to confirm the accuracy or to correct errors from the written notes of the previous meeting.

1.0 Progress to Date: **Action**

1.1 The project is on course for completion as programmed for handover on 1st May 2007. Foundations and ground beams are completed, 50% of the structural steel columns have been erected.

1.2 Details of the roller shutter doors have been issued to the contractor as requested, and the suppliers drawings for the steel cladding fixings for the roof have been received.

1.3 The revised delivery date for the remainder of the steel frame is now confirmed as 1st April 2006. ALL

2.0 Matters Arising from meeting No 3:

2.1 DW confirmed that the steel framework has now been approved by the Building Control Officer.

2.2 GT issued revised layout drawings for the staff toilets.

2.3 RG provided details of requirements for the external signs.

3.0 Any other business:

3.1 Fire Officer to be consulted re. means of escape. GT

3.2 Amendments to spray painting shop to be considered. RG

3.3 It was agreed to change the specification of the roof cladding. GT to issue instructions based on client approval. GT & RG

4.0 Date and time of next meeting:

4.1 Thursday 26th April 2006 in GT Design Office.

Distribution to:

Robert Green, David White and Ron Blake

Author:

Gavin Tunstall **10th March 2006**

The issued minute can also be used as an agenda for the next meeting with the addition of items for discussion under the any other business section (if known) to be circulated in advance.

Minutes as a way of monitoring action

The issued minute has a column down the right-hand side headed **Action**. This is a simple technique for highlighting the parts of the minute which specified individuals need to attend to in order to maintain progress. Whilst the content of meeting records will be of general interest, it is helpful to direct the individual to issues which are their particular responsibility. In a lengthy meeting, when numerous points are discussed, there is a risk that the individual fails to act as required because although an important statement is included, its relevance is not appreciated.

The action column also prompts the writer, or the project manager to take action to see that progress is maintained. By referring each item for the attention of an individual, further meetings or contacts can be arranged to check that action is being taken. In larger organisations, this method enables senior management to see that their own employees fulfil obligations and provide information in accordance with the programme.

For example, after discussion at a meeting, the contractor requests full details of all the steel reinforcement required for the concrete ground floor slab. The record minute could simply say:

Action

- **'details needed for steel reinforcement to ground floor slab'**
  On reading this, it is not clear who is involved. Some members of the design team may fail to appreciate the implications for their own work and consequently at the next meeting the matter is still not fully resolved.

It would be better to write:

- **'details needed for steel reinforcement to ground floor slab'** **Structural Engineer**
  The issue is highlighted for the engineer, who may be reminded of the need to co-ordinate the matter by issuing details to the contractor and ensuring that the design is approved by the Building Control Officer. If cost is an issue at this stage, the Quantity Surveyor (QS) may need to check proposals, comparing the actual design with provisional allowances and considering if any cost adjustments are required. The minute may therefore also include:

- **'details needed for steel reinforcement to ground floor slab'** **Structural Engineer & QS**
  Depending on the complexity of the issue, the minute should be more comprehensive, spelling out each of the actions associated with resolving the steel reinforcement.

Consider the implications of discussions about facing bricks.

- The contractor advised that the specification of the facing bricks for the site boundary wall must be confirmed within the next 5 days in order to maintain the programme. **Contractor**
- The client wishes to approve the choice before instructions are issued. **Client**
- The designer is to investigate alternatives. **Designer**
- The designer will liase with the Planning Officer to ensure that the selected bricks are approved. **Designer**
- The QS will confirm that the cost allowed in the BQ is sufficient for the selection. **QS**
- The designer will confirm the specification to the contractor. **Designer**

All of these might have been statements from members of the meeting and all must be resolved before the specification can be confirmed. Colour coding the action column is useful so that individuals can look for their own colour. There are other ways to monitor action, including interviews, inspections, site visits and tests, but the formality of regularly organised meetings generally obliges the membership to focus on progress. At times throughout the design and construction process, the building designer may be required to present information to the meeting, or even to larger groups of interested people.

3.9 Formal presentations

A presentation or 'giving a presentation' is normally a relatively formal event, an opportunity for an individual or a group to put forward ideas or proposals for the benefit of an audience, who may simply be interested in the content, or who may be charged with selecting or approving on the basis of what they see and hear. The presenter(s) will summarise a topic or a situation endeavouring to secure audience confidence perhaps to persuade them to adopt a course of action. As well as speaking, the presentation may contain visual aids, photographs, drawings and/or models to help the audience. For the building designer such events can be critical to the progress of the project as ideas are put forward for consideration. In a sense, the presentation is a performance demanding planning, organisation and self-confidence if it is to be received well. The two principle considerations are preparation and delivery. The following brief points may help with both tasks.

Preparation

- *Timing*
  A formal presentation given at a prearranged time by 'appointment' is rather like a job interview. Never be late for the start, and if time is limited, plan the presentation to suit; it is usually better to finish early rather than over-run. It is good idea to rehearse timing and sequence before the event. On-the-spot improvisation can be catastrophic.
- *Research your audience*
  If possible, find out who will be in the audience; individuals, groups, employers, clients, experts, students or the general public. Both content and delivery may require variation according to who is being addressed.
- *Content*
  Be objective, relevant, prioritise level of importance, avoid over-simplification and excessive elaboration. Decide what to include such as words, drawings, photographs, graphs, checklists and get all material into the order in which it is required so that the presentation flows smoothly. Try to anticipate audience questions on completion and have a response in mind.
- *Mechanical aids*
  Decide how to display material using projectors, slides, computers, PowerPoint, display boards or handouts and find out in advance how to operate unfamiliar equipment. Check the room lighting to see that audience can see clearly.
- *Visual aids*
  Simplify visual aids so audience listen rather than spend time reading. Select image and font sizes to suit size of screen and distance to audience. Choose colours which are readable and that do not clash with one another. Avoid typographic errors on slides.

Delivery

- *Introduction*
  Introduce yourself, your group or company explaining intensions. The old adage that you '*tell them what you are going to tell them, tell them, and tell what you have told them*' is a sensible way to try to ensure that your message is received. It can be very helpful to issue handouts to the audience summarising complex details so that they spend time listening to the presentation rather than desperately writing notes.
- *Positioning*
  Try to relax and be normal but do not adopt a casual, sloppy stance, avoid excessive hand waving which can be irritating and do not move about too much. Address the audience rather than talk to screens or projectors and do not obscure visual aids from your audience by standing in the way. Make eye contact with members of your audience, involve them, force them to listen to you, but do not intimidate or pick on individuals. Project to the back of the audience as well as those closest at the front.
- *Confidence*
  Cover each issue individually and comprehensively, one point at a time. Convince your audience that you understand your subject; do not read a prepared script as though it had been written by someone else and you have no idea what it means. Use cue cards or bullet points to aid memory; but do not read script directly off screen; the audience will already have read it before you finish. Do not be embarrassed, distracted, giggle or make 'in-jokes'; humour is fine but only if the audience laughs. Speak clearly and slowly enough for the audience to hear, but do not shout or speak too quietly and do not rush or mumble so that ideas are wasted. Pause periodically at key points for emphasis, varying pace and volume. Avoid 'tailing off' at the end of statements as if running out of ideas.
- *Conclusion*
  Watch for audience response to see when to move on or stop. Briefly recapitulate the content then encourage or invite questions, respond professionally, politely and honestly; only bluff if your audience cannot possibly know otherwise. Give positive answers rather than negative excuses. Enjoy yourself, engage the audience, make them want to listen to what you have to say. Be confident, enthusiastic and interested in your own chosen subject; if it seems that you are not, then it will be of no surprise if the audience lose interest too.

3.10 Discussion points

(1) Grammar and spelling is an important part of written communication. Do grammatical and spelling errors diminish the impact of a message? Could abbreviations associated with e-mail and text messaging become common practice in all forms of writing?

(2) Every line on construction drawing has a meaning. What might be the significance of lines being misunderstood? Does CAD eliminate this kind of problem? Can drawings become too complicated to be understood?

(3) Electronic communications are not always foolproof. The speed and volume of transferring information can be overwhelming. What are the risks of important messages being missed or lost? Do the 'senders' of messages automatically assume the recipients have received them?

(4) Is style more important than content? Can a professional presentation make a 'silk purse out of a pig's ear'? Does a poor presentation mean that ideas are easily dismissed as worthless?

3.11 Further reading

Brown S (2001) *Communication in the Design Process.* London: Spon Press.
Edwards BW (1994) *Understanding Architecture Through Drawing.* London: Spon Press.
Edwards BE (2000) *The New Drawing on the Right Side of the Brain.* 2nd Edn. London: Souvenir.
Chappell D (2003) *Standard Letters in Architectural Practice.* 3rd Edn. Oxford: Blackwell.
Ching F (1998) *Design Drawing.* Chichester: Wiley.
Ching F (2002) *Architectural Graphics.* 4th Edn. Chichester: Wiley.
Coles D and **Naoum** S (1998) *Dissertation Writing and Research for Construction Students.* Boston: Butterworth Heinemann.
Goodale M (1987) *The Language of Meetings.* Language Teaching Publications.
Huth M (1999) *Understanding Construction Drawings.* London: Van Nostrand Reinhold.
Jay R (2000) *How to Write Proposals and Reports That Get Results.* London: Prentice Hall.
Laseau P (2004) *Freehand Sketching: An Introduction.* London: Norton.
LightWork Design (2005) *3D Computer Modelling.* www.lightwork.com
Mays P (1997) *Construction Administration: Architect's Guide to Surviving Information Overload.* London: Wiley.
McFarlane R (1997) *3D Draughting with AutoCAD.* London: Arnold.
Moore T (2005) *Spanish Steps: Travels with My Donkey.* London: Vintage.
Oliver P (1999) *Writing Essays and Reports (Teach Yourself).* London: Hodder and Stoughton.
Pandia (2006) A search engine tutorial. www.pandia.com
Reekie F and **McCarthy** T (eds) (1995) *Reekie's Architectural Drawing.* 4th Edn. London: Arnold.
Sampson E (2003) *Creative Business Presentations: Inventive Ideas for Making an Instant Impact.* London: Kogan Page.
Search Tech Target (2006) Electronic references. www.searchtechtarget.com
Styles K and **Bichard** A (2004) *Working Drawings Handbook.* Oxford: Elsevier.
Thompson A (1993) *An Introduction to Construction Drawing.* London: Butterworth Heinemann.
Uddin MS (1997) *Composite Drawing: Techniques for architectural design presentations.* London: McGraw-Hill.
Waterhouse M and **Crook** G (eds) (1995) *Management and Business Skills in the Built Environment.* London: Spon Press.

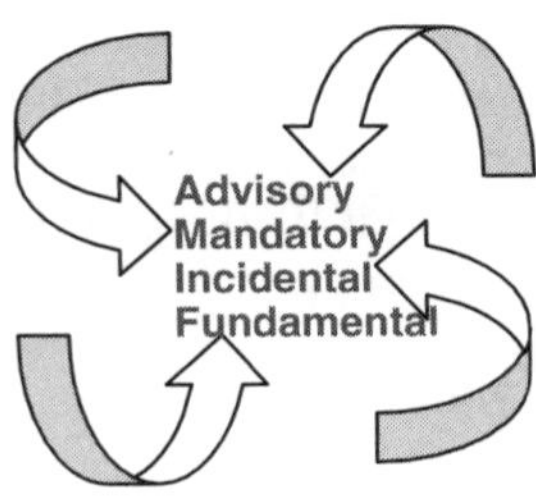

4

Permissions and approvals

4.1 Introduction

It is common practice in the UK for any sphere of professional activity to be regulated by independent authorities, who control or are required to approve the behaviour, performance and potential achievement of those involved. Generally, this acts as a safety valve, attempting to ensure that customers or clients receive services which are appropriate, suitable or legitimate in the circumstances. Such controls range from compliance with mandatory statutory law applicable to any activity to very specific areas of influence or guidance pertinent to individual projects. This chapter will outline principle areas of legal constraint and identify interested authorities, briefly summarising their involvement with design and construction as the necessary approvals are secured. It will include discussion about the responsibilities of all members of the design and construction teams.

4.2 The significance of legal constraints

The design team's principal aim is to create the new building to satisfy the needs of their client, acting as far as is practicable in their best interests. This does not mean, however, that either designers or clients are free to do whatever they wish, or that contractors can build in any way they think fit. The development of the design and the way that it can be built are both subject to the interests and concerns of a number of independent organisations, bodies or **authorities** who are empowered to scrutinise and control both design and construction as work proceeds. The reasons for doing this can be broadly summarised as follows:

- Protecting the existing natural and built environment, minimising potential damages.
- Promoting and supporting high-quality design and management.

- Planning for long-term sustainability of the built environment.
- Limiting unnecessary waste and attempting to reduce energy consumption.
- Constructing new buildings to appropriate standards for their purpose.
- Developing more healthy and harmonious working and living conditions, including directing behaviour and influencing attitudes.
- Minimising health and safety risks.
- Complying with relevant building and common law.

The authorities manage these issues, concerns and interests by offering guidance, making recommendations, imposing restrictions and enforcing standards which may be:

- *Advisory*
  Matters which may be discretionary, or optional, but which are sensible and regarded as good practice, intended to help designers, contractors, clients and members of the public.
  How to create a building which is reasonably secure from burglars.
- *Mandatory*
  Issues which are or can be enforced through legislation, and which cannot be contravened.
  Where to position fire barriers between compartments.

For the design team, compliance with these interests has the following implications:

- *Incidental*
  Compliance can be achieved or negotiated as work proceeds without damaging design ideas or adding significant additional costs.
  Selecting locations for burglar alarms on external walls.
- *Fundamental*
  Noncompliance will prevent further progress at all at any cost, or may add extensive additional costs to correct if neglected or omitted.
  Discovering that fire barriers have been overlooked at the design stage.

The consequences of legislation and advice must be incorporated into proposals at various stages throughout the design and construction periods so that necessary approvals are secured. In some cases guidance is available in the form of printed standards but as these change from time to time, there is a risk that the information contained in them may no longer apply. It is important to keep up to date with current legislation and practice, and to this end specific issues relating to each building and its site should be resolved by consultation and discussion with the appropriate authority at the time. Some issues are open to debate and interpretation and the authority dealing with the project may adopt its own 'likes' and 'dislikes'. Consequently, it is not always possible, or sensible to presume the nature of any response to a particular design idea. Intensions or proposals should be made available to any concerned authority for early, preliminary comment, so that the design team can be confident that their approval will ultimately be forthcoming. There is little merit in proceeding to develop ideas if there is a possibility of later rejection. It is self-evident that this is a critical issue in a commercial context if the designer is to avoid misleading clients and other professionals about possibilities which may be incapable of being achieved. There is of course a design dilemma here; a potential conflict between caution and adventure, between 'tried and tested' and innovation, between safety and risk. There is a definite possibility that both 'designers' and 'approvers' narrow their sights and ambition to follow the rules.

However, each authority will normally be prepared to discuss matters relating to the project giving guidance on how to apply or interpret their concerns accordingly. Incorporating compliance or advice will affect the design of the layout and appearance of the building, and can have significant cost implications. For example, the Fire Officer will require a variety of safety measures to be included in the design, such as fire doors, escape routes and protection of structural elements. If the designer is involved in cost planning, or working with a quantity surveyor (QS) to agreed target budgets, then these items must be identified at the right stage, allocating sufficient sums of money for their provision. It would be unwise to assume that other consultants will make due allowance as it may be the designer who has to advise the client of additional costs as a result of failure to anticipate requirements.

4.3 Statutory controls and regulations

There are many areas of law applicable in the UK which affect the **designers**, the **design** and **construction** as new buildings are created.

The principal sources of law in the UK are as follows:

- *The European Court of Justice.*
- *Parliament through empowering Acts.*
- *The Courts developing criminal and civil common law.*

Sources of law form the basis for other *substantive areas* of legal constraint including a number of regulatory controls applicable to the designer's work. In general terms, the sources of law can only be challenged by appeal to the Courts, whereas the substantive areas allow for discretionary interpretation at the point at which they operate. Common law continues to develop by establishing precedents through test case judgements, where either the point of conflict is not covered by legislation or where there is a loophole in its drafting. In general terms, criminal offences attract punishment whereas civil offences attract compensation.

The following is a summary of areas of law affecting building design.

Acts of Parliament including, for example:

- *Town and Country Planning* to control principles of land use and development.
- *Highways* to control use of publicly accessible roads, paths, squares, etc.
- *Health and Safety at Work* to control working practices and arrangements.
- *Disability Discrimination Act (DDA)* to make buildings accessible to those with disabilities.
- *Fire Precautions* to protect human life and property.
- *The Part Wall Act* to resolve disputes between adjoining building owners.

Regulations including, for example:

- *The Building Regulations* to control performance of construction and use.
- *Construction Design and Management Regulations (CDM)* imposing a duty of care in the design and construction processes.
- *Workplace Regulations* to ensure good practices in operating and managing commercial premises.
- *Statutory Service Provider Regulations* to ensure safe, economical distribution of electricity, gas, water and telecommunications.

Property law relating to the site and design proposals including:

- Ownership, boundaries and party walls.
- Easements conferring third-party rights for access, drainage, light and support.
- Restrictive and positive covenants affecting future use.

Environmental law including controls relating to pollution, clean air, noise, storage and disposal of waste, and remediation of contaminated land.

Contract law liability based on conditions defined by specific agreement confirmed within exchanged contracts. A contract exists when one party offers goods or services in exchange for payment.

Negligence essentially failure to provide advice which can be reasonably expected in association with the offer of *professional* services including a duty of care, breach of duty and consequential damages. Negligence may arise through:

- omission of or incorrect specification,
- failure to comply with statutory regulations,
- failure to consider possible consequences,
- negligence of consultants if nominated by the designer,
- knowingly causing damage to third parties.

Note: **Indemnity insurance** *is now mandatory for building design institute members, and must be maintained for a period beyond ceasing trading.*

Trespass, the unlawful invasion of property, land or buildings owned by others, including entry causing damages, building on or overhanging a boundary, swinging a crane over adjacent property or allowing waste to cross a boundary.

Other forms of legal control include the following:

Arbitration is impartial third-party resolution of disputes.
Bond, a deposit against failure to provide services or goods.
Business or commercial law, including registration, bankruptcy, company accounts, limited liability, taxation, shareholders, partnerships and sole practitioner arrangements.
Collateral warranty is an assurance that the completed building will be 'fit for purpose'.
Copyright including the title to documents and design work, unlawful copying of the work of others.
Employment law including behaviour and relationships between employers and employees.
Nuisance interference with the use or enjoyment of a third-party's property by a public organisation or a private individual.
Public liability is the responsibility for injury and damage to third parties as a result of actions.
Quality assurance is covered in more detail in Chapters 11 and 12, where the quality or performance of individuals, groups, materials and systems can be defined by, or supported by an independent form of authority.

For the design team, any (or all) of the above may have significance for progress on development of ideas about the new building. What is certain is that Planning and Building Control will influence the appearance and workings of almost every building design project.

4.4 Planning

Planning concerns the control and management of the built environment. Archaeological remains in UK from the stone, bronze and iron age periods show what early primitive human settlements must have been like. Many of them were in relatively remote parts of the country and subsequently abandoned as their usefulness declined. The built environment which we live in now has its origins in the works of successive civilisations from the Roman occupation around 2000 years ago, through the Scandinavian and Norman conquests as well as the efforts of the resident population. Scattered rural settlements evolved into villages, towns and cities; an urban expansion generally around churches, castles, natural resources such as quarries and mines or at strategic points of communication along roads and rivers. Although relatively haphazard, development was dense and compact, retaining clear divisions between buildings, agricultural cultivated land and the open countryside.

During the Industrial Revolution, from approximately 1800 up to the start of the First World War, technological advance and commercial expansion spawned rapid, uncontrolled building development. The speed and scale of this process in order to accommodate industry, commerce, workers houses and transport links, changed the appearance of many parts of the country, inevitably encroaching into the open countryside. In some parts of the country, the leading industrialists took a philanthropic view about their workforce. Cadbury, Salt and others created 'new towns' for their workers, and indirectly established the foundations of organisations for the benefit of all (the beginnings of the welfare state which we take for granted today).

Notwithstanding their efforts, development was generally cramped with stress and pollution creating 'slum' conditions leading to a serious decline in health, particularly of young people. After the First World War, there was some political impetus to 'provide homes fit for heroes' (Lloyd George) and the first Planning Act (1919) gave local authorities the power to build low-cost housing at public expense. The Garden Cities were created to give the new generation a fresh start, but business interests and conservative politics sustained intensive expansion, and conditions continued to decline. The ravages of the Second World War brought matters to a head following widespread urban damage through bombing, which left some very poor housing conditions. The labour government, elected in 1945, established the Welfare State (an idea commissioned by the previous conservative government and summarised in the *Beveridge Report*) and the 1947 Planning Act required every local authority to produce a local plan, defining by consultation its land-use policies and means of protecting the countryside from inappropriate development. It also importantly introduced the concept of controlling all new construction within its remit.

This system of control has developed to the **Town and Country Planning Act** in force today. The powers of the Act are established by Parliament and managed by a Secretary of State and are implemented by each Local Authority at County, City, Borough and District levels. The actual title of the government department tends to be changed quite regularly by successive Prime Ministers. An earlier version of Secretary of State for the Environment, quite apt it would seem, is currently Department for Environment, Food and Rural Affairs (DEFRA). At a national level, direction is disseminated in the form of Planning Policy Guidance Notes (see PPG3, for example, concerning urban design) applied by each area's Chief Planning Officer, which informs attitudes to individual building proposals delegated to the project **Planning Officer** (PO) in whose area the development lies.

Most new building development is controlled by the current Planning Act, and must be illustrated and approved by the Local Authority prior to commencement of construction on site. The aim is to control development within the Local Authority's area in the following ways:

- Reconciling the location, siting, use and appearance of new development so that it does not adversely affect the benefits which the locality already enjoys, and that will, wherever possible, enhance the existing environment.
- Forward planning its own area, co-ordinating land use and building construction in consultation with and in the best interests of the whole community. This function includes publishing general planning guides and specific development briefs for how sites are expected to be developed, and some authorities are even acting in an entrepreneurial way working towards the assembly of sites for redevelopment, generating ideas through sponsored design competitions and proactively engaging with developers to create opportunities.
- Confirming ownership and entitlement to develop.
- Protecting or replacing features of special significance or merit.

With these general aims in mind, the following specific issues will be considered:

- Any change of existing use should be appropriate for the building, the site or the locality.
- The need to restrict development density, size, storey numbers and height.
- Development maintains conservation area and improvement area status.
- Alterations and extensions to existing listed buildings are appropriate with regard to protecting original or existing qualities.
- That sub-urban development does not spread onto green belt and agricultural land.
- Protection of the existing landscape from the effects of mining and/or extraction.
- That trees, wildlife and sites of special scientific interest (SSSI) are suitably protected.
- The location and design of advertisements, large and/or illuminated signs are controlled.
- Access arrangements and economical transport systems are co-ordinated.
- A variety of conditions or agreements associated with proposed development. Typical section agreements include arrangements for adopting public highways and drainage systems. Planning Gain negotiations may concern publicly accessible open-space and children's playing areas often associated with new housing schemes.

Figure 4.1 identifies some of these issues in the context of the car-dealership development. Permission to develop takes two principle forms.

Outline Planning Permission

There are occasions when it is useful to establish the principles of development in situations where there is doubt as to whether or not it would be acceptable. For example, a farming land owner considering building a car-dealership on a piece of agricultural land next to a motorway, would like to find out if the local authority might be supportive or not. The proposal can be tested by submitting an outline application for the change of use of the site based on limited definition with little or no detail design. The car-dealership could be shown as a simple rectangular block on the site plan. This indicative explanation minimises detail design costs which would be lost if the application was turned down. However, should outline permission be granted, detailed design proposals will be required eventually in the form of a full application.

Full Planning Permission

In most cases, proposals for new buildings are fully illustrated showing construction intentions. Rather than just the simple block rectangle of the outline idea, this application will include

Figure 4.1 The city of Nottingham looking north, south, east and west. The planning system helps to create and control development of the built environment, a complex mix of different functions and building types.

plans, elevations and sections showing the full extent and appearance of the building, although some details such as the colour of facing bricks may be delayed for future approval when selection is finalised. The Planning Authority may negotiate with the applicant if it feels that amendments would be appropriate.

4.5 Planning Applications

The formal Planning Application, made to the local authority, currently requires the applicant (the client) or their agent (the designer, the contractor or the project manager) to submit the following:

- Forms which include brief statements about the nature and extent of development proposals.
- Drawings (illustrations) showing the site layout, the building floor plans and all its elevations.
- A location plan (1:1250 Ordnance Survey) showing all the adjacent property around the site.
- Confirmation of the client's ownership of the site and that no part of the site is an agricultural holding. It is possible to obtain Planning Permission for development on a site which is not

owned by the client, but the owner must be informed about proposals and the application include confirmation that this has been done.
- The appropriate application fee.

The PO will normally be available to discuss proposals in advance of the formal submission to confirm if work is subject to control and to attempt to highlight and resolve problems wherever possible, but this is no guarantee that the application will be successful. Decisions are normally given within 8 weeks of registration, but this period can be extended by agreement. If the Local Authority decides to reject the proposals, the decision can be challenged by appeal to the Secretary of State, who will appoint an independent inspector to examine the situation and decide if the authority has acted correctly.

Approval or rejection is given in the name of the Local Authority's Planning Committee, which is made up of elected counsellors. They are advised by the authority's employed officers, who scrutinise applications, carry out research and consultations, and if necessary, refine proposals with the applicant or their agent. Some authorities delegate powers to the PO to determine routine decisions, but important or controversial proposals will be put to the Planning Committee, or even a full meeting of the Council. In most cases the Committee will follow the PO's recommendations.

The process is as follows:

- The application for Outline or Full Planning Permission is submitted to the PO.
- The PO checks and registers applications.
- The PO consults with other agencies and interested bodies, and contacts all neighbouring properties in writing to notify them of development intentions, inviting their response. Site notices are erected and sometimes descriptions are placed in local newspapers.
- Discussion takes place between the PO and the applicant or agent to resolve problems or address any comments or objections received through consultations.
- If amendments are made, further consultations may be necessary. If the PO is satisfied that the development proposals are satisfactory, an approval notice will be issued to the applicant or agent, or a recommendation will be made to the Planning Committee that they should approve the scheme. In most cases, the consent will include conditions such as the requirement to submit selected external materials for further approval or to undertake planting at the first practicable opportunity.
- Once work has started on site, the PO may make occasional visits to check that development is in accordance with the approved drawings, and that any imposed conditions have been met. Minor amendments can be made during the construction period at the discretion of the PO.

In almost all cases, Planning Approval must be obtained before work starts on site. Consent cannot be assumed to be forthcoming, and any construction work done without consent is at risk.

4.6 Building Control

The idea of controlling building construction can arguable be traced back to 1666, the Great Fire of London, where a small, local difficulty rapidly spread into a memorable disaster. Subsequent local bye-laws from the mid-seventeenth century onwards, endeavoured to significantly reduce health and safety risks, and promote comfort conditions for the occupants of buildings. National requirements in force today (quite recently introduced in the 1970s) have the intention of securing common standards for all new building work. Once again, the government of the day develops a legislative framework under the direction of a Secretary of State (currently the Office of the Deputy Prime Minister). The Building Act 1984 governs the **Building Regulations**, periodically revised to incorporate lessons from failures and/or changes in policy. They are administered by the Chief Building Control Officer for each Local Authority, who delegates responsibility to the **Building Control Officer** (BCO) in whose area the specific development lies. He or she will check that design proposals comply with current regulations, and confirm by inspection that construction is appropriate and satisfactory. The main aims of the Regulations are to ensure that construction and occupational use achieve the following:

- Secures the health, safety, welfare and convenience of people in, around, affected or concerned with buildings.
- Improves the conservation of fuel and power.
- Controls consumption of fresh water, and disposal of contaminated water.

Figure 4.2 summarises the practical issues related to a simple building. Current regulations are divided into **15 Approved Documents**, each dealing with a different aspect of construction and occupational use. Requirements are stated briefly at the beginning of each document together with an indication of how they can be achieved. Design proposals are **deemed to satisfy** the regulations if they are in accordance with illustrated examples, or if they use criteria which falls within defined ranges. In some circumstances, a tradeoff or relaxation can be obtained to compensate for noncompliance. A mechanism for appeal to the Secretary of State against rejection (determination) is available, if in the circumstances it is felt the Local Authority's interpretation of a regulation is unduly severe.

4.6.1 The Building Regulations

The Approved Documents currently deal with the following issues.

A. Structure

A1: *Loading*: the combined forces generated by the dead loads (self-weight of the materials), the imposed loads (weight of occupants, furniture, machinery, snow, etc.) and wind loads must pass safely to the ground without causing elements to deflect or deform, or causing ground movement affecting the stability of other buildings.

A2: *Ground movement*: construction must resist instability through swelling, shrinking or freezing of sub-soils, and any reasonably predictable land slip or subsidence.

A3: *Disproportionate collapse*: for certain forms of construction, the effects of accidental damage which remove some elements of the structure must not lead to excessive collapse of the remainder.

B. Fire safety

B1: *Means of warning and escape*: in the event of fire, there must be effective means for occupants to reach a 'place of safety' outside the building, available to them at all times.

B2: *Internal fire spread (linings)*: materials for walls, ceilings and internal structures must be capable of resisting the spread of flame over their surfaces and have 'reasonable' rates of heat release.

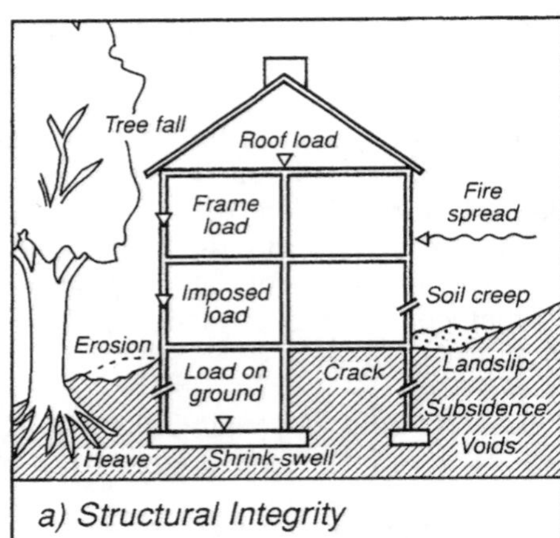

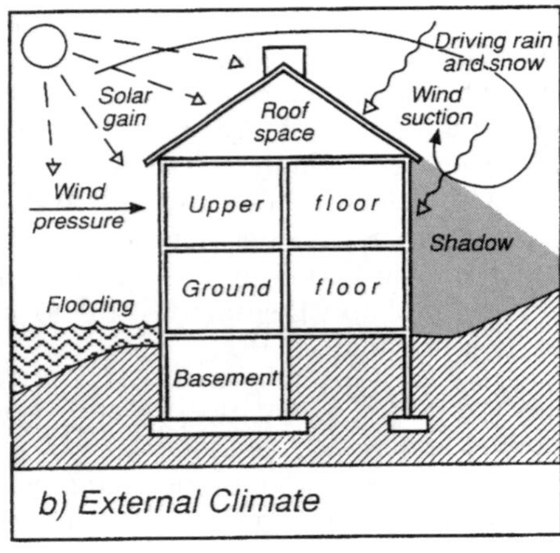

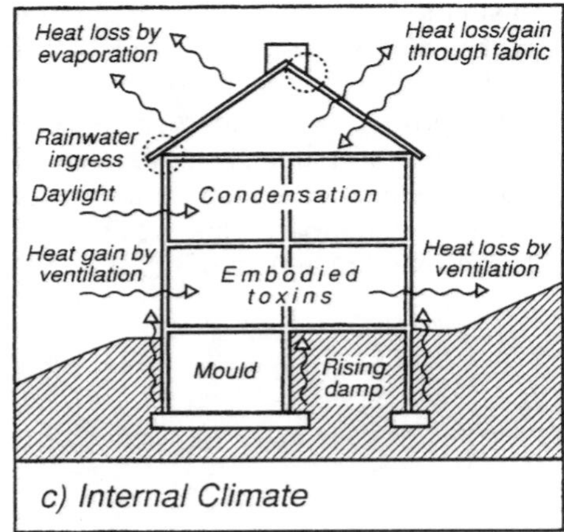

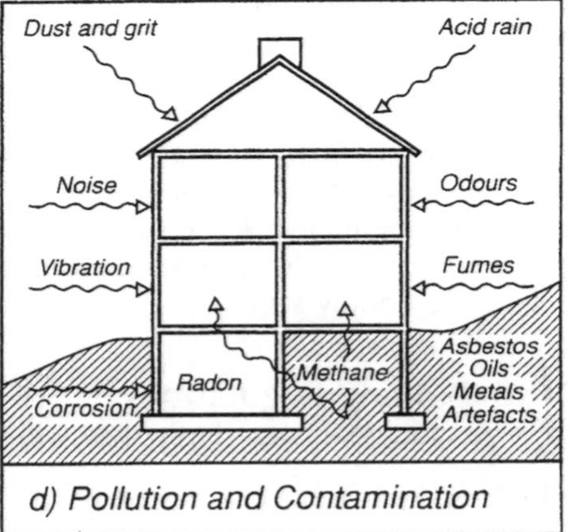

Figure 4.2 Building Control constraints. Diagrams courtesy of Andrew Golland, taken from *Housing Development: Theory, process and practice*, Edited by Andrew Golland and Ron Blake.

B3: *Internal fire spread (structure)*: in the event of fire, stability must be maintained for a 'reasonable' period of time; resist spread of fire through common walls with adjacent buildings; limit fire spread within the building by creating separate compartments and prevent unseen spread of fire and smoke through concealed spaces.

B4: *External fire spread*: external walls and roofs must resist fire spread from one building to another.

B5: *Access and facilities for the fire service*: design and construction must incorporate facilities to assist fire fighters in the protection of life, and ensure fire appliances can gain access to the building from within the site.

C. Site preparation and resistance to moisture

C1: *Preparation of site*: the ground covered by the building must be 'reasonably' free of vegetable matter.

C2: *Dangerous and offensive substances*: precautions must be taken against health and safety risks from harmful substances on or in the ground covered by the building.
C3: *Sub-soil drainage*: sub-soil drainage must be provided for sites subject to flooding.
C4: *Resistance to weather and ground moisture*: floor, wall and roof construction must resist the passage of moisture to the inside of the building.

D. Toxic substances

D1: *Cavity insulation*: material inserted into the cavity of an external wall must not lead to harmful toxic fumes permeating into the building.

E. Resistance to the passage of sound

E1: *Protection against sound from other parts of the building and adjoining buildings*: design and construction should resist the passage of airborne sound between dwellings, another occupancy or another building.
E2: Protection against sound within a dwelling house, etc.
E3: Reverberation in the common internal parts of buildings containing flats or rooms for residential purposes.
E4: Acoustic conditions within schools.

F. Ventilation

F1: *Means of ventilation*: requires provision of 'adequate' ventilation to dwellings, multi-occupancy dwellings, toilets and bathrooms.
F2: *Condensation in roofs*: measures to minimise condensation in a roof or roof void.

G. Hygiene

G1: *Sanitary conveniences and washing facilities*: guidance on 'adequate' sanitary conveniences and washbasins; designed and installed to enable effective cleaning; provision of hot and cold water to washbasins; separation of facilities from places where food is prepared.
G2: *Bathroom*: guidance on bathroom's fixtures, and hot and cold water supply.
G3: *Hot water storage*: this must be installed by a competent person, prevent stored water exceeding 100°C and convey hot water safely to outlets.

H. Drainage and waste disposal

H1: *Sanitary pipework and drainage*: guidance on the disposal of foul water from sanitary conveniences, basins and sinks.
H2: *Cesspools and tanks*: guidance on capacity, construction, ventilation, access for emptying so as to avoid contamination of underground water and avoid health risks.
H3: *Rainwater drainage*: guidance on rainwater disposal from roofs.
H4: *Solid waste storage*: guidance on storage; access from building to store position, and from store to the street, described as 'adequate' for the buildings use.

J. Heat-producing appliances

J1: *Air supply*: supply of fresh air must be adequate for combustion of fuels, and the efficient working of flues and chimneys.

J2: *Discharge of products of combustion*: fumes must be transferred directly to the outside air.
J3: *Protection of building*: guidance on construction to prevent risks of fire to the buildings fabric.

K. Protection from falling, collision and impact

K1: *Stairs, ladders and ramps*: building users must be able to move safely between different levels.
K2: *Protection from falling*: accessible areas to have safety barriers to prevent risks of falling.
K3: *Vehicle barriers and loading bays*: barriers to prevent risk for cars parked within buildings.
K4: Protection from collision with open windows, skylights and ventilators.
K5: Protection against impact from and trapping by doors.

L1. Conservation of fuel and power in dwellings

(a) *Limiting heat loss*: through the fabric of the building, hot water pipes and ducts, and hot water storage vessels.
(b) *Space heating and hot water supply controls*: guidance about energy efficient equipment and controls.
(c) Energy efficient lighting systems and controls.
(d) Providing information to users on energy efficiency.

L2. Conservation of fuel and power in buildings other than dwellings

(a) Limiting heat loss and gain through the building fabric.
(b) Limiting heat loss from hot water pipes, ducts, storage vessels and service pipes.
(c) Energy efficient space heating and hot water systems.
(d) Limiting exposure to solar overheating.
(e) Energy efficiency of air-conditioning and ventilation systems.
(f) Limiting heat gains by chilled water in air-conditioning systems.
(g) Energy efficient lighting systems.
(h) Information for occupants on energy efficiency.

M. Disabled access to and use of buildings
Interpretation: definitions of nature of disability.

M1: *Access and use*: guidance on access into and movement around buildings.
M2: Access to extensions to buildings other than dwellings.
M3: Sanitary conveniences in extensions to buildings other than dwellings.
M4: Sanitary conveniences in dwellings: facilities for disabled use.

N. Glazing

N1: Protection against impact.
N2: Manifestation of glazing.
N3: Safe opening and closing of windows, skylights and ventilators.
N4: Safe access for cleaning windows, etc.

P. Electrical safety

P1: Design, installation, inspection and testing.
P2: Provision of information.

Regulation 7: Materials and Workmanship (1985)
Materials and workmanship: to be 'appropriate' for the intended use.

The details within the Regulations highlighted here need not be memorised as they are periodically revised. It is important to understand what each covers, be aware of the issues which may apply to proposals, and know where to look for advice when problems are encountered. Other areas of interest include the following.

The Disability Discrimination Act – Empowers the Local Authority to ensure that all publicly accessible buildings cater for the needs of disabled visitors and users.

The Part Wall Act – Provides a framework for preventing and resolving disputes in relation to party walls, boundary walls and excavations near neighbouring buildings. Anyone intending to carry out work of the kinds described in the Act must give adjoining owners notice of their intentions.

European Directives – Codes of Practice for activity at a European level largely concerning structural work at present, but likely to develop in the future in an attempt to harmonise standards.

4.6.2 Building Control applications

Almost all new building work is subject to current Building Regulations which must be considered when design proposals are put forward. Of even greater importance is ensuring that work constructed is in accordance with the regulations. Whatever the drawings or specifications state in principle, it is the finished article on site that really counts. Consequently, the BCO has the dual functions of liasing with the design team to approve proposals and inspecting on site to approve construction. There are currently three ways of dealing with Building Control applications.

Full Plans application

- Detailed drawings and specifications are deposited with the BCO.
- BCO checks and registers applications.
- BCO advises the applicant or their agent of any discrepancies, inviting amendment.
- BCO issues notice of approval, including any conditions to be satisfied in due course, such as supplying supporting structural calculations for steelwork when they are available.
- The BCO is notified when construction will commence on site.
- BCO visits the site to inspect work at various stages to check that it is satisfactory.
- BCO makes a final visit on completion and issues certificate of compliance.

The formal approval can be obtained after work has commenced on site in co-operation with the BCO, providing that the BCO is able to carry out necessary inspections. This is a common situation, but there is a risk that completed work may not comply.

Building Notice

A Building Notice is a registration of intent to start work, and does not require the deposit of detailed drawings and specifications. The BCO will liaise with the contractor on site to ensure compliance with the Regulations. A certificate of compliance will not be issued. This is a method of dealing with Building Control normally applicable to relatively small projects and where the

BCO is familiar with the contractor. It would not be acceptable for a large development like a car-dealership which must be formally approved by the Fire Authority. Also, the complexity of construction would place a considerable burden on both BCO and contractor increasing the risk of errors.

Self-certification
It is possible to submit written assurance that proposals comply with the current Building Regulations. However, the consequences of incorrect self-certification could lead to problems if mistakes or misinterpretations are not discovered before construction starts, and even worse if they are found after construction has been completed.

Formal applications, which attract fees must be made at least 48 hours prior to starting work on site, but BCOs encourage advance discussion to confirm if the work is subject to regulation, to offer advice and help to resolve any problems.

For both Planning and Building Control, failure to notify the Local Authority of an intention to build, or carrying out construction which does not comply with the requirements of legislation can result in prosecution, attracting penalties. Legislation also enables the local authority to serve an enforcement notice requiring the client to demolish the new building or reinstate alterations to existing buildings if the regulations have been contravened.

4.7 Other agencies

Planning and Building Control are the principle forms of control for all building development, but there are a number of other agencies who have interests depending on the type of building proposed and its location:

- *Fire Officer*

Ever since the Great Fire of London in 1666, methods of reducing risks to life and property in the event of fire in a building have been regularly revised and improved. Subsequent catastrophes such as the Ronan Point tower block collapse and the Summerlands leisure centre fire have seen the introduction and refinement of further safety measures. The Fire Authority's principal duty in the event of a fire is to attempt to save life. The means of escape from a burning building, detectors and alarms, fire-fighting equipment, and access for rescue and fire fighting are their main concerns. Measures to protect structure and finishes, together with prevention of fire spread to adjacent property, is dealt with by the current Building Regulations, administered by the Local Authority.

The issues relating to design proposals which comply with the Fire Officer's requirements are outlined in Part B of the Building Regulations. The BCO will scrutinise proposals for compliance, and forward them to the Fire Officer for confirmation that they are acceptable. The Fire Officer will inspect the building before it is occupied to confirm satisfaction and will issue the appropriate certificates of compliance. The Fire Officer will also have a relationship with the client once the building is occupied, and may make regular visits to check that the building is maintained in a safe condition.

- *Police Authority*

Advice is available on security measures that can be achieved by design. 'Secured by Design' booklets describe sensible measures for different building types. At present, there is no statutory control in this area.

- *Environmental Health*

The local authority's environmental health officer will be interested in pollution and fumes, contamination, control of storage and disposal of industrial effluent and waste, control of noise, working/operating times and general nuisance to adjacent property owners, and various health and safety issues.

- *Service Providers*

Electricity, gas, water and telecommunications companies operate to statutory controls to provide efficient and safe services installations.

- *Insurance Companies*

For approval of some methods of construction, selection of finishes and installation of security measures such as detectors, alarms and sprinklers which may limit damages or loss to property or contents.

- *Health and Safety Executive*

The Health and Safety Commission (HSC) and the Health and Safety Executive (HSE) are government sponsored bodies responsible for the regulation of risks to health and safety arising from work activity in the UK. The authorities advise, inspect and prosecute in cases of contravention.

- *National House-Building Council (NHBC)*

The NHBC is the standard setting body and leading warranty and insurance provider for new homes in the UK. Its main customers are homebuyers and registered house builders. The NHBC registers builders, regulates standards, promotes sustainability, inspects construction work and offers Building Control services.

- *Local Authorities*

For advice and approval of matters including highways and drainage, traffic control and temporary arrangements during construction. Local Authority departments also procure new buildings; education, leisure and social services for example, and may employ expert staff to brief designers.

- *Parish Councils*

For input about issues associated with the built environment responding to local needs through consultation with the community, voluntary and business sectors.

- *The Environment Agency*

The Environment Agency is a recently established authority based on the previous Inland Waterways Authority with powers to control development close to existing water courses, and where construction activity and subsequent use may affect water table levels, and introduce risks of pollution and contamination.

The Environment Agency provides environmental protection and improvement working with businesses and other organisations to prevent damage to the environment by providing education, guidance and, where necessary, enforcing regulations through prosecutions.

- *Magistrates*

For licences approving certain types of commercial use.

- *English Heritage*

Manages, maintains and cares for many ancient monuments, historic buildings and landscapes throughout the UK.

- *The National Trust*

Founded in 1895 by three Victorian philanthropists, Miss Octavia Hill, Sir Robert Hunter and Canon Hardwicke Rawnsley, its interest is to act as a guardian for the nation in the acquisition and protection of threatened coastline (approximately 600 miles), countryside, including forests, woods, fens, farmland, downs, moorland, islands, archaeological remains, nature reserves, whole villages and buildings (presently some 200 buildings and gardens of outstanding interest and importance), all generally held in perpetuity to secure their future protection.

- *Governmental Agencies*

Including Department for Transport, Local Government and the Regions (DTLR), Department of the Environment, Transport and the Regions (DETR) and the Building Research Establishment (BRE).

- *Other Specialist Activities*

Independent or government-sponsored organisations such as The British Railways Board, The Strategic Rail Authority and UK Coal.

4.8 The Construction (Design and Management) Regulations 1994

Traditionally, the welfare of construction workers on sites had been largely the concern of the general contractors, responsible to the HSE for compliance with legislation. According to 'working well together' approximately 2 million people work in construction, the country's biggest industry. Since 1980, some 2800 people have died from accidents at work, and many more have been injured or made ill. The main causes of injury were falling through fragile roofs and rooflights, falling from ladders and scaffolds, being struck by moving equipment or overturning vehicles, and being crushed by collapsing structures.

Notwithstanding advice, supervision and fines for noncompliance, the building industry had a poor record for site safety. Risks had not always been recognised or treated with sufficient respect and the dramatic annual figures recording injuries and deaths far exceeded other comparably dangerous activities such as underground mining. The designer's contribution was limited to general clauses in contract documents requiring *the contractor* to be responsible for issues which might cause injury, and to protect employees and visitors to the site.

Many accidents resulted from poor pre-construction research and planning, failure to identify areas of concern and inform those at risk. The proposition that *all parties* involved in the construction project should have *a duty to reduce accidents* by promoting a *safe working environment* was the basis for the new legislation. CDM regulations in force in the UK are an interpretation of European Directive 92/57/EEC relating to minimum health and safety requirements on construction sites. They are enforced by the HSE for most projects involving construction work although certain small works are exempt.

The regulations introduced a new member to the development team, the **Planning Supervisor**, appointed by the client to co-ordinate the requirements of the legislation. The designer can fill this role, but for larger projects the Planning Supervisor may be an independent specialist. The Planning Supervisor's responsibility is to ensure that good management practices are followed from inception to completion so that health and safety implications are considered for the *lifetime* of the building, including future use, maintenance and potential demolition.

The regulations impose **duties and responsibilities** on the following parties to any building development:

- Clients
- Planning Supervisors
- Designers
- Principal Contractors
- Subcontractors and self-employed workers.

The legislation focuses on the following key issues:

- **Competency**: appraisal of both design and construction personnel.
- **Risk assessment**: for the handling of existing and potential hazards.
- **Co-operation**: sharing of critical information during development.
- **Communication**: recording and passing on information relevant to future activity.

The regulations require the production of the following:

- **Tender Health and Safety Plan**, compiled by the Planning Supervisor identifying all the risks associated with the works, together with the designer's risk assessments.
- **Site Health and Safety Plan**, compiled by the Principal Contractor showing how risks highlighted in the Tender Health and Safety Plan will be addressed.
- **Health and Safety File**, co-ordinated by the Planning Supervisor and handed to the client on completion, explaining how to use the building safely, with 'as-built' details.

4.9 CDM responsibilities

As the designer may act as the Planning Supervisor, the duties and responsibilities of all the parties will be of concern. The following lists summarise the essential requirements.

4.9.1 The client

Competency

- To allocate sufficient time and resources to enable compliance with health and safety legislation as design and construction work proceeds.
- To appoint a competent Planning Supervisor and to notify the HSE of the appointment.
- To appoint competent designers and a principal contractor.

Risk assessment

- To freely provide, make available or obtain information about existing and potential conditions relating to the site, buildings and processes, where there is a health and safety implication for design and construction work.

Co-operation

- To contribute relevant information to the Tender Health and Safety Plan.

Communication

- To ensure construction work does not commence until the principal contractor has prepared a satisfactory Site Health and Safety Plan.
- To arrange for descriptions of construction and instructions for use to remain available after completion for those concerned with future occupancy, maintenance, repair, refurbishment and possible future demolition of the building.

4.9.2 The Planning Supervisor

Competency

- To advise the client on the competence of designers, and to ensure designers fulfil their duties, co-operate with each other, and avoid or reduce risks to agreed acceptable levels.
- To advise the client on the competence of all contractors.

Risk assessment

- To agree method statements for works which present unavoidable risks.
- To advise the client on the resource implications of achieving a safe working environment.

Co-operation

- To co-ordinate health and safety matters between design and construction personnel.

Communication

- To inform the client of his or her responsibilities under the CDM regulations, and to advise the HSE about development intentions and to notify the HSE when construction is to commence on site.
- To prepare the Tender Health and Safety Plan, identifying areas of risk.
- To advise the principal contractor and approve their Site Health and Safety Plan.
- To advise the principal contractor on the appointment of their own designers, subcontractors and suppliers with respect to competence and risk.
- To ensure that the principal contractor complies with the Site Health and Safety Plan during the contract period, and to incorporate additional safeguards, measures and strategies to cover unforeseen problems or discovered risks as work proceeds.
- To ensure that the Health and Safety File is prepared as work proceeds and is delivered to the client on completion.
- To arrange for training, information and instruction to be provided by the contractor for the building's users where operations may have health and safety implications.

4.9.3 The designers

Competency

- To fulfil the responsibilities of the Planning Supervisor if acting in that capacity.

Risk assessment

- To identify and consider hazards and risks resulting from design proposals.
- To develop proposals avoiding or minimising risks as far as is reasonably practicable.
- To consider methods of safe working where risks cannot be eliminated.

Co-operation

- Co-operate with all parties to achieve safe working conditions to provide information about health and safety for inclusion in the Tender Health and Safety Plan, and clearly identify risks on drawings and in specifications.

Communication

- To ensure that information is readily available on site and updated to take account of design changes or unforeseen eventualities.
- To contribute to the Health and Safety File, including as-built drawings and specifications.

4.9.4 The Principal Contractor

Competency

- To employ competent domestic construction personnel and subcontractors, and to allow adequate resources to minimise risks.

Risk assessment

- To take account of the specific requirements of the project when preparing tenders.
- To obtain subcontractor's risk assessments and agree proposed construction methods.
- To ensure all subcontractors are aware of, and have information about risks on site.

Co-operation

- To take over the Tender Health and Safety Plan, and to develop the Site Health and Safety Plan, ensuring that all site work personnel adhere to it.

Communication

- To co-ordinate site personnel and operations, consulting regularly with employees.
- To ensure that all workers on site have been given adequate training.
- To monitor health and safety performance, to maintain a register of any accidents and ensure that all workers are properly informed and consulted.
- To control access to risk areas to authorised personnel only.
- To comply with legislation, liaise with the HSE and display relevant documentation.
- To contribute to the Health and Safety File so that it is comprehensive on completion.

4.9.5 Subcontractors and self-employed workers

Competency

- To demonstrate their competency and compliance history.

Risk assessment

- To advise on risks arising from their work and proposed methods of minimising them.

Co-operation

- To manage their work in compliance with the Site Health and Safety Plan, and to respond to any directions from the Principal Contractor.

Communication

- To notify the principal contractor about injuries, ill health or dangerous occurrences.
- To keep their own employees fully informed about risks.
- To contribute information to the Health and Safety File.

4.9.6 The Health and Safety File: Contents checklist

Site information

- Project information and description; previous Health and Safety Files.
- Site surveys, investigations and photographical records.
- Contact details for all parties involved in the project.

Design criteria

- Details of design concepts and risk assessments.

Drawings and specifications

- Pre-tender information, construction information and as-built record information.

Methods used in construction

- Method statements produced prior to construction.
- Amendments in the light of experience.
- Details about construction sequences to aid reversal in the event of demolition.

Materials used in construction

- General descriptions of materials used and details of hazardous materials.
- Details of hidden materials hazardous to those undertaking demolition.

Utilities and services

- Location and nature of services including depths, sizes, capacities and details of the risks if exposed or altered.

Maintenance procedures

- For structure and finishes, including manuals from subcontractors and suppliers.
- Manuals for operation and maintenance from suppliers of specialist equipment.

The Tender and Site Health and Safety Plans

- Copies of the original documentation developed throughout the history of the project.

4.10 The client

It goes without saying that the client's agreement at all stages is important if the project is to proceed smoothly. The client is the customer, entitled to get what they want from the advisers and contractors that they select. As far as they can, the advisers and contractors attempt to meet the client's expectations, but is important that everyone should understand that there are specific occasions when the client's formal approval is required to avoid future difficulties. Some of these moments are referred to as 'freezes', points in the design and construction processes which determine the course of future activity. The order in which freezes arrive depends on the client's chosen method of procurement. For designer lead projects, the client will be invited to approve the choice of consultants, the selected contractor and subcontractors during the course of development. Contractor or project management lead projects will generally involve the client in these issues at a much earlier stage. The main issues requiring the client's approval include the following:

- *Conditions of appointment of development personnel*

The client may appoint a building designer, consultants or contractor/manager to begin the process of designing their building.

- *The initial concept and costs*

The client's brief and target costs for design work and construction.

- *The outline proposal*

The agreed scheme which is to be developed into a realistic proposition. This stage may also require agreement to external constraints such as the demands of a Planning Brief, or the satisfaction of neighbours.

- *Elements within the proposal*

Any individual details and requirements.

- *Final design and costs*

This is referred to as the 'design freeze' after which further changes will seriously affect the programme. A fundamental point arrives when the Planning Application is submitted, formally determining the design proposals.

- *Selection of form of contract*

The client's procurement method will influence the specific forms of contract that they wish to enter into (see Chapter 12).

- *Selection of tenderers and appointment of general contractor*

This is applicable to the traditional procurement method where the client agrees which contractors should be asked to tender and approves the selection of one of them to undertake construction work.

- *Standards of materials and workmanship before work commences and before it is accepted as completed*

The client should be involved in agreement about quality standards for all aspects of the work.

- *Selection of fixtures, fittings and finishes*

The client may wish to personally choose any of these items rather than accept recommendations from advisers.

- *Handover of the finished building*

The client should be satisfied that work is completed to their satisfaction as defined in any agreements entered into. See snagging and defects in Chapter 14.

- *Settlement of the final account*

The client may be personally involved in negotiations and arrangements about costs, but may need to approve the recommendations of advisers when all the works have been completed.

4.11 Discussion points

(1) What are the criteria for determining the suitability of land use? Who should be involved in deciding where the Greenfield/Brownfield boundary is located? How should land recycling policies be established?

(2) What responsibility does the government have for co-ordinating the development of the built environment? What private interests might conflict with public policy? What is the relationship between privately owned land and public access?
(3) Imagine a situation where there was no Planning or Building Control. What would a free-for-all system be like if land owners and developers could do exactly as they pleased? Who is responsible for remediation of contaminated sites? How can vehicular traffic and pedestrians use urban spaces together?
(4) Do regional development pressures lead to variations in the way in which the built environment is managed? How important is vernacular tradition when development expansion is urgent? Should today's construction industry take responsibility for tomorrow's built environment?

4.12 Further reading

Adams D (1994) *Urban Planning and the Development Process*. London: UCL Press.
Collingworth JB and **Naden** V (1994) *Town and Country Planning in Britain*. London: Routledge.
European Union and International Regulations. www.odpm.gov.uk
Greenstreet B, **Chappell** D and **Dunn** M (eds) (2003) *Legal and Contractual Procedures for Architects*. London: Architectural Press.
Health and Safety Executive. www.hse.gov.uk
Health and Safety Executive: Designer's can do more. www.hse.gov.uk/construction/designers
Joyce R (2001) *The Construction (Design and Management) Regulations 1994 Explained*. London: Telford.
Law Commission. www.lawcom.gov.uk
Law Society of England and Wales. www.lawsociety.org.uk
Office of Public Sector Information. www.opsi.gov.uk
Office of the Deputy Prime Minister. www.odpm.gov.uk
Perry P (2002) *CDM Questions and Answers: A Practical Approach*. 2nd Edn. London: Telford.
Safety in Design. www.safetyindesign.org
Secured by Design. www.securedbydesign.com
Speaight A and **Stone** G (2004) *Architects Legal Handbook*. 8th Edn. Oxford: Elsevier.
Stephenson J (2001) *Building Regulations Explained*. 6th Edn. London: Spon Press.
Stranks JW (2001) *Health and Safety Law*. 4th Edn. London: Prentice Hall.
Stranks JW (2002) *Health and Safety at Work*. Oxford: Butterworth-Heinemann.
Summerhays S (1999) *CDM Regulations Procedures Manual*. Oxford: Blackwell Science.
The Building Regulations (1995) *Complete Package*. London: HMSO. See also www.odpm.gov.uk
Turner C (2002) *European Law*. London: Hodder and Stoughton.
Uff J (2002) *Construction Law*. 8th Edn. London: Sweet and Maxwell.
UK Fire Service Resources. www.fireservice.co.uk
UK Parliament. www.parliament.uk
Working Well Together. www.wwt.com.uk

SECTION 2 The Design Period

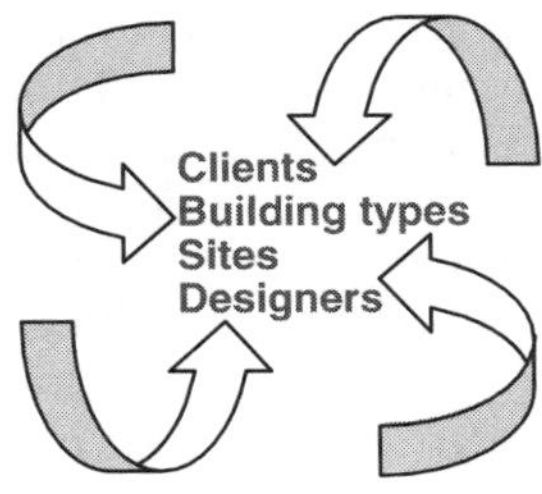

5

Inception

5.1 Introduction

The first section of this book examined the nature of design, communication skills and the extent of external statutory constraints, reviewing in general terms the roles, essential skills and professional responsibilities of the members of the design and construction teams. Section 2 reviews the design process as the new car-dealership is created, starting with collection and analysis of information about the client, the building type and the site, and follows the development of possible realistic responses to the demands of the project through to the point where construction could begin. This section describes the way in which physical and aesthetic issues influence design potential, how project planning affects progress and how ideas are developed, defined and described so that they can be turned into reality.

The idea of creating a new building usually comes from the client, seeking to establish or expand their business activity. The building designer, and others in the design development team need to understand the client and their needs in order to respond in a useful way. This chapter includes a preliminary analysis of the client and explores their ideas or *brief* in two important contexts. Firstly, a dispassionate review of their situation as it stands, the 'status quo' of their current business, and then secondly, the potential of their business and the relocation and development of their new *site*. The constitution of the design team is considered in light of the need for specialist advice and an assessment is offered of the merits or values of initial design ideas.

5.2 In the beginning

The starting point in the creation of a new building is called *inception*, when the client first approaches someone for professional advice about the possibilities of creating their new building. This *someone* may be a contractor, a project manager, a building designer or anyone else with an interest in the client's situation. In the traditional procurement process, the building designer is the client's first point of contact, commissioned to manage the design development process leading to the appointment of a builder. To the client, this method represents a relatively lengthy and costly design period, possibly leading to a tender figure higher than the budget target. In many cases, today the client's first discussions are with contractors or management consultants with whom they negotiate matters relating to costs and programming in advance of detailed design work. Design and build or management contracting procurement methods help the client to establish the feasibility of their requirements quickly on the basis that the risks of making outline design assumptions are manageable once detailed design work takes place. This method requires both parties to incorporate design risks (escalating costs or diminishing design quality) into their negotiations.

Whichever procurement method the client adopts, and at whatever point the building designer becomes involved with the project, the client is likely to already have some idea of their requirements, although this is not always the case. Sometimes, the first contact with the client is a relatively informal meeting, perhaps on the site, simply to consider possibilities. The client may be looking for an opinion or alternative ideas about the viability of any development at all. This is referred to as the *initial consultation*, preliminary discussions perhaps held before the client has even formulated their own outline brief. The client should always be encouraged to give as much general information as they can about themselves, their desires, needs and preferences. They may have a very clear idea of what they want, which can be extremely helpful, but it is not uncommon to find that their ideas are both unrealistic and inappropriate for their circumstances once all the issues have been examined. One of the principle duties of any advisor is to 'act in the client's best interests' to see that what they get is, as far as possible, what they actually need. This can be a drawback in any procurement method which minimizes detailed analysis at the beginning of the process because adopting a specific design solution too quickly may mean that a better solution is overlooked, or is unachievable because of subsequent or consequential commitments.

Inception is the time to establish the framework for the project, so that the client understands what is going to happen. The client needs to be made aware of the design and construction processes so a suitable design team can be assembled (or appointed) and that a programme can be devised allowing sufficient time to complete each stage, including compliance with the demands of statutory authorities. The need for specialist consultants must be considered for matters which are beyond the scope of the building designer's work, such as structural design and air conditioning systems (early advice on design potential or limitations may be invaluable), and the client will look for advice on the total projected cost of the development, including all professional fees and incidental costs.

Even if the client approaches the building designer before anyone else, it is useful at the earliest possible time to consider involving contractors, subcontractors and suppliers in the design process. For example, if the new building is to include some special form of construction or material, or a complex piece of machinery then the advice of builders, manufacturers or suppliers will be useful, or possibly essential in order to ensure that associated elements of the building design are appropriate. The desired speed of construction may dictate the need to take advantage of the experience of contractors at the design stage so that methods of construction are devised with which they are familiar and which are safe, practicable and economical. This may determine the particular form of contract that would be most suitable when the time comes to formalize employment arrangements between the client and the contractor. Above all, communication links must be quickly established between the client and all the other members of the development team. Briefing and approval arrangements must be clearly understood by all parties so that work can progress smoothly.

5.3 Personal organization

One of the most significant factors when working as a part of any team is the way in which good working relationships are formed and maintained, so that everyone has confidence in the way that contributions are being made. The client, or any employer must be persuaded that their advisors or employees can competently undertake the work that they are to be paid for. As well as formal qualifications, this can be achieved through marketing, a prospectus of previously completed work or a traditional curriculum vitae (CV). But confidence is also earned through punctuality, care and attention to detail, fulfilling commitments as promised and responding quickly to requests and instructions. In practice, members of the design team rarely have the luxury of being able to concentrate on a single project. Responsibility for a number of projects simultaneously creates pressures and demands which are difficult to control without strict planning of work and time.

As design work progresses, there is a need to record names, places, appointments and many notes for future reference. Important documents go into the project file, but it is useful to have a system for recording scraps of information such as scribbled notes following meetings and telephone conversations which are easily misplaced. The consequences of forgetting or overlooking important information can be serious and expose the designer to accusations of negligence. At best, it can be very embarrassing to ask for it to be repeated.

The following aids will help personal organization.

A diary

A diary is essential for organizing both business and personal meetings, for arranging appointments and for confirmation of when previous meetings were held. A diary can be used to compile contact names, addresses and telephone numbers, and to record the time spent working on projects, together with associated out of pocket expenses. Some commercial offices keep a group diary to plot the movements of all staff so that a record can be arranged to cope with absences and holidays.

A day book

Faced with many urgent tasks, there is a risk that some of them are ignored or undertaken in the wrong order. It is easy to become depressed by having so much to do that it is difficult to decide what to do first. One solution is to make a list of the tasks at the beginning of the day, highlighting the ones which are most important. The diminished list of tasks can be moved forward to the next day (with new ones added, of course) and new priorities targeted. The daybook helps to plan action for the day, and also records events during it.

A project book

Important information can be collected into a project book, recorded in date order as it is received. If the designer is working on a number of projects, each one can have its own book so that details do not become confused. Together with the diary, the project book can be taken to meetings and visits to record relevant findings.

A database of contacts

Details of the names, addresses and telephone numbers of the specific project personnel can be kept in the project book for easy reference, but building up a larger database of contacts is useful for future reference. For example, groups of builders, subcontractors, suppliers and authorities can be easily updated and are readily available for swift selection or communication.

However information is handled, whether on scraps of paper, in note books or on electronic personal computers, a systematic approach in recording it will help to retrieve it quickly later on. Fee income is essentially based on time, which should not be wasted through inefficiency.

5.4 The client

Although most clients are unlikely to be fully familiar with the process of design and construction, they do have experience of buildings and of their own specialist activities and processes. It is worth noting that approximately 80 per cent of clients are inexperienced, perhaps never acting in this capacity before. Therefore, many clients have little or no background to prepare themselves to evaluate design, or to offer complete briefing. There is a need for strategic front-end advice as to what they should expect of their professional advisors. It can be argued that this is a fundamental problem with the whole process as highlighted by a succession of reports or enquiries into the construction industry and procurement methods.

For the car-dealership, the client's understanding of the arrangement of spaces for car sales and repairs is a good starting point, but it will inevitably be based on their own previous experience, which may or may not have been successful and efficient. In the first instance, the client's assumptions should be questioned to see if there is any better alternative meriting consideration. Undoubtedly, decisions about some elements of the new building will be dictated by the client irrespective of the strength of any counter arguments. In this respect, there is a need to tread a delicate path between passive acceptance of instructions and active promotion of alternatives. To some extent, this depends on the confidence that both the client and their advisors have in one another, and how well the advisors get to know and understand the client and their representatives.

Whether the client is an individual or a group, good communication with an authority figure is essential. Liaising with an individual is relatively straightforward but when the client is a group it can be difficult to obtain accurate information and gain the necessary approvals. It can be especially complicated when members of the group have individual responsibility for certain parts of the development and take different views on the way in which the building should be used, or how the new design should accommodate changes from previous experience. This can be further complicated when important individuals are unable to attend meetings, or fail to supply critical information, or worst of all, where there is a change of personnel altogether, so that there are radically different views part way through.

For example, the personnel in the client's management structure for the new car-dealership could include the following:

- Owner or owners
- Chief executive or managing director
- Sales manager
- Parts manager
- Service manager

- Administration manager, company secretary or personnel manager
- Departmental deputies or assistant managers

Other employees could include receptionists, sales and administrative staff, counter and petrol pump attendants, service and valeting staff, mechanics, drivers, security guards, cleaners and general maintenance staff.

Each person will have their own interests and concerns which may not be compatible with one another. When the needs and requirements of the project are discussed at briefing meetings careful attention should be paid to the debate in order to understand the reasons for variations, which may be unarguable matters of fact or simply the expression of preferences or desires. For example, the parts manager 'requires more floor space for storage' than the managing director would like, who prefers 'to invest in showroom space' instead. The service manager is interested in 'new "state of the art" hydraulic ramps' which are five times as expensive as the standard ones or the sales manager prefers 'private sales offices', but the managing director insists on 'open plan work stations' so that the staff are visible and accessible to customers at all times and so that staff activity can be effectively monitored. There may also be an element of 'empire building' to contend with, where resistance to change may place obstacles in the way of genuine briefing where personal or institutional bias obscures real requirements.

Each view will have its merits requiring due consideration before decisions are made. Advice can be offered about practicalities, but it is more important to prompt the best decisions with regard to satisfying everyone's needs, or at least determining priorities. As this may not be possible, and to avoid embarrassing arguments in meetings, it may be best to give the client the opportunity to confirm requirements later when their team have resolved any contentious issues.

During the inception period, discussions with the client will determine the following.

The client's preliminary intentions

It is useful to meet the client on site to receive their brief and relate their preliminary ideas to the reality of the location. This is the first opportunity for an assessment of the broad feasibility of the client's brief in relation to development potential.

The extent of the designer's involvement

At the outset, the client may simply be looking for feasibility ideas which the designer can illustrate with sketches. Assuming that a scheme is acceptable to the client, the designer's involvement could continue in stages developing outline design drawings suitable for a Planning Application, co-ordinating consultants in the production of fully detailed working drawings for Building Control, tendering and construction. The designer may be asked to undertake full supervision of construction work or act as Planning Supervisor under Construction Design and Management (CDM) legislation. Alternatively, the designer may be employed by a contractor or project manager and will need to agree with them exactly what is required. Each is a separate stage which when completed, could mark the end of the designer's involvement. The designer could be employed stage-by-stage, or appointed to undertake all of them.

The nature of the designer's work and the need for other specialist consultants

The designer must clarify the elements of design work for which he or she does not have the necessary expertise and advise the client of the need to employ other independent advisors. The need for specialist advice is no different with other forms of procurement, but simply transfers responsibility for the employment of consultants to the design and build contractor or project manager.

The statutory controls which may be applicable to the development
The client should be made aware that at various stages in the design and construction periods proposals will be subject to the approval of a range of statutory authorities, who charge fees for their services. Satisfying their interests may have implications on the way that the building can be designed, and will almost certainly be reflected in the cost of construction. The time required to obtain approvals must be included in the design programme. The client should be made aware of the implications of CDM legislation.

The preliminary time scale for design work and for construction completion
The client will want to know how long it will take to reach certain stages, and when their building will be finished and ready for occupation. In a commercial context, they will need to know when they can plan to commence their new business operations.

Costs
The client will have a budget for the development which may or may not be realistic for their intentions. A project of the size of a car-dealership is likely to require the services of a specialist financial consultant or quantity surveyor (QS) who can advise the client and the design team on all aspects of cost planning and budgeting. The client will want to know as soon as possible that their intentions can be financed. Extensive consideration of costs and financing are beyond the scope of this book, but it is interesting to consider the relationship of costs to design quality. One of the biggest debates in the UK construction industry concerns the value of 'good' design, and will clients be 'prepared' to pay for it?

In his 2003 Povey Lecture to the Joint Contracts Tribunal (JCT), Richard Saxon quotes from the Royal Academy of Engineering study of the long-term costs of owning and using buildings, referred to as the 1:5:200 rule; 1 unit is the initial cost of the building, 5 units equates to 20 years of operating costs and 200 units is the extent of 20 years of business staffing costs. He suggests that the business value created could approach 2500 units or more. Bearing in mind that designing and managing the building project is typically 0.1 of a unit, it can be easily appreciated that a small investment in improving design quality can represent enormous lifetime value to the client, building users and the community as a whole.

5.5 The client's brief

The client's brief sometimes consists of a written explanation or list of requirements which forms the basis or framework for discussions about design possibilities at the first meetings. The content of the client's brief generally takes one of three forms:

(1) A simple statement of *their* perception of *their* needs with no notion of design possibilities or how to proceed.
(2) A set of principles identifying *some* limiting factors with a concept of possibilities, but seeking advice on practicalities and options.
(3) A *detailed summary* of requirements with a clear idea of the solution to their needs.

In each case it is sensible to take a cautious approach and not assume that ignorance obscures understanding, or that comprehensiveness precludes alternatives. This is the first of many 'crunch' points in the process of creating the new building and it is a very important one, because it is at this point that aims and objectives can get misdirected by either under or over-valuing priorities so that the whole direction of the design is incorrectly influenced, limiting the chances of producing the best result. Initial ideas should not be adopted too quickly without full consideration of implications as fixed preconceptions about some elements may be incorrect or unhelpful. This statement applies to the designers as well as the client who should both resist the temptation of 'jumping to conclusions' too soon. At this stage possibilities should be evaluated cautiously, based on the elements of information which are available. It is only through a process of collecting and clarifying additional information that one realistic possibility may eventually prove to be worth pursuing.

For the car-dealership, the following is a typical summary of discussions at the first meeting with the client.

Minutes of meeting No 1
Monday 15 August 2005
Present: Client and designer

1.1
The client is the owner and managing director of a small car-dealership business on the fringe of the city in a predominately industrial and commercial area. The existing buildings are in need of extensive refurbishment, but too small to cater for expansion intentions to justify the expense of upgrading the existing premises or constructing a new building. In any case, the client would be reluctant to cease trading for the period of time that would be needed to undertake construction works. The existing business, started in 1966 has developed from repairs and servicing, and presently has up to 25 second-hand cars on display at any one time. In all there are approximately 20 staff employed in the business, although space and amenities for customers and staff are recognized as worse than poor.

1.2
Currently business is buoyant, and the lack of space and facilities is limiting expansion. The client is interested in moving 'up-market' into new car sales and has an opportunity to take a franchise from a major manufacturer. To match the company's 5-year business plan, larger premises are required, together with more external display space. Whilst the manufacturer would normally dictate the style of development, the client has negotiated the rights to remain independent, continuing with the company identity that has become firmly established over the past 30 years.

1.3
The client's intention is to double the size of the present operation, including parts and incidental sales with consideration of the possibility of fuel sales. Assuming targets are achieved, there could be a need for the design of the buildings and external spaces to be able to cater for a further 20 per cent expansion in approximately 5–10 years time.

1.4
Initial research has identified a suitable site, close to the city centre on a major feeder road. See Figures 5.1–5.3, which illustrate the site at different scales. The site is actually made up of two separate parcels of land, both containing redundant buildings. The site is subject to development grants and de-contamination funding. Informal discussion with the Local Authority has suggested

that development of a car-dealership on this site is likely to be acceptable in principle, subject to usual conditions. The status of the existing Victorian pumping station is currently under debate, and may merit some consideration of retention in any possible redevelopment of the site.

1.5
The client has secured a right to purchase the site, subject to a full-site analysis proving it to be viable, and is prepared to appoint a designer to offer full design and supervision services. It has been agreed that a contractor will be selected following competitive tendering. The programme requires design and construction to be completed as soon as possible. Early occupation of the new building is critical for maintaining the financial forecasts identified in the 5-year plan. The client's financial status is such that the new development can be funded, subject to completing a sales contract with an interested buyer for the existing premises.

1.6
The client appreciates that their current organization and arrangements do not match their ambitions. They have built up their business within the constraints of available space with equipment which is now out-of-date. They have a clear understanding of how they wish to conduct their business but would like to re-examine all their processes so that any design solution fully satisfies their practical needs and creates a building that they can be proud of. The client would like to arrange further meetings to discuss possibilities and establish a design and construction programme.

This is a distillation of a lengthy conversation, summarized in the form of a minute. It will be sent to the client who has the opportunity to confirm or change any of the points of information, and is the basis for commencing design work. As well as the general points outlined in the brief, the client may have specific information about some aspects of the development, which will help the designer from the outset. They may have a clear idea of the minimum floor area needed for the activities with which they are familiar, such as the space needed for repairing a car on a ramp or the storage space for the parts that they stock. They will almost certainly have an idea of the maximum amount of money that are prepared to spend on the development. For every building type, it is possible to establish an approximate budget cost per square metre, which is a useful initial guide as to how large any new building may be, and what accommodation can be contained within it.

5.6 Analysing the client's brief

The example of a client's brief given in Section 5.5 is a useful one as it gives simple information about the client, their historical background and the status of their current business activity. It defines requirements in general terms, but accurately enough to understand the company's ethos and ambitions, realistically related to a well-thoughtout business plan. It gives sufficient information to appreciate the potential size of the new development, including the need to consider future expansion possibilities.

The content of the client's brief can be examined in detail to see what further clues are being offered and to assess the implications of each statement.

1.1

- The client is an individual person, successful and ambitious, and as managing director is likely to be personally involved in the new development. The extent of involvement may

range from a general overview of proposals, concentrating on costs and programming to detail concern about every design issue. Design presentations may need to be targeted at the individual, or if the managing director is the chairman of a board of directors, at a group who may make decisions collectively.
- The experience of the existing staff will be useful but there may be differences of opinion at briefing meetings. It is essential that all decisions are approved at the appropriate level of the client's management team.
- Although physical conditions in the existing buildings are poor, business is buoyant, suggesting that customers appreciate the services currently offered, supporting the ambition to upgrade and expand operations and confirming confidence in the viability of the new project.
- The client has little or no experience of selling new cars and may be unable to provide suitable briefing information for this aspect of the development.

1.2

- Although the client has negotiated independence, the manufacturer may be the source of sensible and useful advice about current good practice, which should not be ignored.
- The client's company identity may not be appropriate after 30 years and they may wish to re-examine their position in relation to more modern practice.

1.3

- With regard to the accommodation to be provided in the new development, the brief is rather vague. 'Doubling the size of the operation' could mean a number of things. From Paragraph 1.1, it may mean the provision of space to display 50 second-hand cars and room for 40 staff, or it may mean a new building and display space twice as big as the existing premises. As the existing accommodation was cramped and inadequate, then on this basis the new development will be exactly the same. It could mean doubling turnover or profit with only a small increase in the size of the new building and the number of staff employed. This could be achieved by selling goods of higher value or trading in a more efficient way. At this stage, these issues are unresolved and need to be examined in much more detail before any decisions are taken about the size of external spaces and internal floor areas.
- In any event, the design of the building and the layout of the external spaces must be arranged to accommodate future expansion. This is a considerable constraint on initial design proposals depending on which elements of the development might expand. Is it likely that every area of the client's business will need more space, or only certain activities? Decisions about this issue could have major implications for the design and positioning of the building on the site.
- Fuel sales require considerable space within the site layout, and the location of facilities with respect to the entrance to the site will be significant. This element of business must be prominent to attract passing trade.

1.4

- The site appears to be in an attractive position but is contaminated from previous use. Although grant aid is available, the extent of remediation is unknown and may well be a significant problem. The Local Authority are evidently interested in a commercial development on this site, which would tend to suggest that they will be sympathetic to proposals with regard to granting Planning consent, but the Victorian building may be listed as being of architectural or historical merit. It is uncertain if it can be demolished or whether it must be retained. If the Local Authority insist on retention, the building must be put to some practical use and refurbishment costs may be high. This issue must be resolved urgently.

1.5 and 1.6

- The client wishes to consider ideas about how their site can be developed. The design proposals are not restricted by any pre-determined ideas on their part.
- Financial arrangements appear to be sound, except that disposing of the existing premises could present problems. This is beyond the direct control of the designer but is a consideration in the client's cost planning for the new development.

The car-dealership is a mixture of precise practicalities and processes which can be researched. It also has elements of the client's corporate identity, style, image, ethos and expectations which will influence all the activities identified in the client's brief. The inner city location may mean that proposals will be sensitive in relation to other surrounding development and use, and the site itself, contaminated from previous use must be carefully investigated to see what direct bearing it will have on the way that the new building can be designed, built and used.

This is a hypothetical example of some of the issues that can be encountered at the beginnings of a new project. There are of course, many more and each project is unique with its own range of questions and answers. Some issues are specific to the building type, some to different kinds of client. The process of questioning and analysing answers is continued in the next chapter as the Design Brief is developed.

5.7 The building type

The exploration of the client's needs and intensions is not an entirely abstract process because all members of the design development team will already have an understanding of what constitutes this type of building; a car-dealership in this case. There are advantages and disadvantages to this situation, both of which concern the building designer. There is a risk that some aspect of a typical car-dealership is accepted without question, but equally a risk that an important element is omitted or incorrectly handled through lack of research into well-established precedent. Figure 5.1 shows a variety of car-dealerships; do any of these images help with understanding of what such a building type actually is? At a more detailed level, it might be assumed that all members of the management staff have their own private office space, whereas an open plan arrangement may be more satisfactory for a variety of reasons. The management style and activities of the people involved merits, some thought about optional design solutions. On the other hand, an understood arrangement for servicing cars, including hoists, pits, tool racks, etc. may be the best solution to adopt, and not worth changing just for the sake of change.

This dilemma, common to all building types, presents a challenge which can be the fine line between good and poor design, between something interesting or mundane, something ordinary or innovative. The design of most building types develops progressively as issues are reconsidered by successive designers. But in order to challenge precedent, the building designer should firstly be familiar with what has been achieved before. The simple example of the question of private offices or open plan workspace only arises because there is an awareness of the need for one or the other from previous examples. The selection of an appropriate floor finish in the workshop is a much more precise requirement which must be based on previous experience of its

Figure 5.1 The built environment contains many different types of building, all of which have special functions and design requirements. Some examples from Nottingham, Keddlestone, Bridgnorth and Birmingham.

suitability. Challenging precedent in this case may not be sensible unless a discernable improvement is available. In the future, there may be other options to consider, but there are generic aspects to car-dealerships which can usefully be researched including the following:

- Relationship between public and private functions. Visibility of activities from public spaces, ways to display goods and information, theory about best practice for sales and customer care.
- Typical space requirements or *accommodation* for likely internal and external activities, practicable volumes, floor areas and wall heights.
- Extent of movements of customers, staff, deliveries, storage and waste.
- Flow patterns for processes, for customers and vehicles.
- Potential materials for structures and finishes.
- The 'image' of a car-dealership; the style of the buildings and their surroundings.
- Environmental services, power, heating and lighting levels.
- Staffing levels.
- Security requirements, boundary protection, occupation and surveillance.

These are very much preliminary thoughts which are explored in greater detail in Chapter 7 as the Design Brief is developed. The building designer, and others in the development team need to begin to consider what is, or might be involved with this type of building.

5.8 The site

Unlike most products, buildings are created in the place where they are going to be used. This location is called *the site* and its characteristics have a bearing on the way that the building can be designed and constructed, affecting cost, future use and commercial viability. In some instances, the client will already own or have an interest in a specific site. This site has not been purposely chosen by the client and its characteristics may very well limit the development potential that the client would actually like. There may be problems with this site that make the new building unsuitable for intended use or uneconomical to build on and operate. Even if the site is identified in the Local Authority Land Use Structure Plan as being suitable for industrial development, and it is surrounded by similar development there is no guarantee that design and construction will be straightforward. Site investigation and analysis in this case explores the advantages and limitations of a specific piece of land.

Alternatively, the criteria for 'straightforward' development could be used in the search for a suitable site for the development intended. Investigation and analysis of a number of possible sites may lead to conclusions that one of them is appropriate for the new building type, which the client then purchases. For the car-dealership, the principle criteria are likely to be commercial viability and site acquisition costs. Many other issues affecting design proposals will still need to be examined after this point as few, if any sites present no difficulties, and all of them have their own characteristics which will influence the design of the buildings.

For other building types, the criteria for selecting a site will be different. New houses for families, accommodation for the elderly, hospitals, and sports and leisure centres are located in response to other requirements and needs. Some are prominent and public, others discreet and private. The client's or developer's view of a site's attractiveness will vary from place to place, and prevailing market conditions is a powerful influence. New development can make an area attractive for other development when its general condition had previously been thought unsuitable. Viability changes as areas improve or decline and fixed ideas are reconsidered. There can be no absolute definition of the right or wrong site which cannot be changed in the future.

For the development of a car-dealership, its location, prominence and relationship with other activities going on around it will affect its potential trade. The siting of petrol sales is a particularly delicate balance of these factors together with issues such as ease and convenience of access and the nearness to competitors. A site for this kind of development must be carefully chosen and justified so that its characteristics are essentially positive rather than negative. This is a matter best handled by specialist research and analysis consultants who would be commissioned by the client to find a suitable site. For the purposes of the case study, it is assumed that the client has been well advised and has purchased a site in the belief that it is appropriate for their development.

For most buildings, ownership and use extends beyond the external walls up to the boundaries of the site and sometimes the building itself occupies a relatively small percentage of the total site area. The relationship of the building to its external spaces is governed by existing site features and constraints, and in many cases today, further complicated by the remnants of previous development. The design and use of many buildings is also influenced by factors outside the site boundary, by conditions in the immediate locale and beyond in the surrounding neighbourhood. They may even be affected by conditions in the area generally, in the town, city, region or country.

Figure 5.2 shows the client's existing premises and the new site in the context of the city. Figure 5.3 relates the site to its surrounding locality and Figure 5.4 illustrates the site in sufficient detail to be able to appreciate the boundaries and the principle features within them. The advantages and constraints for development of this site will be examined in some detail in the next chapter, but the effect of the site's characteristics may be as follows:

- *Dictate the form and layout of the building, and its location on the site*
  For example, the need to retain attractive, mature trees identified and protected by a Local Authority 'Tree Preservation Order' (TPO) may determine the position of the building on the site and its maximum possible size. Even if the trees are removed, it may be decided to avoid building over areas of disturbed ground where the roots once were.
- *Increase the cost of construction*
  If sub-soil investigations show that the make up of the ground below the top surface comprises weak, loosely compacted material then the foundations for the new building will be relatively complicated and more expensive than a simple, traditional form of construction. This fact does not necessarily influence the location of the building, but unavoidably adds to its cost.
- *Enhance or limit subsequent use*
  An attractive view away from an elevated site can be a considerable bonus for the development of some building types such as residential accommodation for the elderly, whereas the noise and health and safety risks associated with a busy railway line at the side of the site may have serious consequences for the quality of life of families with young children. For any building type, design may minimize hazards for future users to the point where conditions are tolerable, but nevertheless are relatively unattractive.
- *Prevent development*
  In some cases, the site's characteristics may prevent development at all. The costs of remediation of dangerous contamination before construction and to make the site safe for its future occupants could be uneconomical. Some forms of legal constraint cannot be overcome such as previously imposed covenants or the local authorities determination not to permit a change of use. For example, it is unlikely to be acceptable to construct a tanning factory in the middle of a housing estate.

Some of these issues will be apparent at the first visit to the site and could well be the basis for expressing opinions about the viability and practicality of development. Mature trees, the size of the site and the significance of the railway line are visible and obvious. But the sub-soil conditions, contamination and Planning Authority attitudes to development are not. They must be researched as possible constraints on development, and their existence can only be anticipated together with any effect that they might have on design proposals.

5.9 Assessing the need for consultants

Offering professional advice in every area of modern day life carries considerable risk in the event that the advice is incomplete or wrong. The professional advisor must be certain that advice is correct, including any incidental implications that could reasonably be anticipated arising from it. The building designer should be aware that there are many aspects of design and

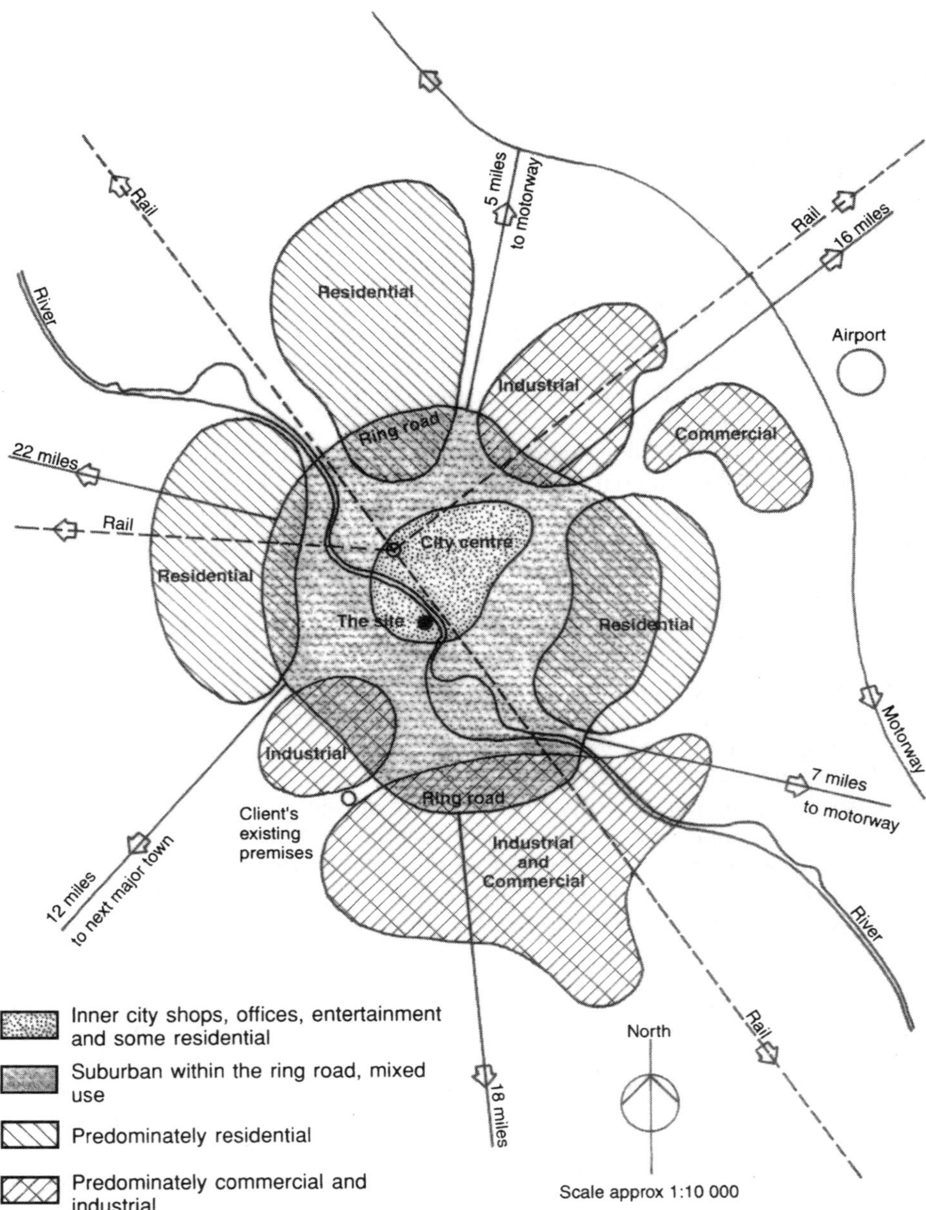

Figure 5.2 The site in the context of the city.

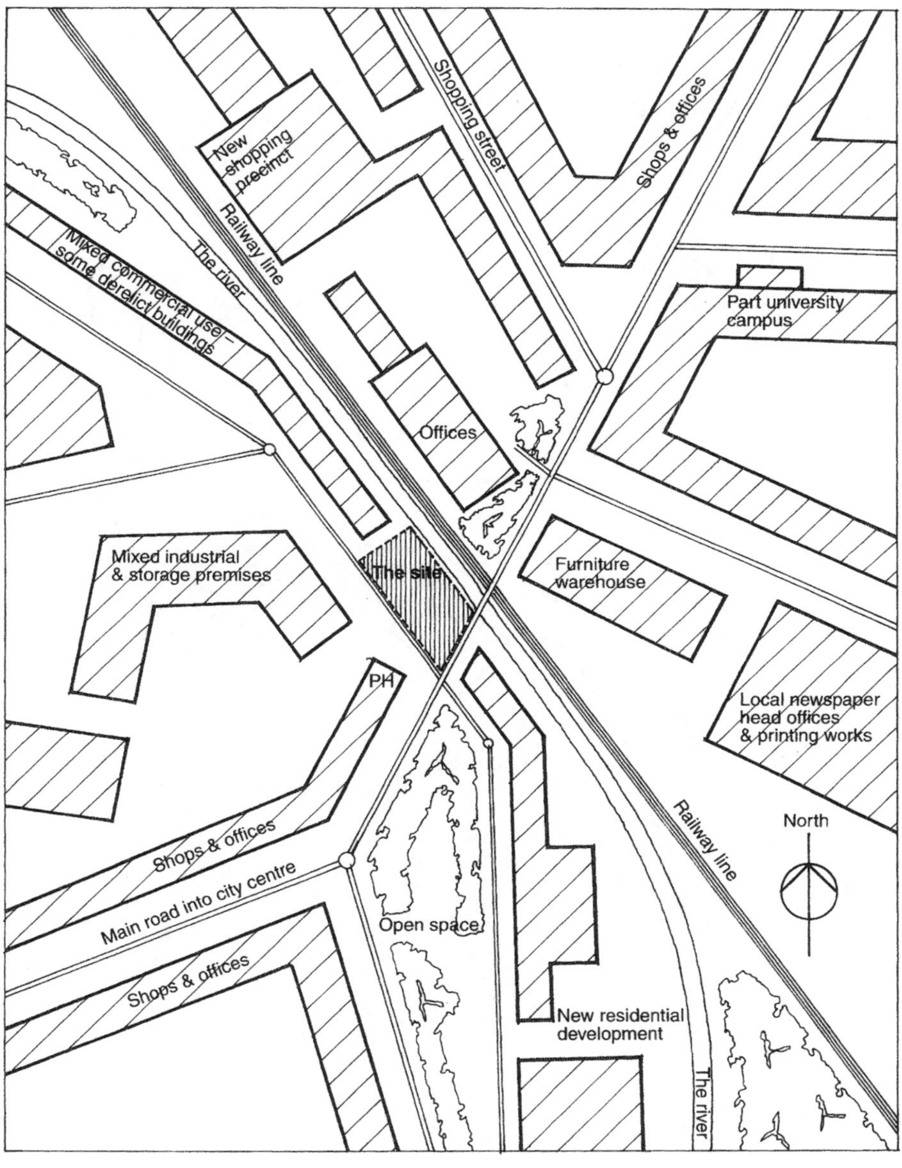

Figure 5.3 The site in the context of its locality.

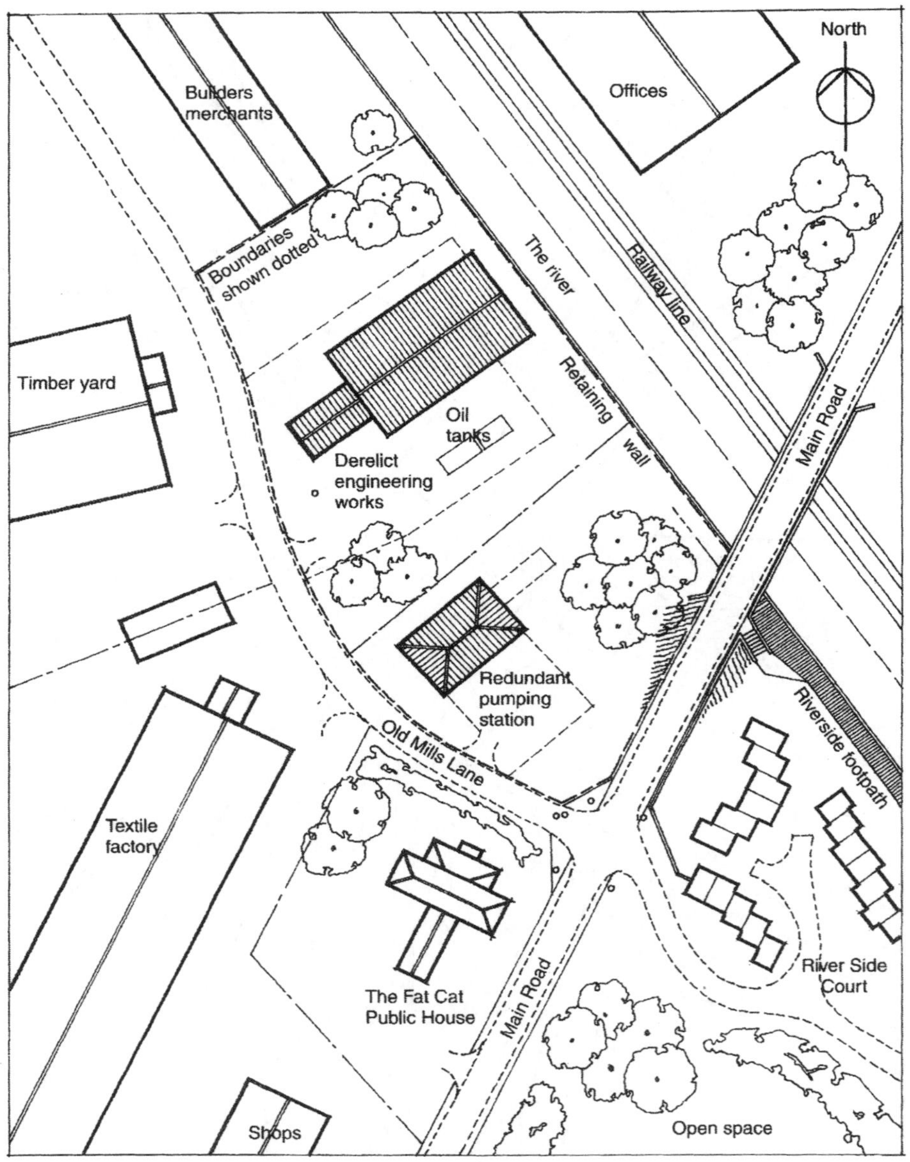

Figure 5.4 The site itself.

construction for which they should not offer professional advice, either through lack of education, training and experience or most importantly of all, that they are not insured to do so. Adequate insurance cover against building failure, death, injury and loss, and expense is essential for the building designer, whose services will be closely defined and limited by their insurance company. The legal implications of offering professional advice can be studied in the many publications already available.

The practical reasons for seeking specialist advice at the design stage can be summarized as follows:

- *To undertake specialist investigation and analysis*
  Some tasks require the use of expensive and elaborate machinery, as for example, undertaking bore holes for ground investigation or video filming the conditions inside underground pipework, which the designer is unlikely to possess. Specialist advice will also be needed to correctly assess the results of findings.
- *To help to develop feasibility ideas into practical proposals*
  Consultants can advise on possibilities and constraints that the designer may not appreciate.
- *To contribute knowledge and experience that the designer does not possess*
  Consultants provide specialist skills such as the complex calculations needed for justifying structures, internal environmental conditions and finance, about which the designer may not have sufficient knowledge.
- *To ensure that responsibility for performance is supported by suitable qualification*
  It is essential that necessary tasks are done properly and that design proposals will lead to a new building which is fit for its purpose. Specialist advice should be independent of the designer, supported by appropriate forms of quality assurance so that in the event of a failure resulting from poor advice about specialist areas of design and construction it is clear where the responsibility lies. The building designer must ensure that risks associated with his or her own work are limited to the extent of his or her ability and avoid being exposed to accusations of negligence with respect to advice given.
- *To meet the timescale imposed by the programme*
  Consultants are also needed to obtain proper advice quickly so that design work is not unduly delayed. Given that the building designer were in a position to do everything, for large projects, it would take too long to complete.

The building designer can recommend consultants for the project but great care must be taken to clearly establish the effect of professional relationships depending on the procurement method adopted by the client. Some large-design firms employ their own specialist consultants and are therefore able to offer the client a comprehensive design service. Design and build and project management organizations will employ consultants directly in order to offer the client a complete package. The traditional arrangement for the specialist building designer is that consultants should be employed directly by the client and be fully responsible for their own work.

The specialist services or advice likely to be required for a complex project such as a car-dealership include the following consultants.

Land surveyor

For small sites, the designer should be able to produce survey drawings showing the size and levels of the land together with existing features and buildings. It may be more cost efficient to use a commercial firm with the necessary specialist equipment and resources for larger, more complicated sites.

Structural engineer
The structural engineer will co-ordinate sub-soil investigations, undertaking or appointing other specialists to dig trial holes or bore holes to analyse the ground conditions in advance of designing the foundations. Further design work may be required for ground stabilization, retaining walls, reinforced floors and superstructure, limitations on sizes of openings and compliance with current legislation.

Quantity surveyor
The QS prepares cost plans and advises on the economy of design, construction and materials leading eventually to the Bills of Quantity if contractors are to be invited to tender for the work. The QS may undertake valuations and payments as construction takes place and agree the final account when building work is finished.

Services engineers
The design of mechanical and electrical installations and the requirements for the installation of specialist equipment will be undertaken by a range of services engineers. They will see that certain performance standards are achieved for artificial lighting, power and ventilation.

Managers
The client may appoint a project manager to oversee design and construction work. This is a possible scenario for a car-dealership where the new building is to be in the company style of one of the major manufacturers. The project manager would ensure that the client's requirements are achieved in relation to the need to harmonize any standard arrangements and finishes.

Planning supervisor
This is a relatively new position required under the current CDM Legislation. The Planning Supervisor's duty is to over-see and co-ordinate all aspects of design and construction in such a way that health and safety risks are minimized.

Planning consultant
If the proposed development is in a sensitive location and there is a possibility that the Local Authority may refuse to grant Planning Permission, a Planning consultant would be able to offer advice on the best approach to the design to minimize the risk of failure. The Planning consultant would more commonly be used to co-ordinate an appeal in the event of rejection by the Local Authority.

Some of the other consultants listed in Chapter 2 may be required immediately or later, when the design work is well advanced depending on perceptions of the relative importance of their work on the design concept. Landscaping and planting, for example may be simply a matter of tidying up external spaces when the building is finished, or it may be fundamental to the site layout and the design of the building itself.

5.10 Development possibilities and limitations

Invariably at the first meetings, the client will ask questions along the lines of 'what do you think' or ' is it possible, and how much will it cost?'. 'Do my/our ideas make sense to you and are they feasible?' To the client, these are the most important issues of the moment, and they want a professional opinion about their ideas or desires. They may have lived with the notions of development for a long while, gradually realizing that something must be done and pondering possibilities. Their needs are often expressed in the abstract, in simple statements such as 'we need a new building' or ' we would like a conservatory' either unaware of, or ignoring the constraints imposed by their site or their building. They want a professional view, linking the two together to be assured that their ideas are realistic.

The practicalities of adding a conservatory to the client's house can be quickly assessed by inspecting the insides and outsides of the house, and the space available in the garden. The risk of giving poor advice about this element of construction should be minimal because there are relatively few constraints that could prevent development and those which might be significant can be easily seen. Actually designing the conservatory is a separate issue, but the principle can be established with confidence.

For a larger development such as a new car-dealership, instant advice is much more problematic. There are many reasons why such a development may or may not be practicable, which as already mentioned, must be researched before a conclusion is reached. An 'off the cuff' opinion about a development on this scale can lead into deep water, if and when assumptions are found to be incorrect. As with the conservatory, the design of the new building is a separate issue, but at this stage, establishing the principle is crucial and there is little point embarking on the design of the new building if there are any fundamental constraints.

In Chapter 2, an early stage in the design process was defined as 'feasibility' which means:

- Firstly, checking the client's brief and the site to see that development is practicable, and the extent to which the site can be developed.
- Secondly, assuming that this is the case, producing ideas about how it can be done.

The second set of feasibility studies will be described in the next chapter, and preparing them may take some time before all the relevant constraints are assessed and a satisfactory outline design emerges. It is the first feasibility study that needs attention at this stage, establishing as quickly as possible the important parameters of the project to justify starting detailed design work. Setting the element of cost to one side for a moment, which of course may be fundamental to the success of the whole idea, there are a number of issues to consider before making any statement.

- *Is it possible to visualize a finished building in principle?*
  If the client's intentions were implemented, do they add up to something that could actually be physically achieved?
- *Are there any obvious constraints?*
  Is there anything about their building or their site which immediately suggests that their ideas cannot be achieved?
- *Could there be hidden constraints which may become significant following survey, research or consultation?*
  Are there any unknown, but predictable factors which once discovered, would alter the advice or opinions?

In giving any opinion, each consideration must be looked at briefly but carefully if it is to be of any value, or more importantly, that the client is not given bad advice or encouraged to believe that something is possible, which cannot subsequently be achieved. For the car-dealership, the essential immediate considerations include the following:

- Is the site in the client's ownership? Have the boundaries been correctly defined and are there any legal difficulties?

- Is the site big enough for the proposal, and could it accommodate expansion in the future?
- If the site is not big enough, can the client's brief be modified so that the maximum possible size of the new building is still satisfactory?
- Is access adequate for proposed use? Can alterations be made to existing roads and footpaths to enable vehicles and pedestrians to get onto the site and into the building?
- Are there any features on or around the site which could affect proposals?
- Is the site contaminated in any way?
- Could proposals affect the amenity value of other developments around the site?
- Could potential development around the site affect the amenity value of proposals?
- Is it likely that the proposed development and use will meet current Planning requirements?
- Could other agencies impose conditions which make development difficult such as the Fire Officer or the Environment Agency?
- Are there adequate drains and services available? How will foul sewerage and surface water be disposed of, and is the high-voltage electricity needed to run machinery available in the locality?
- Is the client's need correct, sensible, economical and practicable? If the client or other sources of finance invest capital in this development, what will be their return? Will the value of the completed development be sufficient to justify the investment? If the client decided to sell the property, could it be a viable proposition to some other user and will its market value recoup the initial investment?
- Are the requirements within the general capability of the designer?

Other issues may be apparent depending on the client, the site and the building type. Findings may reinforce or question the client's brief. Given the constraints which are discovered, elements of the client's brief may be wrong or impossible to achieve. The determination to have a building of a certain size and height may be limited by Planning considerations. There may be insufficient space to include fuel sales and the cost of constructing a new electricity sub-station may be prohibitive. A public right of way may exist, running through the centre of the site, which could prevent construction unless it can be modified and repositioned.

Some of these issues will be confirmed quite quickly but others will take time to resolve. Further meetings must be arranged and more information collected to clarify the issues which are fundamental to the principle of development and which must be assessed before design work can commence. This process together with the subsequent work of all the members of the design team must be planned to define and plot the progress of their work as ideas become approved proposals ready for construction. The next chapter considers design planning and the preparation of a design programme.

5.11 Project File content

General aims at this stage are as follows:

- Meet the client and explain the process leading to the creation of the new building.
- Outline the professional services being offered.
- Understand the client's business criteria.
- Establish the broad brief, costs and timetscale.
- Assess the need for specialist consultant advice.

- Select, interview and appoint members of the design team.
- Check that the client is able to finance development.
- Compile and exchange documents confirming agreements with the client including fees for work up to an understood stage of design development.
- Set up administration and communication systems.

File material could include the following:

Record notes, letters and minutes

- Notes detailing informal telephone conversations with the client, specialist consultants, representatives of local authorities, builders, subcontractors, suppliers and manufacturers.
- Copies of letters asking for or confirming points of information, and copies of replies.
- Minutes of meetings formally circulated to confirm information and agreements.

Briefing information

- The client's initial intentions, business criteria, requirements and preferences.
- Information about other similar developments gained from visits, with details of typical commercial operations, general accommodation requirements and staffing levels.
- Assessment of Health and Safety issues associated with the client's operations and the site itself.
- Commencement of the CDM Tender Health and Safety Plan File.
- Preliminary thoughts about the site and its surroundings, including photographs, maps, and and current and historical survey plans of the area, which may be particularly useful as visual aids in meetings to discuss specific design constraints.
- Initial advice from the Planning Authority or guidance in the form of a Planning Brief for the site and its development potential.
- Market research justifying development potential, including land and building values, local construction costs, proximity to competitors, population, employment trends and area status.

Appointment details

- Confirmation of services required, acting as the client's agent or as part of a development team.
- Consideration of costs and design programming.
- A check list of activities to be undertaken once design work has commenced.
- Project personnel and contacts, including names, addresses and contact numbers of organizations and their representatives who will be directly involved with the project, including a summary of external bodies likely to have an interest in the development.
- Blank application forms which will be required for submission to authorities for later approvals with the likely fees payable (e.g. Planning, Building Control and the Environment Agency).

Preliminary thoughts

- Constraints identified using existing maps, plans or other drawings.
- First ideas and freehand sketches of possible design solutions overlaid on the same existing maps, plans or drawings.

- Information from books, magazine articles and catalogues which may provide guidance for any aspect of the design in the future.
- A project reference number.

5.12 Discussion points

(1) Do clients, customers, developers, authorities have fixed attitudes towards old buildings and community areas? Do members of the general public prefer older buildings? Why is there hostility towards newly designed buildings sometimes? Are the qualities of old buildings only recognized after many years of use?
(2) What does it mean to say that a development proposal is viable? Whose point of view should be given precedence? To what extent should buildings be designed to last?
(3) What is the significance of designer/client friction throughout the development of a new building? How do their expectations vary or coincide? Does the designer 'know better' than the client? Is the 'customer' always right? Should the designer always do what the client wants?
(4) Can a building designer understand how all types of buildings work? Should he/she limit their activity to specific types only? What are the best forms of reference for designing buildings for the future?

5.13 Further reading

Baden Powell C (2001) *Architects Pocket Book.* 2nd Edn. Oxford: Architectural Press.
Boothroyd C (1996) *Risk Management: A Practical Guide for Construction Professionals.* London: Witherby.
Brawne M (1992) *From Idea to Building: Issues in Architecture.* Oxford: Butterworth-Heinemann.
Buchan R, **Grant** F and **Fleming** F (2003) *Estimating for Builders and Surveyors.* (2nd edition). Oxford: Butterworth-Heinemann.
Cox S and **Hamilton** A (1997) *Architect's Guide to Job Administration.* London: RIBA.
Department of Trade and Industry: Constructing Excellence. www.constructingexcellence.org.uk
Emmitt S (2002) *Architectural Technology.* Oxford: Blackwell Science.
Graham P (2003) *Building Ecology: First Principles for a Sustainable Built Environment.* Oxford: Blackwell Science.
Green R (2001) *The Architects Guide to Running a Job.* 6th Edn. Oxford: Architectural Press.
Hackett M and **Robinson** I (eds) (2003) *Pre-contract Practice and Contract Administration for the Building Team.* Oxford: Blackwell Science.
Kernohan D *et al* (1992) *User Participation in Building Design and Management: a Generic Approach to Building Evaluation.* Oxford: Butterworth-Heinemann.
Morledge R and **Sharif** A (1996) *The Procurement Guide: A Guide to the Development of an Appropriate Building Procurement Strategy.* London: RICS.
Salisbury F (1998) *Briefing Your Architect.* 2nd Edn. Oxford: Architectural Press.
Salisbury F (1990) *Architect's Handbook for Client Briefing.* London: Butterworths.
Saxon R *Povey Lecture Writings.* www.richardsaxon.co.uk
Thompson A (1999) *Architectural Design Procedures.* 2nd Edn. London: Arnold.
Willis A and **Chappell** D (2000) *The Architect in Practice.* 8th Edn. Oxford: Blackwell Science.
Worthington J and **Blyth** A (2001) *Managing the Brief for Better Design.* London: Spon Press.

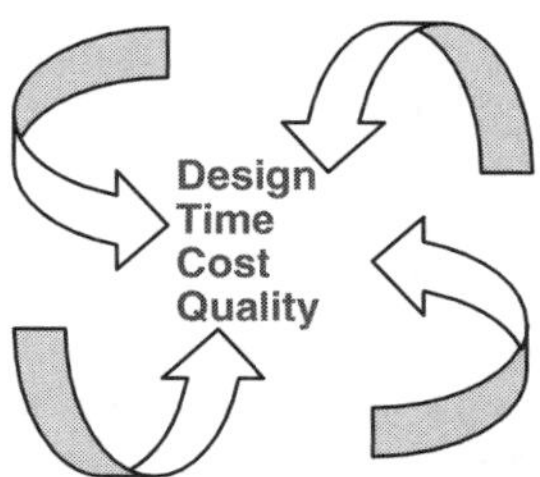

6

Design planning

6.1 Introduction

Building design is not an isolated activity where proposals appear suddenly as if generated by a flash of inspiration. They develop quite slowly in response to the availability of information, the demands of external agencies and the need to co-ordinate critical decisions. This is a *process,* which needs defining in terms of its length from start to finish. This chapter considers the concept of programming and time management, and shows how an activity can be managed through a step-by-step approach to undertaking and completing all its constituent elements. The completed diagrammatical programme is like a map, showing how to get to the end product. In a commercial context the programme is a key document, perhaps determining both practical and financial success.

6.2 Time passes

The client's decision to commission a new building reflects *their* perception of shortcomings in their present situation. Numerous factors may have contributed to this conclusion, based on lengthy research and analysis of their current and future requirements, but it is the availability of sufficient finance which is often the key factor. The decision to invest their own capital, or finance development through borrowing is clearly a serious commitment which they will have considered for some time. However, once the decision has been taken, most clients will be anxious to proceed as quickly as possible. For the car-dealership, the transition from 'old' to 'new' has other commercial implications, affecting amongst other things investment in machinery, personnel, advertising and even stationery. There may also be a significant loss of trade as one site runs down and the other builds up. Running a business through this transition can be not only expensive but traumatic as well, as the client and their staff try to keep operations going under pressure.

The client must, therefore, decide which is the most appropriate procurement method to suit their circumstances. The traditional design lead method focuses on costs. It is essentially a linear process starting with sketches and progressing to detail, measurement, competitive tendering and construction supervision. Its aim is to create a building representing the best possible value for money in terms of its capital cost, but it extends the time scale of the development because the design team must complete a large percentage of their work before tenders are invited. There is a risk of further delays because it is a system of instruction and approval for every stage of construction; it does not allow the construction team any flexibility, it tells them what to do. It is inherently confrontational with the design team, representing the client on one side against the contracting team on the other, both taking all the necessary steps needed to protect their positions.

Other forms of procurement, whilst still linear in nature, permit the time scale to be compressed in some areas because the stages at which competitive pricing is agreed are repositioned, and much of the confrontation between the design and construction teams is removed altogether. Consequently the focus of these methods includes time as well as capital cost. However, it does mean that the client themselves must take a much more active part in the process in order to understand what is being offered. For construction lead methods based on negotiated costs, the client must assess 'value for money' and determine the suitability of design proposals and the specification of materials. This is not to say that the client would not do this anyway with the traditional procurement method, but the element of advice given by *independent* consultants is no longer available to them because they are acting for the contractor or project manager.

The point at which the designer starts to make a contribution to the project depends on the client's chosen procurement method, which determines responsibility for management of the design process. Whichever method is selected, the process must still be organised or programmed by *someone* so that the necessary work is carried out in an appropriate, economical sequence.

The word 'programme' is a familiar one in every day life used in a variety of contexts. It can be a collective noun for the content of a television or radio feature such as 'Coronation Street' or 'I'm sorry I haven't a clue'. A printed programme is a statement of what is going to happen at an event over a period of time, prepared by the event's organisers for the benefit of participants. A concert programme advises the audience of what they will hear or see, and a football programme confirms the selection of players in each team. These programmes are purely informative, requiring no audience or spectator participation, other than that they should read them if they want to know what is going on. Programmes for an agricultural show or a primary school Summer Fayre on the other hand, directly involve their visitors, who must choose which activities that they wish to take part in from the many that may be going on at the same time. This programme is an essential plan for the organiser's work, but enables visitors to decide their own priorities. All of these events are relatively straightforward to document because they either take place one after another in fixed periods of time, or they take place independently of one another.

The design process can be regarded as an event, occupying a considerable period of time, but unlike the previous examples, it is difficult to describe the content as precisely because the timings and the interactions between events are not so clearly understood at the outset. It can be stated with confidence, for example, that the time needed to play the first movement of Beethoven's Moonlight Sonata is, say, 6 minutes and 55 seconds, give or take a few seconds, but the time required to obtain information or come up with a feasibility idea for the new building is unknown, and difficult to quantify. It could take 10 minutes, 3 days, 2 weeks or 6 months.

Nothing interferes with the pianist, who can play steadily until the work is completed, with no pressure to finish in 5 minutes and 45 seconds so that the concert hall can be closed on time. But at different stages throughout the design period, specific activities and contributions from all the members of the design team must be finished as intended if the time limits are to be met. In this respect, design programming essentially operates in reverse, fixing time limits at which points must be completed, so that, for example, a period of 3 weeks is allowed for collecting information, and a period of 6 weeks to prepare a sketch scheme. The numbers are a prediction based on the demand or hope that the work can be completed within the time available so that the next stage is reached as planned. The programme is a working document illustrating how everything must be co-ordinated to arrive at a finish point.

6.3 The concept of programming

The process of design and construction involves the manipulation of many elements, one of which is time, influencing both efficiency and profitability. It is therefore essential to establish the organisation and co-ordination of all the members of the design and construction teams so that their contributions can be made practically and economically at the right times.

To appreciate the commercial significance of programming, consider the following simple example of specific stages of design and construction work:

- One designer working alone may take 12 weeks to prepare sketch proposals for presentation to the client. This can be quantified as 12-unit weeks of design work.
- Two designers working together on the same project may only take 5 weeks to reach the same point, which is 10-unit weeks of design work, but they arrive at the finish point 7 weeks earlier.
- Once construction has started on site, four steel erectors could complete the installation of the steel frame for the new building in 8 weeks, which is a total of 32-unit weeks of construction work.
- The same task undertaken by six erectors may only need 5 weeks to complete the work, which is a total of 30-unit weeks of construction work, but completed 3 weeks earlier.

In this example, the allocation of more resources for both design and construction actually saves the cost of 4-unit weeks of work, and gives the client the substantial benefit of 10 extra weeks of commercial operations because the building is made available for earlier occupation. A similar approach to each stage of work may show advantages, although there will be optimum conditions where further improvements are impossible. For example, the employment of 20 steel erectors would not result in *pro rata* savings if there is only sufficient work for six. Alternatively, of course, the extra time made available through careful planning could be used to improve design and construction quality, or be retained in the programme as a contingency against unforeseen delays. In any event, considered planning of time and effort will show advantages to someone which can be easily missed by lack of forethought.

There is some overlap between the design and construction period programmes with any method of procurement, as once construction has commenced on site it is not uncommon to find that detailed information is still being prepared by the design team, but for the purposes of explanation, this section will deal with the factors to consider when developing a pre-contract programme for design work up to the point of commencing construction on site. The programme is like a map, establishing a route from start to finish, or more significantly showing the best or shortest path to completion.

The objectives of design programming are as follows:

- Identification of all the stages of work needed for the development.
- Determination of all the necessary resources needed for the client and the development teams.
- Planning work in the appropriate sequence, co-ordinating the contributions of all the members of the development teams.
- Fixing target dates for the achievement of activities to meet an overall agreed finishing date.
- Establishing critical dates after which it becomes difficult, impossible or very expensive to make major alterations to design work.
- Assisting the client in cost planning and establishing a realistic target date for commencement of their new business operations.
- Providing a basis for checking that design work is progressing as predicted.

6.4 Time management

All the activities associated with design process can be quantified diagrammatically in blocks representing a number of hours, days, weeks or months required for each to be completed. The starting and finishing points can be highlighted, technically described as nodes. A simple programme could show the blocks in line one after another, each one starting when the previous one is completed. This would add up overall projected development period but not be particularly efficient, failing to account for waiting time while others are working, or when various activities could be going on at the same time. This arrangement would be unnecessarily long and incur unacceptable costs and delay to everyone involved.

A more useful approach, when different activities can be moved forward together at the same time, is to overlap the blocks showing the relationship between dependent activities which can be commenced when the earlier ones are sufficiently advanced. For example, at some stage in the period allocated for site investigation, sketch design ideas can be considered, even though site investigation continues on into the future. Similarly, once sketch designs are presentable, specialists can begin to contribute, which they could not do without the sketch drawings. *Activities* and *event nodes* can be analysed as a *network* showing how they are interrelated. The shortest possible route from start to finish is called the *critical path*, which links those event nodes which are absolutely dependent on one another. The earliest starting time and the latest starting time for these activities are the same and cannot be varied. However, other activities, which are not critical, can start sooner or later without affecting the overall programme.

Illustrated graphically on a bar chart, the critical path can be drawn as a straight line at an angle linking the start and finish points. The nodes at the start or finish of individual stages within the programme will straddle the critical path. The critical starting nodes will be on the line, whilst those offering flexibility can be on either side. If everything goes to plan, all the activities will be completed by the final completion date. However, should any critical starting nodes slip onto the right-hand side of the line, then subsequent activities will be delayed making it impossible to achieve the target. The subsequent activities must be completed in a shorter period of time than had been planned if the overall programme is to be maintained.

Some stages in the design process, like the period needed to process the Planning Application have fixed time blocks which are relatively easy to determine. Others such as thinking of design concepts are much more indeterminate, but they cannot be left entirely open ended.

The phrase 'open ended' is not particularly helpful, as the familiar saying that 'work expands to fill the time available' nearly always applies. Faced with a multitude of activities or the need to find inspiration for design, it is often helpful to have target dates to aim at rather than to hope that inspiration will arrive eventually. Apart from the obvious commercial implications of time management, it can also be useful to focus attention on problems and concentrate on finding solutions. Pondering the rights and wrongs of any particular element of the new building by searching through manufacturer's catalogues for alternative roller shutter doors or facing bricks, for example, could carry on indefinitely. However, constrained by the need to complete a sketch drawing to a deadline means that decisions must be taken, which may be right or wrong but have merit by keeping the project moving.

It is interesting that sometimes quick, intuitive decisions are proved to be correct. It is also perhaps worth mentioning that a quick, wrong decision is not always a mistake, providing that it is circulated straight away to the other members of the design team. This enables them to return their observations about the 'incorrectness' of the decision so that the matter can be examined again. If a problem remains stationary at one point for too long, then it will inevitably affect subsequent progress by delaying the work of others. This is one of the primary functions of having a programme to work to, obliging all the members of the design team to take responsibility for their work in relation to the work of others to see that problems are resolved and information is supplied at the correct times.

Preparing a programme incorporating deadlines requires an analysis of activities in reverse order so that it is possible to see how much flexibility there is to complete open-ended activities in order to arrive at the desired finishing points. As an example, consider the following scenario where the client asks the design team to produce a preliminary sketch scheme for discussion within 2 weeks (10 working days). Starting on day 1 and simply working forward, it may be day 10 before enough information has been collected, and there is no time to prepare the sketch. The important factor in the arrangement is being able to make a presentation on the target date. To achieve this means planning in reverse to work out what must happen in order to present the sketch proposal on day 11.

What could be involved during this period of time? It may include the following:

- An examination of the preliminary brief.
- A site visit to assess constraints and measure significant elements.
- Meeting(s) with the Planning Officer to discuss attitudes to development and determine any constraints.
- Meeting(s) with other statutory bodies if necessary.
- Deciding if specialist consultant input is necessary and seeking initial advice, maybe holding further meetings on site to inspect conditions.
- Consideration of the building type for historical reference and undertaking research.
- Considering design options and sketching out a possible solution.
- Preparing a drawing or a number of drawings to a suitable level of quality for presentation.
- Printing drawings and arranging presentation material.
- Organising the presentation meeting.

For this activity, illustrated in Figure 6.1, there are overlaps and nodes, and a critical path. Time is needed for communication; telephone calls to people who may be unavailable or who fail to return a call as promised. Time is needed to arrange appointments. The Planning Officer, for example, is not sitting in an office waiting on the off chance that someone will call for advice. It may take days before the responsible Planning Officer is available, in which case the sketch

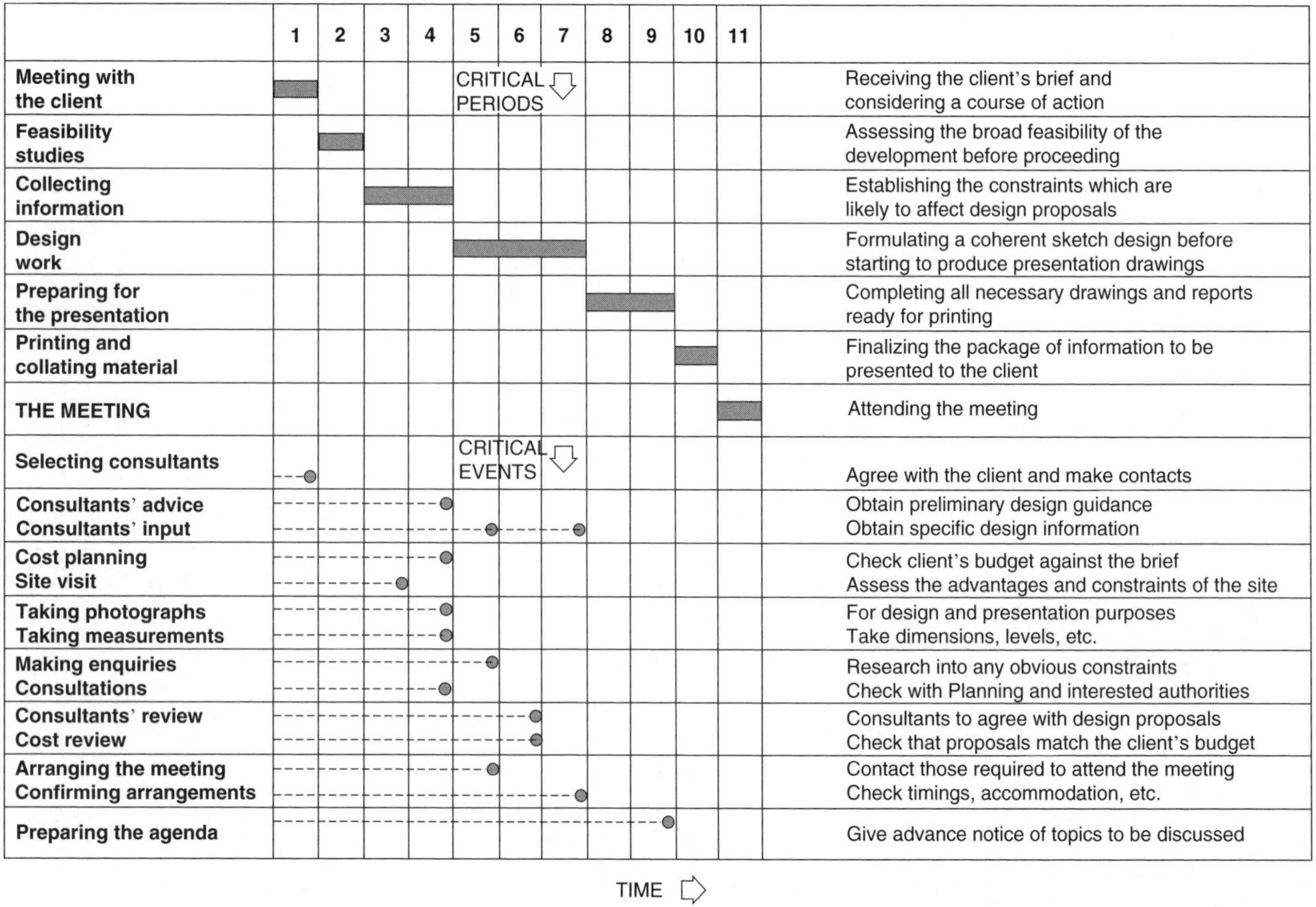

	1	2	3	4	5	6	7	8	9	10	11	
Meeting with the client					CRITICAL PERIODS							Receiving the client's brief and considering a course of action
Feasibility studies												Assessing the broad feasibility of the development before proceeding
Collecting information												Establishing the constraints which are likely to affect design proposals
Design work												Formulating a coherent sketch design before starting to produce presentation drawings
Preparing for the presentation												Completing all necessary drawings and reports ready for printing
Printing and collating material												Finalizing the package of information to be presented to the client
THE MEETING												Attending the meeting
Selecting consultants					CRITICAL EVENTS							Agree with the client and make contacts
Consultants' advice **Consultants' input**												Obtain preliminary design guidance Obtain specific design information
Cost planning **Site visit**												Check client's budget against the brief Assess the advantages and constraints of the site
Taking photographs **Taking measurements**												For design and presentation purposes Take dimensions, levels, etc.
Making enquiries **Consultations**												Research into any obvious constraints Check with Planning and interested authorities
Consultants' review **Cost review**												Consultants to agree with design proposals Check that proposals match the client's budget
Arranging the meeting **Confirming arrangements**												Contact those required to attend the meeting Check timings, accommodation, etc.
Preparing the agenda												Give advance notice of topics to be discussed

TIME

Figure 6.1 The 10-day programme.

proposal cannot incorporate the constraints that he or she may require. Other factors may take even longer to resolve, like negotiating a wayleave with the electricity company, or confirming boundary positions. Unrealistic assumptions can be dangerous, causing difficulty later on when they are fully identified.

Remember too that the print machine may be broken or the journey to the client's office is disrupted by road works. Life itself is never easy!

6.5 Stage-by-stage programming

The work of the consultants involved with the design of the new car-dealership must be managed or co-ordinated. Mention has already been made of the wide range of potential professionals and specialists typically involved including architects, structural and services engineers, quantity surveyors, landscaper designers and specialists for equipment. There are specialists subcontractors and suppliers too who's contribution must be planned and controlled and fitted in with other parts of the contract, ensuring that targets are met in time in line with anticipated costs.

The role of design manager has traditionally been filled by the project architect or engineer, but for projects with large design teams, extensive detailed engineering works or for contractor/management lead procurement, there can be advantages in someone else acting as the co-ordinator, able to take an overview of the whole process.

The following section highlights programming issues at each stage in the traditional procurement process, with reference to the implications of other procurement methods wherever relevant.

Inception – meeting the client

This is an open-ended stage, but may be limited by the client's determination to establish working relationships so that meaningful design work can begin. The client must decide which form of procurement to adopt, possibly after seeking independent advice about the costs and timings of alternative methods. If the client's first relationship is with the designer, the programme proceeds through detailed design work to the appointment of a contractor by competitive tendering, or to the production of initial sketch ideas forming the basis for negotiations between the client and a builder. If the first relationship is with a builder or project manager, negotiations will take place immediately leading to detailed design work and commencement of construction. Whoever is the lead figure, meetings and visits will begin to clarify the client's brief, and determine a programme and costs for design work and construction. The critical date for the client is likely to be when the new building could be ready for occupation.

Feasibility – formulating the design brief, considering basic options

This is also an open-ended stage, depending on the complexity of obtaining sufficient survey information and the extent of research and consultations needed before being able to think about possible design options. Specialist advice about site conditions, structural design and cost planning may be essential at this stage. Consultations with interested authorities will reveal further constraints. Checklists can be circulated as an aid to gathering information about the client and the building type. Time will be needed to organise and present sketch ideas.

Outline proposals – establishing a practical proposal in principle

The design brief becomes an increasingly important working tool, informing the work of all the members of the design team. Constraints and limitations for possible structural and environmental services designs are explored and costs examined against the emerging idea. Sketches will be prepared and sent to authorities for initial advice. Formal presentations are made to the client for approval to proceed. The design team must be certain that the idea can be developed, and that proposals match the client's budget.

Scheme design – developing an agreed idea into a coherent working proposition
In terms of the overall programme, there will be a fixed point at which the sketch idea is agreed as the basis for more detailed work. The outline idea may only illustrate parts of the building like the front elevation, or simplified plans with little detail. Continued analysis and further information may alter the initial ideas, but at some stage the whole building must be illustrated showing, for example, all the floor plans and all the elevations.

Design work focuses on the development of a specific proposal rather than continuing to search for alternatives. When the design drawings are sufficiently advanced, the scheme can be submitted for Planning Approval. The Planning Process can be lengthy, so a decision must be taken on the appropriateness of proceeding with detail design work. This is a difficult point in the programme because the client may be anxious to get on, but detailed design work beyond this stage is at risk if consent is not forthcoming. Even if the scheme is eventually approved, changes may occur as a result of further consideration, or following negotiations with the authority.

Consultations continue with the authorities about building control, fire safety, highways and drainage design and services installations. Health and safety issues and the requirements of Construction Design and Management Regulations (CDM) legislation are discussed with the Planning Supervisor and the client, and the design team must consider the preparation of the Tender Health and Safety File. It is at this stage that the important elements of the design brief must be formally agreed and Plan that projected costs are fully understood. The client must be satisfied that this is the scheme that will be developed. Fundamental changes to the design concept after this point will be significant in terms of costs, and will inevitably affect subsequent programme timings.

Detail design – finalising the scheme
The detail design stage leads to a full understanding of all the parameters relating to the site layout and the building. Regular exchanges of drawings and information will refine the layout and appearance, finalise methods of construction, selection of materials, fixtures, fittings and details for installations and the operation of machinery. The specific requirements of statutory authorities are incorporated. The aim at this stage is to be certain that the framework for the new building is designed in sufficient detail to enable the design team to move into the next stage and begin preparing the necessary precise descriptions of the building and all its elements of construction.

Production information – working details and practical considerations
The design team begin to prepare working drawings to illustrate the arrangement and assembly of the detailed parts of the new building with detailed specifications describing requirements for materials and construction. The design may change in some respects as a result of detail considerations at this level. For example, the exact size or height of spaces may be changed to suit specific items of equipment or machinery. The dimensions of a roller shutter could be altered to suit a particular manufacturer's range or additional windows be added into a previously blank wall. Minor revisions such as these can normally be incorporated without too much difficulty, but changes by one member of the design team can cause considerable inconvenience to others who may need to alter designs and revise drawings which they had thought completed. Changes which are not essential may result in the client facing unwelcome additional design costs, and

an extension to the design period. Some changes may also affect the Planning Consent. Minor amendments may be acceptable by negotiation, but major changes can result in the need to make a fresh submission. The Building Control Application can be made and the Health and Safety arrangements described under CDM Legislation can be finalised. The QS may begin the preparation of the Bills of Quantity (BQ).

Measurement – BQ

The BQ is a summary of all materials and labours involved in construction of the building and any other associated costs. It is used in the traditional procurement process to obtain tenders from contractors, control the costs of development at pre-contract and post-contract stages and helps to ensure that everything needed has been accounted for. With other forms of procurement, it may not be used for tendering purposes, but for large projects may still be necessary as a means of cost control. It is essential that the QS is provided with sufficient, correct information as the basis for measurement, demanding close liaison between the members of the design team. If the measurement process starts too soon, there is a risk that sections of the BQ could be incorrect as subsequent changes may not be incorporated. Late changes in specification may not get included in the BQ if it has already been printed. This is an important factor because if expensive elements are not included in the BQ, the money that the client is expecting to spend will be exceeded by legitimate claims for additional costs once construction commences.

Tender action – obtaining competitive prices

This stage is only relevant in the traditional procurement method. The design team arrange contract matters with the client and put together a package of information which is sufficiently comprehensive to enable contractors to understand the potential cost of the work. Selected contractors will be approached to check their interest and time must be allowed for them to price the work and submit their offers to the client. The drawings included in the tender package are indicative only, so that contractors can see the extent of works required. Throughout the tender period, the design team will continue to prepare information to complete the description of the building. For example, a single typical section through an external wall will show a tendering contractor the principles of construction, but many more sections will be needed to show specific points of detail.

Project planning – tender analysis, appointment of contractor, confirmation of construction methodology, CDM considerations and construction programming

In the traditional procurement method, the returned tender figures are assessed and recommendations made to the client leading to the formal appointment of a contractor. After appointment, meetings are needed with the successful contractor to discuss CDM requirements, the construction programme and site establishment matters. The contractor will need time to appoint staff and arrange for site accommodation and machinery to be made available. The design team issue sets of information to the contractor. The information required immediately such as setting out dimensions, foundation details and specifications for hardcore, concrete and reinforcement must be complete so that the contractor can start work on site. Fundamental statutory approvals must be in place.

With other methods of procurement, project planning and contractual issues may be discussed and resolved at a much earlier stage, possibly during the inception period, but essentially at any time after the client and the contractor or project manager have agreed to work together to create the new building.

6.6 The completed programme

This example of a bar chart programme, shown in Figure 6.2, has been arranged to meet a notional target start date for commencement of construction on site 30 weeks after inception. The emphasised deadline dates are the points that have an affect on subsequent work. If the deadline dates are altered, because the task is completed sooner than expected, or takes longer, it is possible to see the implications at a glance. For example, a request from the client for an earlier start on site or an extension of time in the Planning Process will speed up or delay other activities. Earlier activities must be compressed, if possible to accommodate the client's request, whereas later activities extended until Planning Approval is assured. The bar chart indicates which, if any particular blocks of time have any flexibility. It is of course always advisable to leave a little float to account for unforeseen circumstances. It is usually better to be ready sooner rather than later.

In practice, preparation of the programme reflects the circumstances of each project. The length of time allowed for the completion of each activity or the point at which it is appropriate to start one activity which depends on partial completion of another, will not be the same every time and cannot be given as universally applicable figures. There are many points in the design process where timings may vary, where individuals need more or less time to make their contribution. Their previous experience is clearly significant, giving them an understanding of how long activities ought to take, but actually achieving the targets can be another matter. The client's chosen procurement method, the management structures of all the members of the design team and the attitudes of authorities can all result in the need for shorter or longer time periods at different stages. Figure 6.3 compares the time implications of three different procurement methods, showing the advantages of earlier completion.

Methods of communication and preparation of information may speed up some stages of work. Computer generated drawings can be revised and circulated sooner than those constructed manually. The use of fax machines and e-mail communication is much more immediate than traditional letters, or even telephone conversation when contact can be difficult if people are not available. The speed of response at points when individuals must approve the work of others can also be critical. For example, the design of a straightforward rectangular steel-framed structure can be progressed smoothly from sketches to detailed construction drawings, but a complex shape take much longer to fully consider. In a commercial situation, individuals in the design team must programme their own work with respect to many other projects that they may be involved with. Reasonable time should always be made available so that individuals have sufficient time to give a properly considered response, which is not so rushed that poor decisions are taken leading to avoidable errors. Whoever is responsible for preparing the programme, all of these issues must be reflected in the time made available to undertake tasks from start to finish.

6.7 Project File content

The general aims at this stage are as follows:

- Determination of the extent of design work required by the client and the consequential professional consultant involvement.
- Production of a programme of the essential work for all the members of the design team.

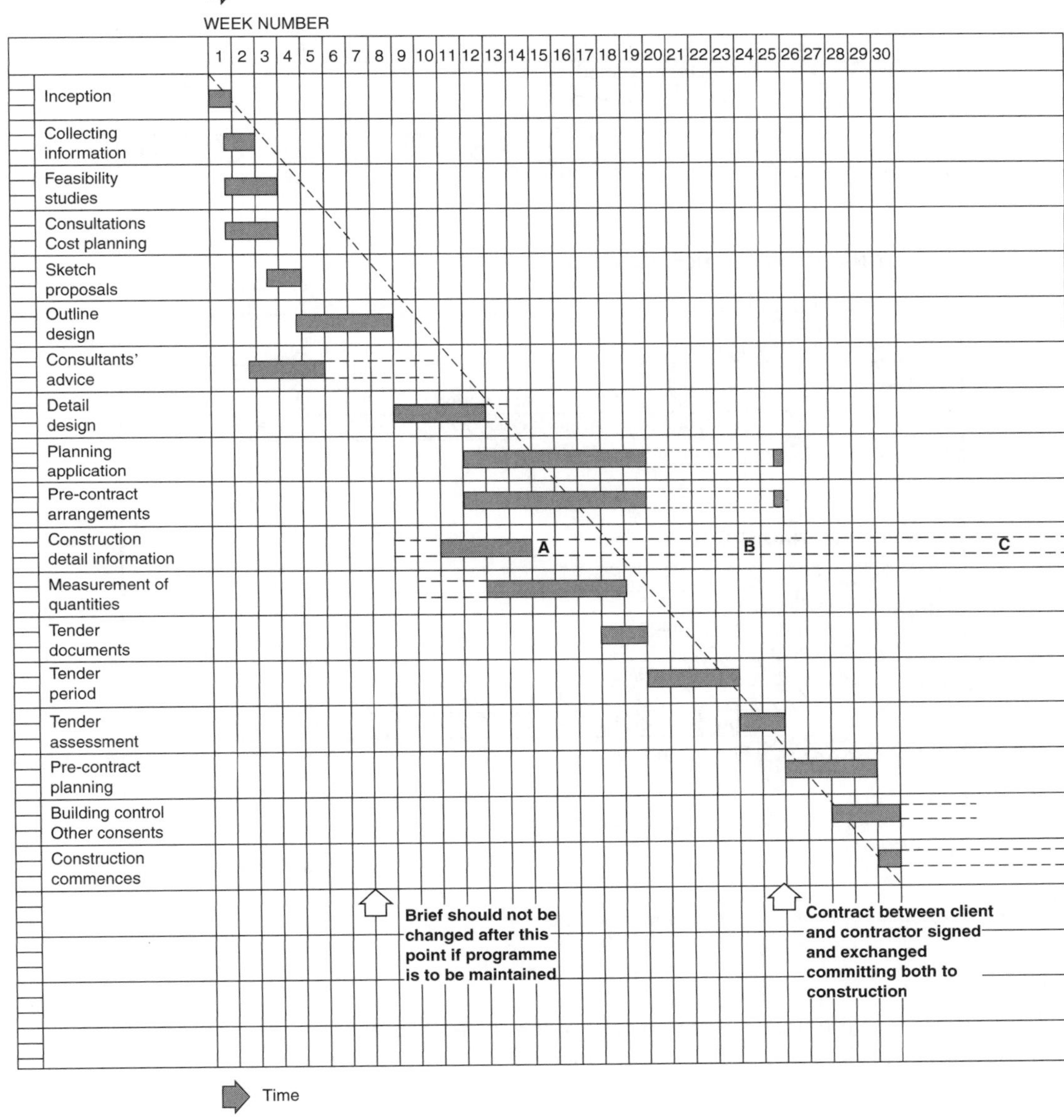

Figure 6.2 A traditional procurement programme.

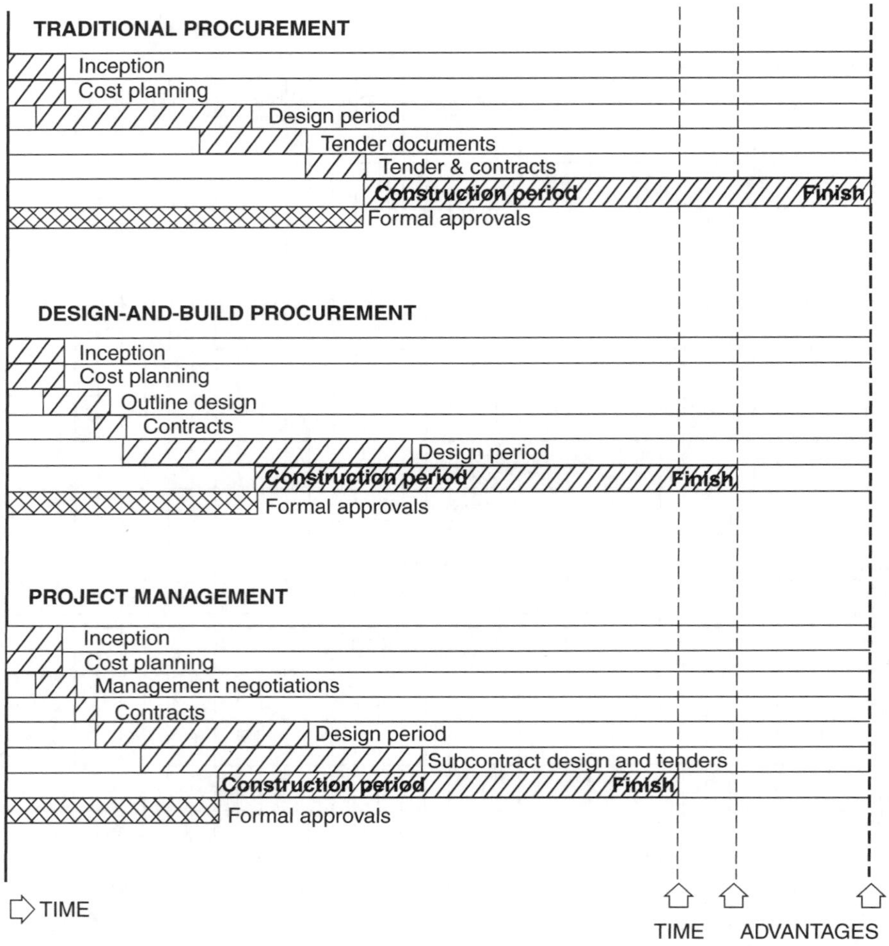

Figure 6.3 Programming alternative procurement methods.

- Establishment of the financial framework of the development.
- Instigation of preliminary consultations and site investigation works.

File material could include the following.

Record notes, letters and minutes

- Details of informal and formal exchanges between the members of the design team.
- Copy of letter advising the client on the economics of type of design envisaged.
- Copies of letters asking for or confirming points of information, and copies of replies.
- Details of arrangements with specialist surveyors or site investigators.

Appointment details

- Further consideration of the procurement arrangements and agreements with client.
- Details of initial costings and budget planning.

Bar chart programmes

- Preliminary exploratory programme charts for discussion.
- Agreed programme forming part of contract arrangements with the client, circulated to all members of the design team.

6.8 Discussion points

(1) Is there a clear relationship between the ability of any designer to create a successful building and the time available for design work? Will the quality of the new building be improved as a result of an extended design period? To what extent is a building designer exposed to risks in pre-empting the results of consultations with statutory authorities?

(2) How long is it sensible to spend on theoretical design before committing to detailing? How can the building designer accommodate specialist elements such as heating systems in advance of consultants input? Why is there always pressure to complete the design process quickly when the new building may be in use for the next 50 years?

(3) 'Fast, cheap, high quality and aesthetically pleasing'. Is this a fair summary of a client's expectations for their new building? Most clients are unaware of their responsibilities for making decisions in line with a necessary programme. Is this a good or a bad thing? How can a client be offered flexibility in the timing of the handover of their new building?

(4) How should health and safety issues by incorporated into a programme? What steps are required to ensure that all hazards are identified at the right times? Is there a risk that speed makes design and construction work more dangerous?

6.9 Further reading

Best R and **de Valence** G (2002) *Design and Construction: Building in Value.* Oxford: Butterworth-Heinemann.

Calvert RE (1995) *Introduction to Building Management.* 6th Edn. Oxford: Newnes.

Cherry E (1999) *Programming for Design: From Theory to Practice.* Chichester: Wiley.

Cooke B and **Williams** P (2004) *Construction Planning, Programming and Control.* Oxford: Blackwell.

Duerk DP (1993) *Architectural Programming: Information Management for Design.* London: Thompson.

Englemere Services Ltd and the **CIOB** (2002) *Code of Practice for Project Management for Construction and Development.* Oxford: Blackwell.

Foster G (1981) *Construction Site Studies: Production, Administration and Personnel.* Harlow: Longman.

Fryer B and **Ellis** R (2004) *The Practice of Construction Management.* 4th Edn. Oxford: Blackwell.

Picher R (1992) *Principles of Construction Management.* 3rd Edn. Maidenhead: McGraw-Hill.

Popescu C and **Charoemgam** C (1995) *Project Planning, Scheduling and Control in Construction: An Encyclopaedia of Terms and Applications.* Chichester: Wiley.

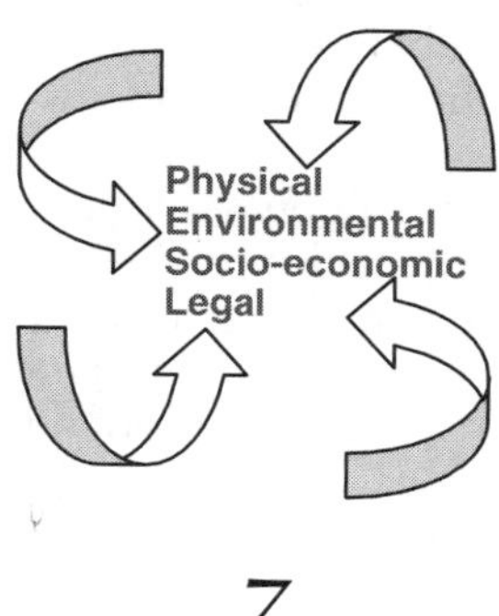

7

The design brief

7.1 Introduction

The initial thoughts about a possible new building discussed in the previous chapters are likely to be idealistic and relatively loose. In many cases they may be no more than expressions of intent, or perhaps a 'wish-list' of preferences. Once design work begins, decisions will be made as information is discovered which effectively create anchors, fixing the way in which the project will or could advance. There is a period at the beginning of design work when ideas and information are exchanged beginning to describe what is required. The needs of the project may be clearly understood and agreed by everyone involved, perhaps based on precedent backed up by research. But more often than not, after analysis some initial preferences may prove to be unnecessary, incorrect, unachievable, or too expensive, but it is right that they should be carefully considered as options. There are three significant areas of research at this stage: the client, the building type and the site; and this chapter will identify the information which can be collected, organised and prioritised leading to the notional design brief which forms the basis for the work of the design team as ideas are developed.

7.2 Developing the client's brief

The client's brief, discussed in Chapter 5 is a preliminary guide to *their* requirements for developing *their* site which may be sufficient to determine a concept for the new building, or at least to suggest a number of options which might be viable. On the basis of known facts, conceptual design work or feasibility studies can commence straight away to see if an idea or ideas are worth developing in greater detail. Assuming that development is practicable in principle, it is possible, and even probable that subsequently revealed elements of information will further change or modify the client's brief in some way. Unforeseen circumstances can mean reconsideration of issues which were believed to have been resolved or which had not been fully appreciated straight away. In these circumstances, alternative concepts must be considered, as the first conceptual designs may no longer be valid.

In any event, parts of the client's brief will be described in general terms only, such as 'showroom', 'workshop', 'parts sales' and 'waste disposal' which will all need careful examination to see what they actually mean. If conceptual design is to have any practicable value, it should be used to explore and define the true functions and requirements of the project by continuous, dispassionate questioning, prompting consideration of alternatives, as far as possible eliminating preconceived ideas. This process leads to the *design brief*, clearly establishing what the elements of the building are, or should be and involves collecting and recording information about them in order to assess their significance.

For example, a concept for the layout of the car-dealership based on available information may appear to be attractive until an enquiry with the local authority reveals the previously unknown existence of a 900-mm stormwater sewer running under the area proposed for the showroom. The restrictive 10 meter wide easement corridor required by the local authority for future maintenance and replacement works, prevents construction in this area unless the client is prepared to accept the major expense of diverting the sewer. This is a possibility, as the line of the sewer could be changed to run under areas of the site which will not be built on such as car parking, landscaping and planting. At this stage the consequences for design proposals are essentially twofold:

(1) further detailed design work on this particular concept will achieve nothing because the showroom cannot be built in this location or
(2) the concept can be developed unchanged, but involves substantial and possibly unnecessary additional costs, which could be avoided with a different design.

Simply attempting to relocate the showroom to a different position within the conceptual layout will be difficult or impossible because the essential relationships with other spaces and areas will no longer be correct or appropriate. As will be shown later, the organisation of activities in the building and around it on the site depends to a large extent on how people and materials move from one position to another. If the correct relationships are based on the showroom being in one position, then changing its position will mean changing the position of everything else. If for example, in the conceptual layout, the showroom is at the front of the site in a prominent location obscuring the more unsightly workshops and storage areas, moving it to one side to accommodate the sewer easement exposes unwanted views to the passing customers and changes the arrangements within the building. The probability is that the original concept is no longer practicable and in order to accommodate the new finding in a better way will require a new concept.

Other elements in the design, such as the number and specification of toilet cubicles are not immediately important and can eventually be determined and integrated into the layout without damaging the concept. In conceptual terms, it is important to try to 'see the big picture' before becoming swamped with details, but as failure to identify and account for fundamental elements like the sewer easement can make initial design work irretrievable, it is *very* important to concentrate on obtaining *critical* information as quickly as possible, taking an overview of the situation to try to see where the problems may arise. It would be unreasonable to expect the designer, or any of the other members of the development team to automatically somehow 'know about everything' but there is an expectation that sufficient, professional care has been taken to try to find out, so that progress is not affected, leading to accusations of negligence.

The design brief is not necessarily a formally produced document. It may take the form of a report identifying the project-specific factors to be taken into account, or could be a series of checklists summarising design data or criteria. It may simply be seen as an attitude, or a means towards an agreement about decisions taken as work proceeds. In whichever way the information is recorded, the design brief should be seen as a stage in the development of design proposals establishing a common understanding of the requirements of the project so that all the members of the design team are working towards the same goals. The objective is not to publish a perfectly complete document at any particular time, but rather to set up a management system for information to which everyone can contribute and can use for reference once design work starts, and which, as far as is possible, eliminates the dangers of verbal misunderstandings leading to confusion and the risk of subsequent problems. It can be expanded and refined by agreement as more and more information is collected. There are three principal areas of project-specific elements to be investigated:

(1) Client
(2) Building type
(3) Site

Information about them can be obtained in any of the following ways:

- Initial meetings, discussion and site inspection (sometimes referred to as a 'walkover').
- Research and enquiry (often referred to as 'desk study').
- Measured survey.

Consequently, the aims of the design brief can be defined as follows:

- Investigate and record everything which is visible or known.
- Research all possible sources for factors which are invisible or unknown, but predictable.
- Focus the attention of the client and the design team on issues which are unpredictable but could apply to the development, but which may not be immediately apparent. The *'what if'* scenario.

7.3 The client

Discussion with many clients about their needs and requirements is often affected by the problem of their familiarity with their current situation. The inadequacies or failures of their existing arrangements are often not recognised and change is resisted because previous experience has apparently been adequate or they have not been able to step back and look at their situation dispassionately. Sometimes, even the good points or advantages of their situation are obscured because they have been used to doing things in their own particular way. It is not uncommon to find that they will perceive some situations as if they had been positively chosen in the first place, completely ignoring obvious deficiencies. In many cases there is a tendency to defend the *status quo* to avoid the embarrassment of admitting to inadequacies or facing up to the expense of correcting them. At best, they work around problems forgetting that things could be different and maybe better. This is very common in an expanding commercial business when the client is too busy to rationalise their activities in the available space. It is entirely understandable that they prefer to concentrate on the quality of the goods and services offered to *their* clients rather than worry about their own conditions.

However, they eventually realise that they must extend their building or relocate to a new one. They will use their previous experience to dictate firm ideas, often proposing design solutions themselves which they expect to be implemented. Whilst experience is obviously very important, existing practices and arrangements may not be the best basis for considering the requirements of the new building. Some aspects of their brief may be incorrect or could be satisfied in a far better way once all the elements have been discovered and accounted for. They may be unable to 'see the wood for the trees'.

The design team can help the client to analyse their own situation, establishing priorities for elements which are as follows:

- Essential
- Important
- Preferred
- Not needed at all

The factors to consider include the following.

The client's existing operations

In many cases a client seeks a new building to replace an inadequate existing one, or to enable expansion of their business activities. There will be some features of the original business that they wish to retain. They have a relationship with their customers based on the style or image that they have already created which they would like to be maintained and enhanced in the new building. They will have a company ethos or philosophy about how they treat their customers and respond to market forces in their own area of trade. They may have a market position to sustain demanding a certain level of quality or prominence. Elements of the building can be dictated by all these facts. Even the most simple and elementary factor can play a major part in design proposals. For example, the colour of the external plastic coated profiled steel cladding for the walls and roof could be dictated by the need to match it exactly with the colour of the client's logo on their recently printed stationery.

The client will already have a management structure with different people responsible for various activities within the building. New activities may mean new staff, bringing in their own experience, ideas, preferences and understanding of their needs. Both existing and new staff will have expertise in their own areas which should never be ignored.

An analysis of the client's current floor areas and volumes can be revealing, perhaps highlighting areas where space is wasted, or could be used more efficiently with an alternative layout. Current staffing levels will help to establish new floor areas and the space needed for each activity. Other issues to consider include processes and sequences, customer handling, sales arrangements, staff facilities, times of occupation and how existing operations interlink or overlap. It is important to see where there are any problems at the moment so that they are not repeated in the new building. They may already own expensive machinery and equipment which must be transferred into the new building. This is a major area of research, questioning existing staff about their perception of current occupation and activity.

Details of needs, requirements, essentials and preferences for the new development

At this stage it can be useful to explore possibilities by asking different members of the client's staff ' what would be the ideal design solution to suit your needs?'. Although the managing director may wish to be fully involved with all briefing arrangements and ultimately may have the casting vote on what is to be provided, it is more likely that individual members of the staff will be able to provide the necessary detail observations. For example, the parts manager will have detailed knowledge of the company's storage requirements, the number of parts carried in stock at any one time, the means of stock control, and even the number and size of shelves or racking needed. The parts receptionist will have experience of a preferred counter layout to serve customers and contain computers and other equipment. However, this can be a time to tread carefully in order to avoid causing internal friction in the client's management structure. Individuals should not be lead to believe that their personal requirements will automatically be incorporated into design proposals. The managing director, or the clients project leader will need to take an overview of all matters within the complex, particularly where decisions affect cost.

The general issues to discuss could include the following:

- Proposed additions or reductions to staffing levels.
- The male/female split in staffing numbers.
- Intended relocation of staff to different areas of the new building.
- Details of working arrangements for different types of staff.
- The numbers of customers and trade deliveries anticipated at any one time.
- Preferred facilities and amenities for customers and staff.
- Changes in the way people or materials are handled or processed.
- Details of storage requirements.
- Details of processes which generate waste.
- Any specific sizes and dimensions of spaces or equipment that are technically essential.
- Desired or required environmental conditions.
- Processes which present hazards or risks to health and safety.
- The style or finish of any elements.
- Costings for any specific elements.

The extent of client's own subcontractors and suppliers

Many commercial clients will have some in-house expertise. Employees or previously employed specialists or consultants will be able to offer advice and guidance about some elements of the building. For example, someone may have intimate knowledge of the heating or electrical systems, responsible for their maintenance with regularly employed subcontractors. They would be in a good position to help with the Tender Health and Safety Plan required under Construction Design and Management Regulations (CDM) legislation, identifying requirements in principle and will be able to assist in the design of the heating and electrical systems to suit their technical requirements in the new building.

Programming

The client can help to establish the programmes for both design and construction by making time available for meetings, responding promptly to requests for information and formalising opportunities when approvals will be given to proceed from stage to stage. They must understand when finance is to be available once construction has commenced and forward their plan their own move into the new building, ordering specialist equipment for delivery and installation at the right times.

Checklists

Checklists can be exchanged in the form of question and answer sheets to confirm information in each of the areas above. The purpose of such checklists can include the following aims:

- Remind the design team to ask all the relevant questions.
- Help the client to rationalise thoughts and make commitments to progress.

- Give the client time to conduct research and consult with staff or other advisors.
- Act as a reference as design work proceeds, confirming the decisions taken.
- Assist with transfer of information between members of the design team.

Checklists can remain 'live' for the duration of the design period, being changed and amplified as more detail is added, kept up-to-date and circulated regularly. Assumptions should not be made about simple issues as something that is obvious to one person may not be to another. The designer's understanding of an issue may not be the same as the client's, and equally, the current technical content of some aspect of design may be quite different from that with which the client is familiar. For example, the designer's presumption that a workshop should have windows to let in natural light could be wrong if the activity in the workshop demands a fixed level of artificial light at all times. The design implications with regard to complying with the current Building Regulations in respect of fire safety may bear no resemblance to conditions in the client's existing building or in relation to their existing practices. The client may need to rethink a variety of activities in the light of the implications of other applicable legislation. In both of these examples, the client and the design team must work together to arrive at the correct answers. The checklists are a useful way to get clear accurate answers to the easy questions so that attention can be focused on the more difficult ones.

7.4 The building type

Buildings are identified by simple descriptive names such as 'house', 'factory', 'church' or 'car-dealership', and inside the buildings there are more names; 'kitchen', 'workshop', 'vestry' and 'showroom'. They imply a purpose or function generally thought to be understood, but the influence of previous experience may make this understanding incomplete or inappropriate. Historical example is not necessarily the correct basis for design requirements or expectations today and elements once given priority may not now be so important, or even have any relevance at all.

For example, the word 'office' once carried with it suggestions of privacy, staff only, type writers and filing cabinets, but is currently seen as open plan, accessible to customers and full of computers. A whole wall of shelves stacked with thousands of pieces of paper has been replaced with a few plastic discs. Computers once occupied whole rooms but are now small tools within each work space. The word 'office' may not mean the same thing to every client, and may contain uses specific to the requirements of their building with which the designer is not familiar. It goes without saying, of course, that today's expectations may not be relevant in the future, and considering the pace of change in the recent past, that future may not be too far away. In this respect, the best approach is to anticipate change by offering flexibility, creating environments which are not so fixed and constricting that their use, once defined cannot be viewed in any other way.

At this stage, it is essential to consider what the building and the site *could* contain in terms of use as it is only through an awareness of actual requirements and possibilities of activity that spaces can be defined and dimensioned. As a simple example, consider an entrance lobby into a public building. It could be designed purely as a means to entering the building, kept to the minimum size needed to open and close the two doors needed to keep out cold air. This is one definition of the function of an entrance lobby which the design has clearly catered for. Alternatively it could be designed to be large enough to accommodate displays or information to attract and direct entrants, or a receptionist to welcome them and maintain security from unauthorised or unwelcome visitors. It could contain access to toilets and a waiting area with seats, tables and a telephone. In the car-dealership, it may need to be large enough to be used to bring cars into the building and at some point in the future, the client may wish to relocate the lobby to another part of the building, bringing customers into the building in a different way. If in the future the car-dealership ceases trading and the building is to be put to some other purpose, the entrance lobby may need to be used in a different way altogether, containing other, unknown functions.

The use of space definitions alone is limiting because it ignores the fact that different activities can overlap in the same space. To say that the car-dealership should have spaces for an entrance lobby, a reception, a showroom and offices can encourage the belief that they are separate, individual and unique spaces, whereas they may be the same space, or can be designed to be in the same space. A checklist of the principle activities and issues which might be associated with them, or influence them will help better understand the needs of the project. Influences can be ideas that may or may not be relevant, to be incorporated or discarded after due consideration, positives and negatives, advantages, disadvantages or conflicts.

In the next chapter, when design work commences, the movements between activities will be investigated as they are assembled in the designed layout, but first it is useful to examine individual activities that can take place in buildings. The words listed here are associated with activities that normally take place in a car-dealership. For other building types, there would be other ideas. At this stage, each word can be regarded as an idea, which can be explored for meaning, possibility and relevance. This is a form of brainstorming, trying to think of anything that might be involved however unlikely it might seem at face value. The list is not structured, and can be added to at any time:

- *Public access onto the site for pedestrians and vehicles*
  Visibility, accessibility, prominence, sight lines, traffic calming, display, advertising, attraction, image, style, deliveries, collection, car parking, turning, shelter, security, facilities for the disabled, safety, lighting, signage and directions.
- *Public access into the building reception and control of visitors*
  Information, direction, advice, service, display, welcome, security, deliveries, storage, ambience, views, orientation, warmth, glare, shade, noise, toilets and telephone
  sales reception for new and second-hand cars, accessories-parts reception for parts, parts storage, display, accessibility, control, service reception, customer waiting, collection, retrieval, courtesy cars, test drives.
- *Private access and entrance for regular occupants, management and staff*
  Security, supervision, control, cloaks, toilets, storage, mail for deliveries, new cars, second-hand cars, damaged cars, fuel, parts.
- *Circulation for public or private movement around the building*
  Entrance porch, foyer, passages, corridors, stairs, lifts, landings, escape routes, communication, directions, signage, control, security, internal and external movements.
- *Work*
  Managing, administering, typing, filing, selling, manufacturing, repairing, distributing,
  records and files, sales, repair, servicing, painting, fuel, air and water, car wash, noise, dust, fumes, catering, washing, cleaning, drying.
- *Display: internal and external*
  New and used cars, parts, accessories, advertising and directions
  exhibiting, informing, demonstrating.

- *Relaxation facilities for customers and staff*
  Toilets, refreshments, information, resting, drinking, eating, sitting, watching, listening, reading, canteen, rest room, first-aid arrangements, training.
- *Meeting space for staff and customers*
  Public, private, large, small, formal, informal, casual, arranged.
- *Study*
  Thinking, reading, writing, teaching, listening, watching, learning, referring, computers, video monitors.
- *Storing*
  Equipment and goods, stationery, catalogues, warehousing, retrieving, security, control, replenishing, accounting.
- *Disposing*
  Unwanted waste, solids and liquids, vapours and gas.
- *Personal hygiene*
  Customer and staff toilets, basins, urinals, sinks, showers, disabled facilities, facilities for children, washing, bathing, showering, dressing.

There are many others including sleeping, sports and recreation, performing, acting, singing, worshipping, being ill, being young or old. For all these activities there are extra considerations for people with disabilities. Of course, every building will not accommodate every activity, and some will take place outside like gardening, sunbathing, eating and cleaning the car on Sunday mornings.

The bubble diagram shown in Figure 7.1 is a preliminary expression of first thoughts, produced without particular regard for location and relationships. It is a form of *brainstorming* to take a look at what issues might be relevant to the design of a car-dealership. The outcome is presented in a rationalised way in Figure 7.2, establishing possible activities and relationships in a more structured manner. All the members of the design team should be invited to contribute to this process of analysis as an attempt to understand needs before allocation of specific space names, because although the space names may describe the principle activities, they may also enclose other activities needing further consideration.

For example, as part of the activity of display, the car-dealership may have a showroom. What is meant by the word 'showroom'; what are its constituents, what does it do, what can be said about it? Some of the issues are identified in Figure 7.3, which can be amplified by considering access, inclusive activities, association with other spaces and activities, prominence, shelter, security, environmental controls, use by whom, display, information, welcome, reception, direction, sales, communications, access, disability, toilets, refreshments, deliveries, waste, maintenance, cleaning, visibility, privacy, finishes, equipment and so on. Each of these words can be expanded and analysed for content, importance, cost and overlaps. Ultimately, this should give sufficient information to determine practical size, shape, volume and location.

Now which of the defined criteria are essential, important, desirable, peripheral, unnecessary? Which ones are anchors in the design? Which ones are secondary functions, enclosed within the anchors? Questions prompt answers that will vary from designer to designer and from client to client. There may be no absolute answers, or there may be a multitude of possibilities, depending on view point, expectation and experience. This does not affect the need for the questions as it is the nature of answers that leads to good design as decisions have a rationale, the result of positive assessment and choice rather than happening by default. Default design will be incomplete, unsatisfactory or poor.

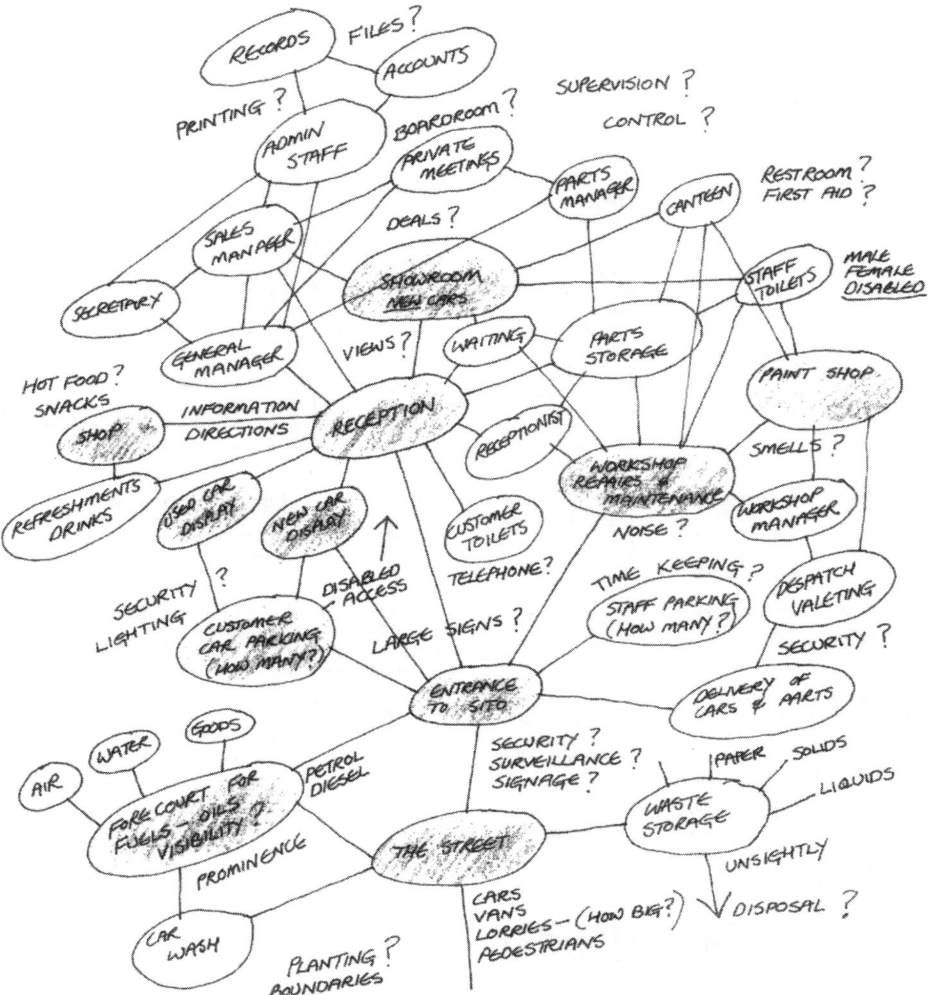

Figure 7.1 A bubble diagram of first thoughts.

As the showroom is a principle part of the car-dealership business, consider some of the cultural issues briefly referred to earlier. Take a look at current car-dealerships and see how the design reflects many of these words:

- *Image*: appearance, prominence, significance.
- *Status*: authority, power, importance, affluence, wealth, focal points.
- *Confidence*: assurance, influence, continuity, stability.
- *Excitement*: thrill, stimulation, inspiration, exhilaration, flamboyance.
- *Meaning*: allusions, reflections, context, ornament, decoration.

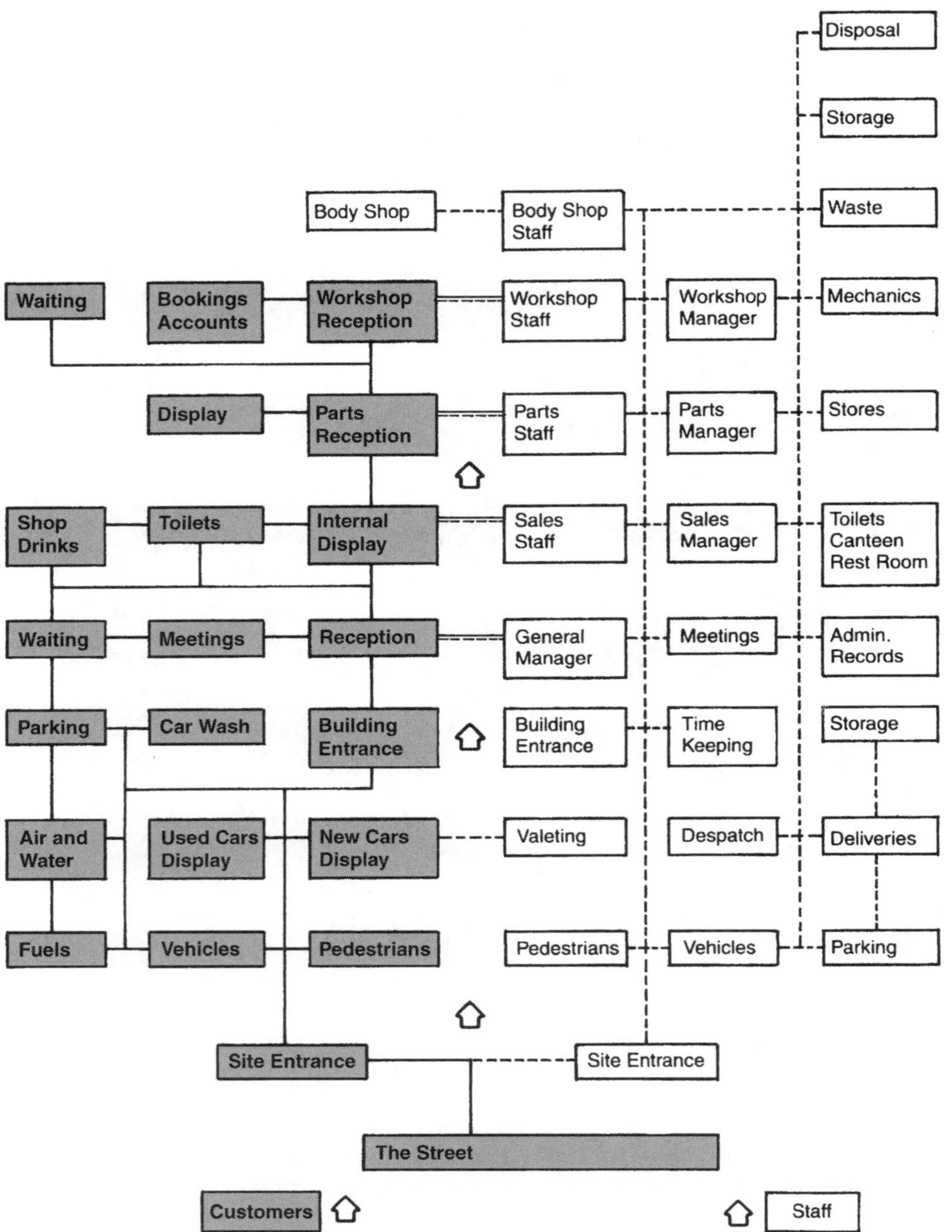

Figure 7.2 A structured analysis of the whole complex.

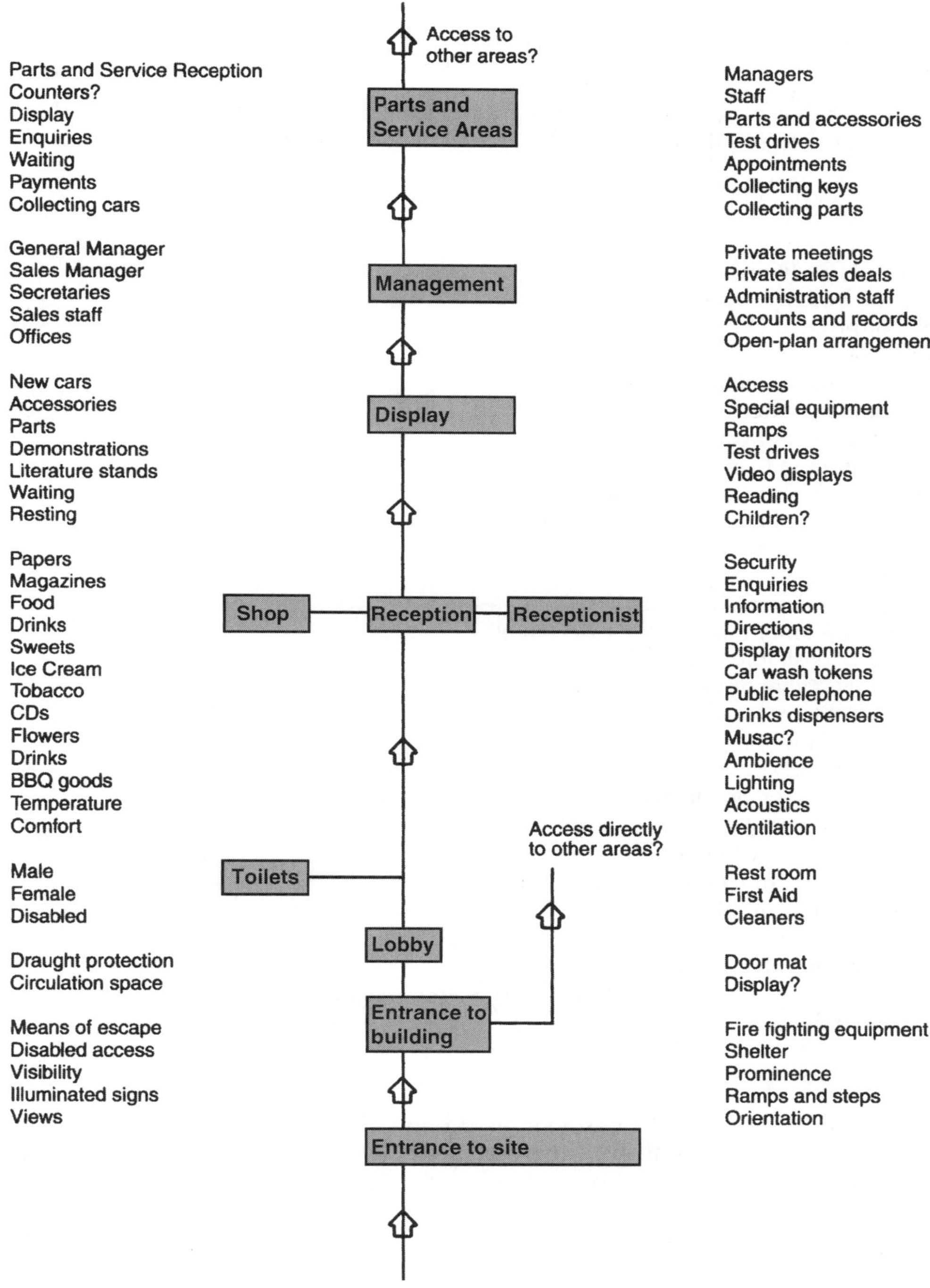

Figure 7.3 A structured analysis of the showroom.

- *Flexibility*: change, redevelopment, expansion.
- *Belief*: value, importance, ritual, symbolism.
- *Politics*: power, control, influence, community interaction.

Questioning and analysis can be organised with detailed checklists for each space. The questions are tailored to suit the building type. For the car-dealership, the following issues require clarification:

- Name of space
- Principle function
- Additional functions
- For use by whom
- Public or private
- Number of occupants, male and female
- Number of visitors at any one time
- Links with other areas
- Space for specialist equipment
- Space for built in fixtures or fittings, racking
- Anticipated or necessary floor-loading capability
- Special construction such as ramps or pits
- Ground floor or first floor location, means of access, floor levels
- Means of escape
- Maximum/minimum size (floor area)
- Maximum/minimum height (floor to ceiling)
- Preferred or desirable shape
- Need for flexibility or future extensions
- Internal or external walls, or both
- Need for windows, how many, what size
- Doors, ironmongery
- Working environmental temperature, preferred heating system
- Power requirements
- Lighting, emergency lighting
- Ventilation and extraction, air conditioning
- Communications installations
- Alarm systems, smoke detection, fire alarms
- Fire protection, fire fighting equipment
- Floor, wall and ceiling finishes; decorations; colour scheme
- Signage, graphics
- Fixtures and fittings, furniture
- Colour schemes

Checklists should be established as soon as possible. Some of the issues can be resolved straight away but others will only be determined once the concept design has been completed. All the items listed above must be defined, selected and costed at some point in the process. Checklists can be very useful not only to obtain information, but to ensure that important and expensive items are not forgotten when the client and contractor commit themselves to a contract to construct the building.

7.5 Evaluation and prioritisation

The factors determining the value or priority given to many elements as the new building is designed are either matters of fact or preference based on someone's perception of their significance. For example, it could be a well-established commercial fact that the new car showroom should be located in a more prominent position than the fuel sales courtyard. Alternatively, it could be the other way round, or maybe the demands of business practice are such that both elements require equal prominence. Whichever is true, it is an over-riding fact which must take priority when considering the layout of the building. On the other hand, there may be no factual basis for choosing one option or the other and the relative prominence of one to the other is unimportant. It may actually be unknown or change over time in response to market forces. The way that the layout is developed in this case depends on experience or preference. Similarly, large areas of glazing may be one approach to the design of the showroom, opening up the interior to display as many new cars as possible to potential customers or, alternatively, solid walls punctuated with smaller windows can create greater interest by limiting views, inviting inquisitive customers to come into the building and explore its contents. In this example there is no 'right' or 'wrong' design solution. The selection of one or another is a matter of choice in the way in which the issue is evaluated.

Assuming that the prominent display of new cars *is* a primary function of a car-dealership, then this activity is a critical anchor in the overall layout and determining its size and location is a primary consideration. Conversely, staff toilets, which are obviously essential are secondary or even tertiary issues, can be fitted in later, once the form of the development has been established. It would be rather perverse to start from the position of 'where shall we put the toilets?'. What can be said though is that '*x*' number of toilets to satisfy the needs of staff and customers is a fact, but their location within the development is a matter of preference, or that there are a number of options which can be assessed later on.

It is impossible to categorise facts and preferences in neat and tidy columns. Every element in the building must be examined to see whether its performance is predetermined in any way. This is rarely a simple task as the influences that bear on any element are complex mixtures of fact and preference which can be interpreted in different ways.

Evaluating and prioritising elements depends on **the interest of the assessor**.

For example, consider a roller shutter door in the external wall of a workshop. The client, designer, engineer, manufacturer, Planning and Building Control Officers will have different interests in the performance of the finished product. Their understanding of cost, construction, appearance and use will not be the same and the criteria that each use to assess the shutter door may not be compatible with one another. The client's demand for a very large opening in a particular location may conflict with the engineers diagonal steel bracings, the manufacturer's ability to guarantee smooth operation, the Planning Officer's attitude to appearance and the effects on adjacent property and the Building Control Officer's application of current legislation. Which requirement takes priority?

First of all, clarification is needed as to:

- *Whether one fact specifically controls another*
  The client's requirements for the shutter door are *paramount,* and the engineers steel bracing must go somewhere else.
- *Whether one fact leads to possible options*
  The client's requirements for the shutter door are *flexible* and it can be located in a variety of possible locations to suit other constraints.

It may help to evaluate and prioritise elements by establishing a hierarchy of importance, starting with principle headings and expanding them into primary, secondary and tertiary sub-headings, until all the elements have been accounted for. For example, *car-dealership* is the principal heading. What are its principle ingredients? New car sales, second-hand car sales, repairs, maintenance, parts, fuels, etc. Once again, a brainstorming session may open up important strands of thought. Concentrating on new car sales, amongst other things does it contain an *internal car showroom*? Following on from this:

- Does the showroom include *reception*?
- Does this mean that there should be *a counter*? If so, what should its *dimensions* be and *what materials* should it be made of?
- How will the counter be constructed?
- What are to be the specifications for its *finishes*?
- What *colour* should they be?
- How will they be *applied*?
- How much will it all *cost*?
- What are the implications for future *maintenance and/or replacement*?
- Is it better to *invest* in 'long life' materials or renewable finishes?
- Which elements are affected by *statutory controls*?
- Should thought be given to *flexibility* to be able to reposition reception in the future?
- If so, what are the *implications* on floor finishes and services if it is moved?
- Could the space be used for *a different function* within the car-dealership if required?
- Could the space be used by *a different occupant* altogether for some other use?
- What are the implications if *business methods change*?
- What are the implications for *incorporating advances* in technology?

And so on.

Evaluation can start from any of these points, whichever is raised first. Some of the answers will look forward to their solutions, others will look back to previous answers. The answers all have a bearing on each other. Is a car showroom necessary, essential, optional or preferable? Should there be a formal reception counter, or an informal area? What are the constituents of reception; how might it work; how might it not work? What are the advantages and disadvantages of any suggested answers? Who will make the decisions or establish priorities? Is it the client, the designer, the engineer, the quantity surveyor? Is cost more important than appearance? Is there a statutory involvement determining some aspects of the solution?

There is not enough space in this book for all the answers, even if they could be given. The point is that every issue involved in the creation of the building can be explored in a dispassionate manner. For some elements of the building it would be extravagant to question everything. But the process can be considered in reverse; that is, can a conceptual proposal withstand critical analysis? If not, the design must be changed as it is probably the wrong solution. At the end the design must as far as possible, rationalise conflicting facts, opinions and preferences. Always remember that design decisions will have a profound affect on the building's users. For example, the height of the parts department counter may be the most important element of the building for the parts receptionist. Everything else in the building may be perfect, but if this element is wrong, then to that occupant the building is a failure.

7.6 The site

In Section 5.6 in the previous chapter, reference was made to the way in which the site's characteristics might affect development proposals. The site's characteristics can be grouped into the following four principal categories:

(1) *Physical features*
Within the site boundary, above and below ground level.
(2) *Environmental conditions*
Affecting the site or immediately around its boundary.
(3) *Socio-economic factors*
Relating to the neighbourhood infrastructure and general area status.
(4) *Legal issues*
Which control development potential and future use.

The discovery of one specific piece of information may have implications under each heading. For example, a preliminary inspection of a site may show the existence of a well-trodden footpath running from one corner to the other. This 'desire line' has developed as a result of local use as a shortcut. It is a visible **physical** entity in existence on the site, and research may show it to be a Public Right of Way, which has **legal** status. Even if it is not, its removal has an **environmental** implication as its users may still attempt to cross the site after completion of the new building, presenting security problems, and its removal could have an effect on the general amenity value of the area by denying an established means of access, upsetting the **socio-economic** balance of the neighbourhood.

As each characteristic is noticed or discovered, it should be recorded and summarised on an accurate survey drawing, building up a picture of advantages and constraints. The significance of any factor is not always obvious at the beginning and its implications for development proposals may not be fully appreciated until design work is well advanced. The aim of site analysis is to discover and understand the site's characteristics in advance of design and construction. It is a systematic process of inspection, research and measurement, and wherever practicable, the building designer should be involved in this process as a means of learning about the site in order to gain the fullest possible understanding of it in relation to development requirements. This will help to ensure that important issues are not overlooked as design proposals are developed. It is not always possible to discover every relevant characteristic or to measure it accurately. However, a rigorous approach to site investigation will minimise the risks of omissions and help to establish contingencies against unforeseen problems which may arise later. The following four sections identify the factors and issues which should be considered during site investigation.

7.6.1 Physical features

At the first visit to the site, the physical features visible above ground level are the most obvious, which in themselves may determine design decisions, or which suggest the possibility of other hidden factors which can only be revealed by further research, exploration or accurate measurement. For example, the remains of a derelict building visible above ground could mean that there are concealed floor slabs, foundations, basements, fill, contamination, redundant drains and services, none of which are immediately visible, but any or all of them could be anticipated as probable. The presence of any of them may affect the positioning of the building on the site and amongst other things, increase the cost of foundations.

Figure 7.4 illustrates some of the specific physical features evident on the car-dealership site, but other general issues to consider could include the following.

Previous activity and use

Whether or not a site has been used previously is not always obvious. Remnants of ceased activity may be visible, but in many cases it will have been covered over and forgotten. Research may reveal its existence, recorded in historic documents. Indicative street names such as Quarry Side, Spring Road or Old Mills Lane can give vital clues to previous activity and use, as can discussions with local residents. 'I remember a few years ago seeing hundreds of lorries go in there full of broken concrete and bricks!' may contain an element of exaggeration, but it could be very important and well worth further investigation. Other indications include:

- *Structures*: buildings, foundations, basements, cellars, demolition residue, walls, retaining walls, machinery and ancient monuments.
- *Workings*: open trenches, quarries, pits, open cast mines, mine shafts, tunnels, spoil heaps, burial grounds, industrial and archaeological remains.
- *Hardstandings*: roads, footpaths and car parks, slabs, setts, gravel, stone, ramps and platforms.

Damages

Visible signs of damage to buildings and land may indicate the nature and strength of the sub-soils. For example, leaning structures, cracks in structures, settlement of buildings and hard-standings may all indicate potential problems. Overburden, landslip, erosion and subsidence may suggest obvious instability of the apparent natural ground levels, requiring special attention to the design of the new building if it is not to suffer in the same way.

Figure 7.4 Site analysis: physical features.

Significant issues affecting development of the site include:

1. The existing derelict engineering building is in poor condition and will be demolished and cleared. There will be existing foundations to deal together with the possibility of basements and cellars, redundant drains and underground pipework. The existing concrete hardstandings and submerged deposit tanks may result in problems with large voids and probable local contamination of subsoils.
2. The existing redundant Victorian pumping station presents similar problems to item 1 although there is a possibility of retaining the structure within any future development of the site.
3. The existing commercial building on the north-west site boundary may have foundations projecting into the site restricting closeness of any new construction.
4. The inclined road leading to the bridge over the river and railway line will have projecting foundations and the sloping embankment limits closeness of any new development and effective use of the site in this area.
5. The piled retaining wall along the river bank suggests possible constraints on the closeness of new development to this boundary.
6. The remains of a previous riverside path indicate the possibility of a desire line used by local people for access along the embankment, perhaps for fishing or for informal play for children onto the derelict land.
7. The existing mature trees will present problems with the construction of foundations to any development in their proximity.
8. Access conditions off the Main Road may present problems such as the operation of the traffic lights, road junction dimensions, radii and sight lines.
9. The existing site levels indicate a fall from one corner of the site to the other with implications for finished building and site levels and drainage connections. Public sewer manholes at 8, for foul and surface water drainage have limited invert levels.
10. The general subsoil conditions must be tested by further exploration, including details of water table level, and the possible spread of ground contamination.

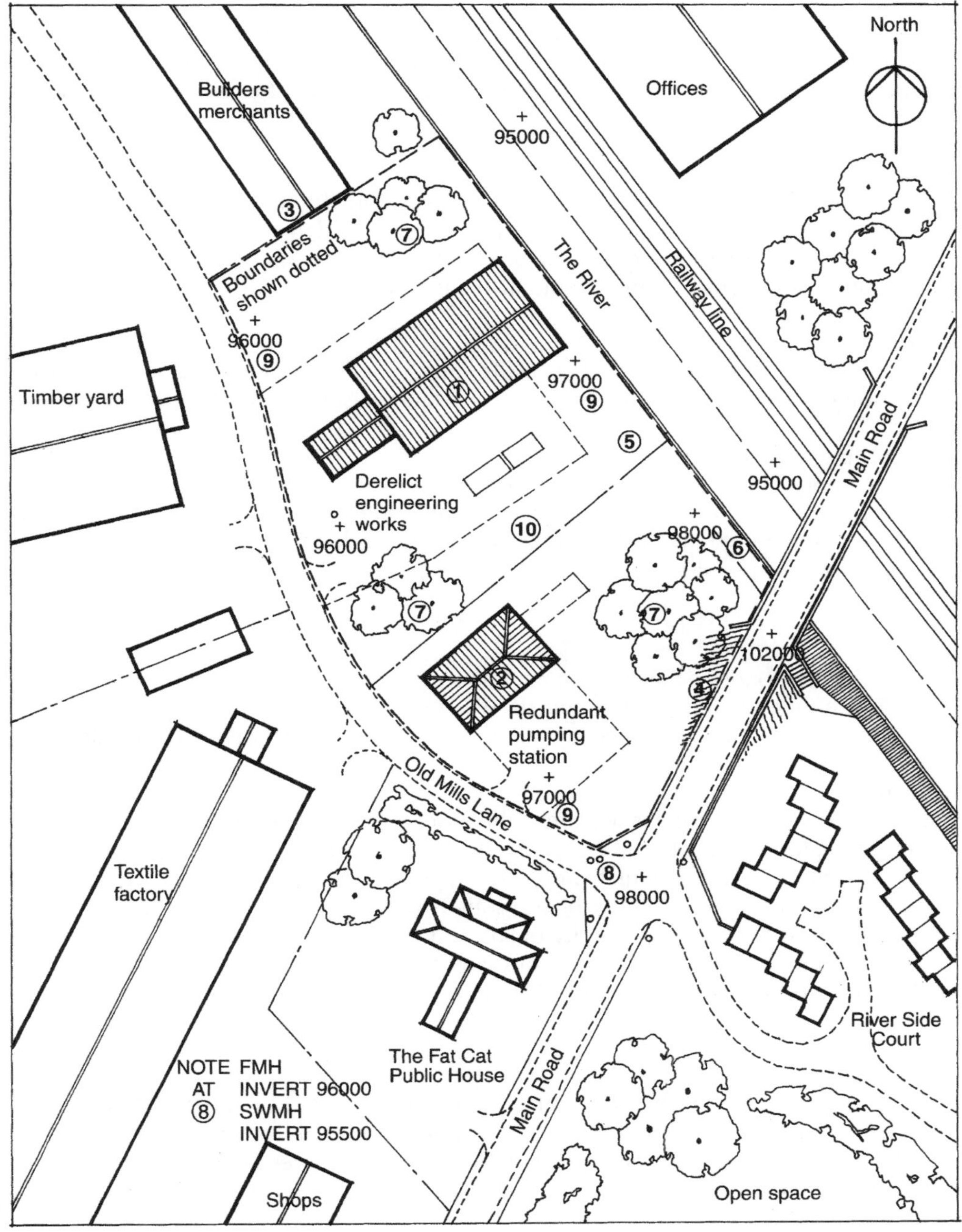

Figure 7.4 Site analysis: physical features.

Topography
The size, shape and levels of the ground, slopes, abrupt changes in levels, cuttings, valleys and embankments are all issues which affect the possible positioning of the new building and landscaped areas around it. There are technical issues regarding the design of foundations on sloping sites.

Vegetation
The existence of trees, shrubs and plants, reeds and lush growth may indicate possible soil type and water content. Note the species, height and spread, signs of disease and evidence of removal of any trees. Building close to mature trees or in areas where trees have been removed may show signs of settlement which could also affect new construction.

Boundaries
Some boundaries will be obvious such as walls, fences, hedges, ditches, bollards, painted lines. However, some boundaries may be unmarked and their existence must be clarified by research.

Access; existing and potential
Check for existing vehicular and pedestrian routes onto or through the site, formally constructed hard surfaces (such as paths and roads) and informal routes (such as desire lines). Look for bus stops, lay-bys, vehicle turning and parking space and check for potential space, width, visibility and sight lines at junctions. Although these are physical factors, they also subject to legal constraint.

Water
Look for culverts, ponds, ditches, streams, canals, rivers, tidal movement, flooding, springs, bogs, wells, run-off and infilled water courses where the possibility of moving water could affect development. The Environment Agency will advise about construction close to water courses.

Water content
The water content of the sub-soils will affect the process of construction below ground level and the effects on materials placed in the ground. The natural level at which the ground is saturated is called the water table level, which varies in different types of sub-soil relative to the prevailing weather conditions. The stability of sub-soils is dependent on its permeability, the rate at which water is absorbed and the acidity of sub-soils determines the strength and resistance to attack of materials such as concrete and brick used for foundations and footings. Information about these factors can only be obtained through testing by specialist survey and analysis consultants.

Soil condition
Possible variations in ground bearing capacity may be indicated by the presence of imported fill, rubbish, cracking, stickiness, softness, sponginess, shrinkage, rock outcrops indicating possible fault lines. Warm ground and smells may indicate the presence of dangerous gases or chemical contamination.

Bearing capacity
The ability of the sub-soils to support the load of the new building is dependent on the geological structure below ground level. This can vary substantially and can only be determined by

specialist survey and analysis. The usual procedure is to excavate trial holes or drill bore holes deeper into the ground to discover the nature of the layers of sub-soils below the area where the building is to be constructed. This information can be used to calculate the bearing capacity of the sub-soils and to design the foundations accordingly.

Contamination

The nature and extent of chemical pollution can only be determined by specialist survey and analysis. Indications of contamination would be fill, spoil heaps, disused workings, coal, industrial waste, domestic waste, chemical tips, effluent, agricultural waste, discolouration and odours. Specific contaminants such as arsenic, methane and radon will not be visible.

Services

A previously used site will have connections to services including power, fuel and water which may be turned off, dead and redundant. However, in some cases they may still be in operation, live and dangerous, or may be connected to and serving other property. The existence and status of all existing services can be discovered by survey and with the use of special detection equipment, or may be revealed through research. As well as the direct influence on design and health and safety issues for construction, the designer will ascertain the suitability of existing services if they are to be retained and the capacity of local main feeds to supply the demands of the new development:

- *Above ground services*: electricity, gas, water, communications, cables, pylons, poles, substations or other services machinery, inspection chambers, sewage or surface water outfalls, stop cock access chambers, drainage gulleys.
- *Underground services*: pipes, ducts, underground cables, sewers, soakaways, culverts, availability and capacity.
- *Health and safety risks*: Identify and record any physical features which could present health and safety risks to the design team, the contractors, the future users of the site and occupants of the new building including, busy roads and railways, potential falls from height or into holes, contamination, soft ground, running water, live services, dangerous structures, etc.

7.6.2 Environmental conditions

The location of every site is a unique environment, situated amongst other buildings and spaces around its boundary. Environmental conditions may have affected how the site was used in the past, maybe sustain uses at present and certainly will have implications on the way that it can be developed in the future. The issues considered in this section concern the development and use of the site for the benefit of any new occupants and the impact of the new development with respect to needs and activities of the neighbouring occupants and property owners. The issues are essentially fixed points which the designer cannot change, unlike a tree, which can be removed and replaced with a new one to suit the layout in a more convenient position. Environmental conditions are inherited 'facts of life' which may be harmful or beneficial, ugly or attractive, dangerous or safe.

Figure 7.5 is an analysis of environmental conditions on or immediately around the car-dealership site, but general issues to consider include the following.

Usage

For any site, there can be a variety of established activities taking place around or beyond the boundary, including industry, commerce, housing, shops, schools, playgrounds, pubs, sports, leisure facilities and agriculture. The consequences of their activities include traffic movement and hazards, heavy vehicles, trains, aircraft, noise, fumes, smells, pollution, dust, waste and effluent storage, vibration, soil contamination, animal noise, health and safety risks, and unsociable working hours. For the car-dealership, it may be advantageous to have shops and pubs in the immediate locality, but the presence of a glue factory next door would not be quite so attractive. The arrival of early morning bread deliveries to the local shop or the 'crack of dawn' coach party of fishermen next door are not good selling points for the prospective purchasers of a new house. Any of these factors may influence the design of elements of the building or their location on the site.

Of equal importance in considering design proposals is the environmental affect that they may have on the neighbours. For example, an increase in traffic movements to and from the site becomes a significant hazard for adjacent residential development and schools. Activities generating noise, fumes and dust contained in a busy workshop or spray painting shop, or which take place at night or during the weekend are likely to meet with resistance from local residents. Special provision must be taken to deal with these issues if development is to be satisfactory.

Boundary conditions

The boundary around the site is rarely a simple fence separating it from the rest of the world. For every site there will be a variety of features, activities and obstacles either along the boundary or just beyond it. It is to be hoped that a part of the boundary touches a public highway, but this is not always the case. Possible access points across the boundary will certainly be limited by adjacent ownerships which can lead to complex negotiations and tradeoffs. The position of existing buildings or structures along the site boundary may limit the location of new buildings and access into the site.

The distance of new construction from existing property along the boundary is governed by the risks associated with undermining their foundations, made worse if the floor level of the building and the space around it is higher than the site. Sharp changes of level at the boundary may require

Figure 7.5 Site analysis: environmental conditions.

Significant issues affecting development of the site include:

1. The main road is a busy route into the City centre with associated noise, fumes and health and safety considerations. However, the elevated highway offers excellent views to potential passing customers, both into the site and into the new building.
2. There are similar problems with the adjacent railway line, with the addition of potential vibration effects; however, the development will also be visible to passing customers from a wide area; maybe from other parts of the country.
3. The river presents health and safety problems, but offers a potentially attractive outlook from the site and the new building.
4. The existing commercial building on the north-west site boundary may cause overshadowing, and any existing openings overlook the new development. There may be security risks from the rear of the building, which is hidden from public view.
5. The remnants of the public footpath under the Main Road bridge may be a well-used desire line along the river bank, or possibly a short cut to the Fat Cat public house. Local people may use this route and expect to continue doing so after redevelopment.
6. The adjacent timber yard could present problems of noise, air-borne pollution and conflicting access.
7. The closeness of the Fat Cat public house may cause difficulties with access, noise and associated security risks, particularly late at night.
8. The existing mature trees may support valuable wildlife, which could have legal protection.
9. Contamination in the ground may present health problems during the construction process, and to occupants once the new building has been completed.
10. The most prominent corner of the site is south facing and may be subject to maximum solar gain at certain times of the year. This is however, an ideal vantage point to attract passing customers.

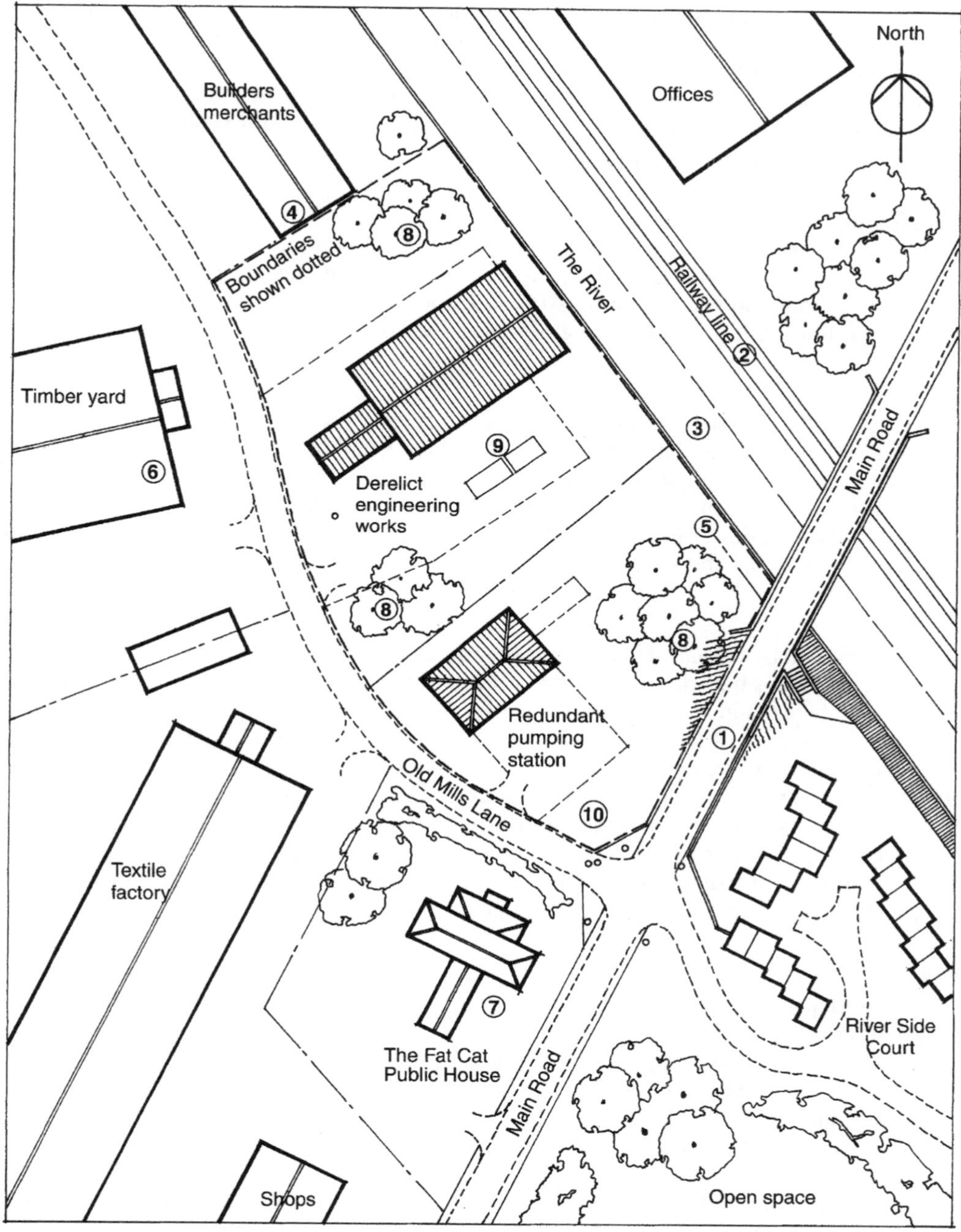

Figure 7.5 Site analysis: environmental conditions.

expensive retaining walls or space consuming embankments to ensure stability. If the level of land next to site is higher, there is a risk of rainwater or pollution running down onto the site.

There may be door and window openings in the walls of existing buildings on or close to the site boundary. Depending on the activity in the building, they can be a source of additional noise or pollution and present fire safety hazards. The extent of natural light reaching existing windows is defined in the current Building Regulations, limiting the proximity of new construction.

Overshadowing

Existing buildings, structures, fences, walls and trees may overshadow areas of the site restricting the penetration of sunlight, and vice versa, parts of the new development will be controlled to prevent overshadowing of neighbours diminishing the use of their property. A row of four-foot conifer saplings will soon become a dense barrier to sunlight, sterilising the land beside it.

Privacy

Existing windows overlooking the site reduce privacy, and existing amenity space in adjacent property is entitled not to be diminished by overlooking from the new building. Views out of the site can be restricted by adjacent development, which in turn restricts views into the site. Privacy for the occupants of the new building and for those living or working around the site is maintained, improved or diminished by the way that the spaces and activities in the new building are positioned within the site.

Orientation

Every site is affected by climatic variations depending on its location. Exposure to wind and rain may influence the positioning and design of the building. Even a gentle prevailing wind can bring unpleasant smells into the area from a considerable distance away from the site. The daily and seasonal movement of the sun affects shading to internal and external spaces, creating cold or warm spots and may cause discomfort through solar gain and glare. Access to the site is often governed by its orientation to feeder roads, and visibility and views into and away from the site are fixed points to be used or excluded as appropriate to the use of the new building, dictating the location of internal and external spaces. Failures of structures and materials along the site boundary or in the immediate locality may be very unattractive. The arrangement of habitable spaces can obscure them or reduce their impact.

Wild life

Increasing emphasis is being placed on protection and enhancement of the existing natural environment and the ecology of plants and animals. The site may contain the habitats of common or rare species. Preserving flowers, shrubs, trees, animals, birds, fish, butterflies, hedgehogs, nests and breeding grounds, for example, can be a significant obstacle to development limiting the extent and timing of construction, and requiring careful integration into design proposals. Many species and habitats now have the benefit of legal protection.

As part of the site investigation the designer should prepare an 'environmental audit' based on the drawn survey plan, recording all the issues mentioned. Highlighting the areas affected by environmental conditions will begin to help to organise the activities within the new building and the site as a whole so that the new development can fit into its surroundings in the best way for its occupants, without damaging the amenity value of its neighbours.

7.6.3 Socio-economic factors

The phrase 'socio-economic factors' is used in this context to describe the prevailing cultural and political influences in relation to the extent and quality of the local infrastructure, affecting the site and its development potential. The way that people live and organise themselves varies throughout the world and designers must respond to the traditions and practices in an appropriate manner in the circumstances. In the UK, England, Scotland, Wales and Northern Ireland have their own history and identity, expressed in many ways including language, art and music, industry and commerce. Some of these factors have influenced their buildings and general land use policies, and within each country, each region, county, city, town and even village it is possible to find distinct differences in the style of its buildings and the way that people use them. The area is said to have 'character', a word often used to describe the visual appearance and coherence of the buildings, but which is really more an expression of the qualities of the environment and the feelings of community that bind the area together. A small market town can be beautiful but in decline, with empty shops and a depressed population. Conversely, a run-down inner-city area can be ugly, but thriving, busy, active and dynamic, an exciting place to be in.

The site itself is just a small part of a small community, subject to layers of influence up to a national level and increasingly today, a European level, even a world level as politicians struggle to control international environmental problems. Some of the issues described will have a direct bearing on the development and use of the site, whilst others will make it more or less likely that development is attractive or successful.

The socio-economic implications on the car-dealership site are indicated in Figure 7.6, but other relevant matters may include the following.

Area status

Previous, current and planned use of the area will indicate the suitability of the new development in the context of its location, which may be either compatible or in conflict with other activities in the area like residential, commercial, industrial and leisure use. In 'run-down' areas, the lack of investment could mean further decline, whereas a planned increase in investment will lead to expansion or regeneration. A proposed by-pass could remove essential passing trade, but a new supermarket will expand possibilities. Employment trends, population statistics for numbers, age groups and income, movement trends and the number of properties 'up for sale' are all issues indicating growth or decline in an area's status.

Neighbourhood infrastructure

For many building types, local facilities and amenities are essential to support the development or make it more attractive. Shops, schools and leisure facilities, a post office, a doctor's surgery or a pub. Immediate proximity, as described in the section about environmental issues can be a problem, but access to the same facilities within a reasonable walking distance could be a significant advantage. The success of a new development may hinge on the local amenities, as for example, a local authorities decision to relocate approximately 1500 council tenants from the city centre to a new estate on the very edge of its boundary, well away from shops, community and leisure facilities and access to employment, a factor which contributed to the estate's decline and demolition barely 20 years after it was first occupied.

Existing and proposed transport facilities and access arrangements can also encourage or deter people from using the area. A city centre site can be difficult to reach by car, with traffic congestion and limited parking, but an out of town site can only be reached by car if it is not served by public transport.

Commercial attractiveness

It may be clear that the proposed development is an attractive proposition because of the general status of the area or that confirmation is needed through market research and consumer testing. Land and property values vary enormously in different areas, as do rates and other commercial taxes. The costs associated with land purchase and subsequent running costs can influence the need to develop to a high density, or determine the type of building that represents an economical investment on the site.

Security risks

A brief exploration of the neighbourhood will confirm the extent of vandalism, litter and dog mess, which may be problems for the new development. The local police authority will supply statistics about burglary, property theft and muggings, and can offer advice on the extent of security measures needed such as the choice of materials for walls, fences and gates, and the use of specialist equipment for locks, surveillance lighting and cameras.

Local style

Consideration should be given to other buildings in the area which may indicate a typical architectural style and vernacular tradition of using certain details or materials in construction. There may be a history to earlier development indicating that some forms of construction are more suitable to the location than others, suggesting useful influences for the style and finishes of the new building. If this were the case, check the availability of local materials and labour needed to achieve similar qualities.

National politics

The demands of national politics through the UK and European Parliaments are becoming ever more significant in the construction industry. Controls on land use, transport policy, energy

Figure 7.6 Site analysis: socio-economic factors.

Significant issues affecting development of the site include:

1. The site is in an attractive location in a busy commercial and residential area. It is surrounded by a mixture of shops, offices and commercial premises as well as some residential development. It is likely that the activity will continue in the area throughout the day and into the evening.
2. The site itself, is in a prominent position, visible to passersby and can become a familiar landmark in the locality.
3. New development in the locality suggests that the status of the area is improving. The new shopping precinct will attract even more people to the area.
4. The amenities and facilities in the area could be a useful consideration in attracting new staff.
5. There are security problems and risks of vandalism and theft from the users of the Fat Cat, from the nearby residential development, the open space and the nearby university campus. Security and measures to reduce criminal activity could be very important.
6. There is evidence of desire line footpaths along the riverside and across the site to the Fat Cat which may mean that redevelopment of the site will interfere with current use.
7. The site is easily accessible to users of public transport.
8. The local building style is a mixture of traditional and new construction and may give useful clues to the form and appearance of the new development.
9. The availability of grants for remedial works to deal with problems of contamination suggests a positive approach to development by the local authority.
10. Closeness to the airport, motorway, main station, ring road and other towns and cities makes the development accessible to a wide customer base.

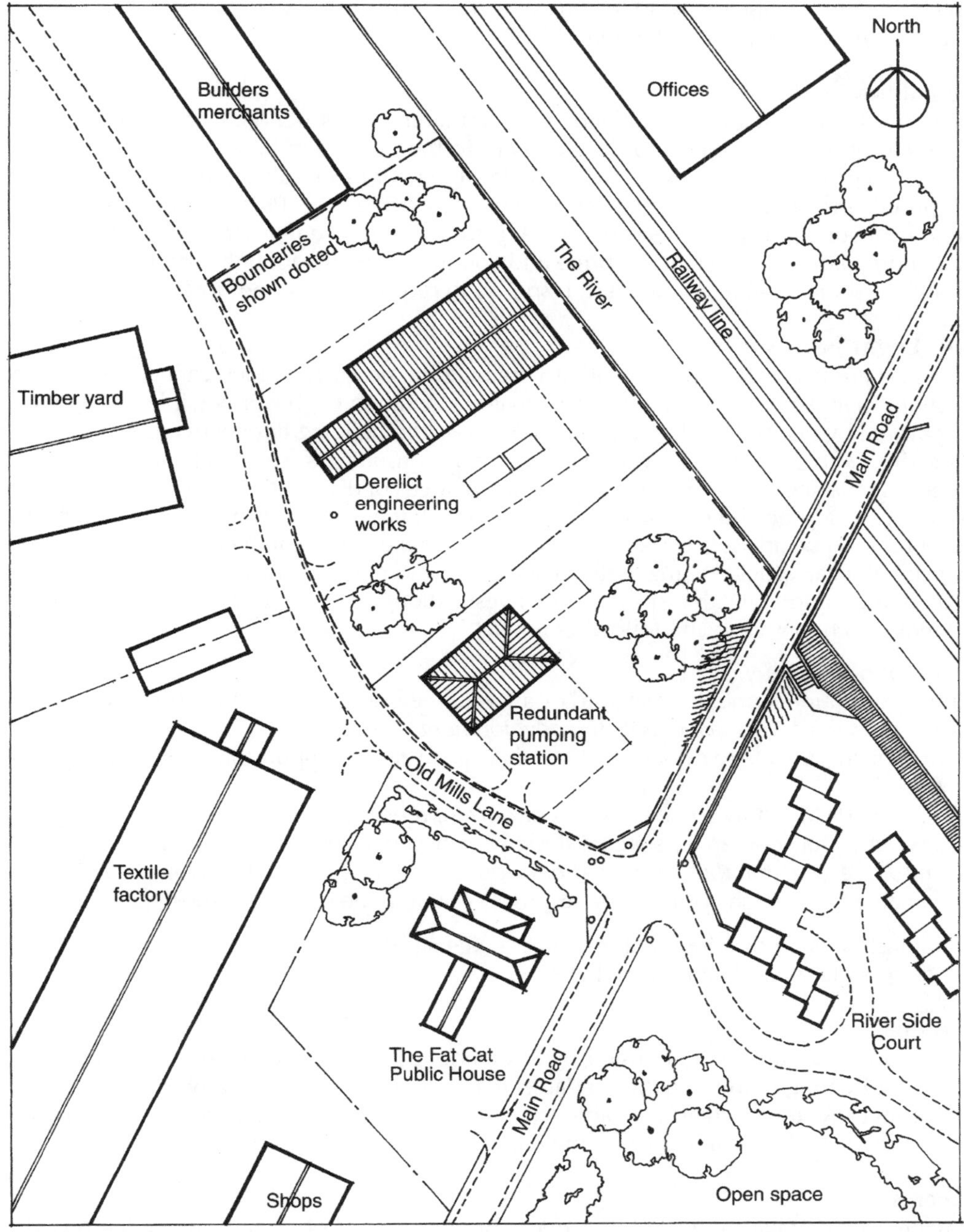

Figure 7.6 Site analysis: socio-economic factors.

conservation and sustainability, for example, will undoubtedly lead to increased intervention in the development of any site in the future.

Local politics

Local politics plays a major part in land use and building construction. It can be useful to know the attitude of local politicians to development in their area and be aware of their support or resistance to proposals. In some cases, new development proposals attract objections from the local population which must be resolved or satisfied before Planning Permission can be obtained. Negotiations may yield assistance to stimulate new employment. Local opinion may offer a good guide about feelings of pride and community spirit, about the stability of the neighbourhood and the general attitude and response to redevelopment proposals.

7.6.4 Legal issues

Legal issues are particularly problematic to the building designer as they cannot be observed by inspection. Their existence can only be anticipated as either probable or possible, to be confirmed or discovered by research and enquiry. In most cases the building designer can personally establish the effect of legal constraints on development by consulting the external authorities who have an interest in the site. This is not always possible for areas of dispute like ownership and boundary conditions, or for specialised aspects of law with which the designer is not familiar, in which case the client should be advised to use a solicitor to undertake a full search before design work starts.

The general legal issues relating to the designer, development proposals and construction were indicated in Chapter 4. With respect to site investigation, there are four principle effects of legal constraint which may have the following impact:

(1) *Prevent development*
If the constraint cannot be removed or altered, then development may be impossible.
(2) *Restrict or limit some aspects of the development*
It may be impossible to develop parts of the site, or design opportunities may be significantly reduced.
(3) *Positively dictate how some elements of development are handled*
Constraints may demand a specific design solution with no viable options.
(4) *Control the way that the completed development can be used when occupied*
Legal directives which positively or negatively establish the criteria for subsequent occupation.

Legal constraints on development of the car-dealership site are shown in Figure 7.7. The main issues to consider include the following.

Ownership

Ownership can usually be confirmed by reference to property deeds for land, buildings and boundaries. Bearing in mind that boundaries may not always be physically marked on site, it is essential that the area of land available for development is legally verified before any design work commences. This is a matter for the client and their advisors to resolve.

Easements

Easements usually take the form of corridors of land on the site, onto which authorised third parties have a right of access for legitimate purposes. Typical easements would include vehicle access to adjacent property by its users, and access for statutory authorities to maintain or replace essential services such as sewers or water mains. The normal effect of the easement is to prevent permanent construction on the strip of land, as for example in the case of a local authority sewer, the client would be required to sign an agreement entitling the local authority to demolish the construction at any time in the future, if it became necessary to replace the sewer. This is not an agreement to be entered into lightly, and in most cases is an uneconomical risk.

Consequently, the layout would be designed to avoid the need to use the easement strip for construction. Whilst an easement sterilises land from the point of view of constructing buildings, it does not generally limit use. The strip of land can remain available for car parking or for open space and planting, but if the local authority are to adopt the service contained within the easement, it is a usual practice to design the layout so that it lies within the areas of public highway. The local authority can then dig up the service as many times as they wish.

Covenants

Covenants are commonly established by previous owners of land, wishing to control future development. They are described in the property's deeds and form part of the purchase agreement at points of subsequent sale. They can be positive or restrictive, limiting the way that the land can be used. For example, a philanthropic land owner may have arranged a covenant on part of the site requiring it to be developed as a playground for children. A residential developer can include a covenant in the terms of sale to the effect that purchasers cannot alter or extend their houses without permission, or that open frontages cannot be enclosed with high fences and walls.

Boundaries

As well as confirming the position of boundaries in order to define the site area, it is necessary to know the ownership of fences and walls to establish responsibility for repair and maintenance, or entitlement to undertake modifications or replacement. Agreement is required for access arrangements between properties to carry out necessary works. There are also statutory controls over development on or close to the boundaries of adjacent property, particularly for development with shared or party walls.

Access

Existing access onto, around or across the site may have a legal status. Public highways and footpaths, pavement crossings and rights of way are protected by law and cannot be altered without approval. The present forms of access to the site may have been acceptable for the previous use, approved under the applicable legislation when they were constructed. For a new development, access requirements must be designed to current legislation which will include all aspects of public highway design for widths, gradients and cambers, visibility at junctions and bends, turning and parking arrangements, traffic calming, quality of construction and materials, surface water drainage, space for fire fighting requirements, deliveries, waste disposal and public transport access.

Some of these issues are details which can be resolved later on, but others are fundamental. If an existing, narrow entrance cannot be widened, or the appropriate visibility sight lines cannot be created, then increased vehicle use will not be permitted.

Services

The suppliers of power, gas, water, telecommunications and domestic drainage operate on the basis of statutory controls and duties applicable to their service. Each must be consulted directly to ascertain their requirements and responsibilities.

Statutory regulations

Briefly referred to in Chapter 4, there are a range of statutory regulations and common law issues constraining development including health and safety issues, interests and effects on neighbours, fumes, pollution, noise, ground contamination, water pollution, works to adjacent structures, party wall regulations and rights of light.

Statutory authorities

The principle statutory authority with respect to site investigation is Planning, administered by the local authority. Apart from a very few minor works exceptions, the development of any site will be subject to Planning approval. The site will be defined in the local authority's Structure Plan and may also have a Planning Brief, setting out the authorities requirements for any development. Special criteria may be applied to the development if it is part of a City Challenge area or in a conservation/improvement area. Development possibilities may be constrained and additional works may be required as 'Planning Gain' negotiated or conditional in relation to obtaining consent, such as improvements to existing highways off-site, provision of play space for small children or a financial contribution to a communal amenity space.

Some elements of construction such as publicly accessible highways and drainage can be adopted by the local authority who take over responsibility for future maintenance. Developers must enter into a variety of formal agreements with local authorities to guarantee compliance with their own standards. Other agencies previously mentioned must be consulted to check on the extent of their legal control over development.

Figure 7.7 Site analysis: legal issues.

Significant issues affecting development of the site include:

1. The local authority Planning Brief requires the construction of a riverside path for pedestrians running through the entire length of the site. It may continue past the site at some time in the future, as part of a continuous public riverside walk.
2. Although the redundant pumping station is not a formally listed building, the local authority would like to discuss the possibility of its retention in any new development, if practicable. They may dictate the extent to which the building can be altered or extended.
3. The Highways authority require improvements at the road junction in association with any redevelopment of the site, taking into account the increased traffic flow.
4. Measures must be taken to deal with contamination prior to commencement of construction works, in accordance with current legislation under the supervision of the Health and Safety Executive and the local authority Building Control Officer.
5. All the trees on the site are protected with formal preservation orders, and cannot be removed without approval. A trade-off may be negotiable by providing sufficient replacement trees if any are removed.
6. The local authority are interested in relocating the entrance to the Fat Cat public house onto Old Mills Lane to reduce traffic problems on Main Road. This may be regarded as a Planning gain issue with consequential costs for redeveloping the site.
7. There is an easement permitting the Environment Agency access to the full length of the riverside retaining wall, to enable future maintenance and possible replacement.
8. The Environment agency will also impose strict constraints on development close to the riverside and with respect to the drainage installations, particularly owing to risks of oil and petrol spillage contaminating the water table.
9. The Planning authority will impose conditions relating to external display and signage.
10. Old Mills Lane must be kept clear and accessible at all times, and no mess must be deposited on Main Road at any time during the construction period.

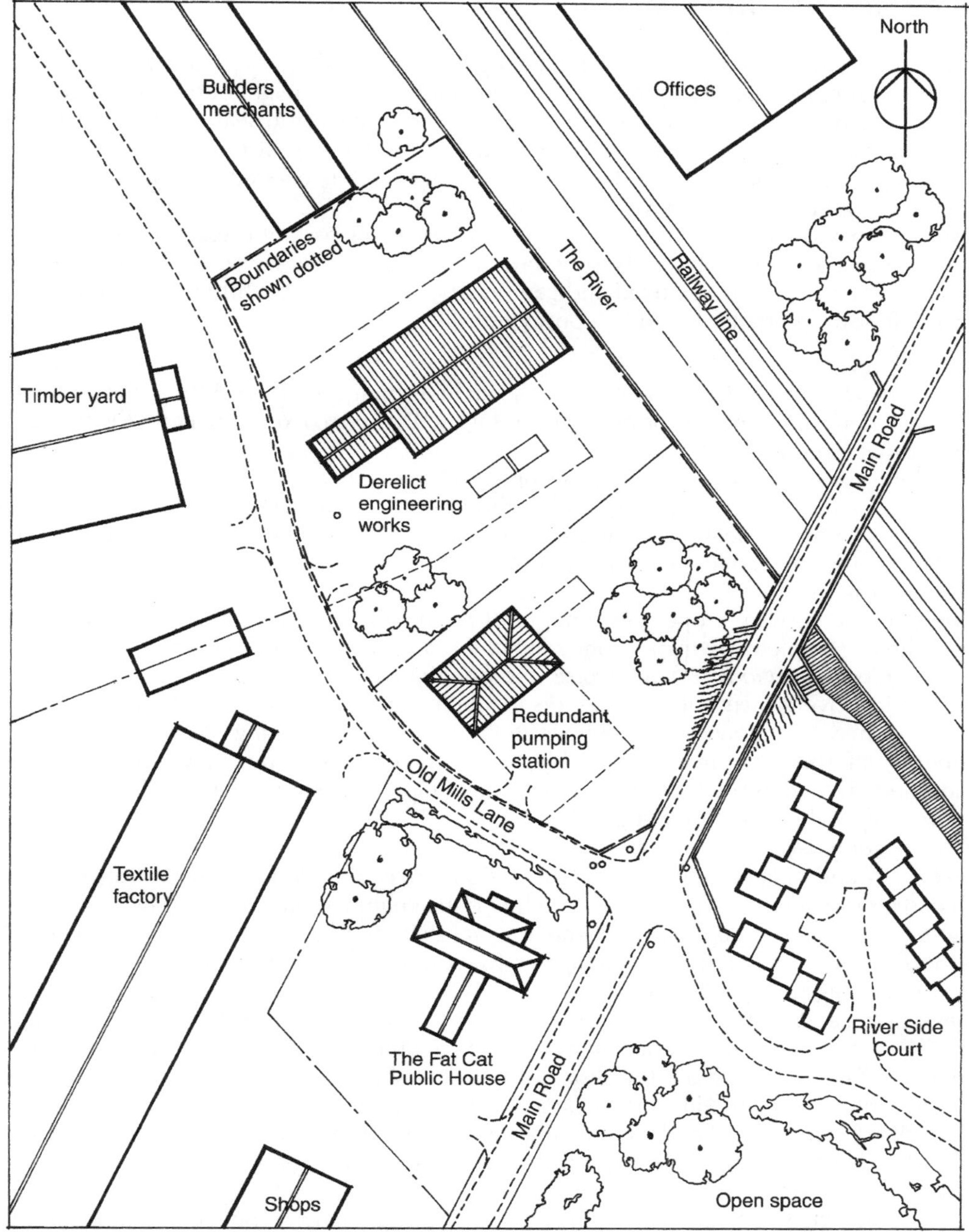

Figure 7.7 Site analysis: legal issues.

7.7 Collecting information

Many of the features and factors which make up the site's characteristics can be easily seen and recorded by simply walking around the site and its surrounding neighbourhood. Others must be accurately measured, scientifically tested or discovered through research. The first step is the initial site inspection or 'walkover' to gain some understanding of the general conditions, to record the obvious points and to decide what further work is needed and which, if any other specialists should be called on for assistance.

As the site may be some distance away from the designer's base, it is essential to be organised and well prepared if the journey is to be productive. Nothing is worse than finding that some important fact has been overlooked and can only be resolved or confirmed by making a further visit. The following points should be considered:

- *Take a map*
  It can be extremely difficult to find a small site in a strange location. Empty pieces of land do not have convenient names or numbers and those that once existed may no longer be there.
- *Check that the site is accessible*
  If buildings and land are likely to be secured with locks, take the necessary keys. If the site is occupied, carry proof of identity and authorisation and always contact occupants before commencing the inspection. If the site or buildings are derelict, be prepared to explain the presence of 'strangers' to legitimate onlookers.
- *Do not trespass*
  Avoid entering private property without prior permission. If it is essential to go onto an adjacent site to complete the inspection, contact the owners or occupiers first.
- *Take copies of previous survey records and photographs*
  A derelict, overgrown site can be disorientating and difficult to understand. Important elements may be hidden from view or simply inaccessible. Existing records can be useful to confirm information already available, and photographs may show previous features subsequently demolished. Existing aerial photographs can be particularly helpful to get an impression of the extent of a large site.
- *Be fully equipped*
  Take tools and aids to help with access, vision, measuring and recording. For example, be able to lift inspection covers, shine a torch into the chamber to inspect it, have a tape or a rod to measure the invert depth and a camera to record its condition. Wear suitable clothes, shoes or boots which can stand up to the aggressive conditions likely to be encountered.
- *Record findings comprehensively*
  Make notes and sketches which can actually be read on return to the office. Very scribbly details, crossings out and crude alterations may be unintelligible later on, and therefore of little value. Take care with abbreviations or personal shorthand, and remember that it is difficult to write or sketch with pen or pencil in pouring rain! Use a dictaphone which can be transcribed in the office, and take photographs so that details can be examined more closely later. A video recording will show much more detail than single shots.
- *Talk to local people*
  Local knowledge can offer vital clues about previous activity, but information should be treated with caution and be confirmed by reference to other sources.
- *Establish contacts*
  Find out the names and addresses of owners of adjacent property who may be affected by the new development and details of the relevant local authorities. Pick up application forms.

- *Locate elements critical to future survey work*
  Check the existing boundaries to confirm the area of the site and find the position and value of the nearest bench mark for levels. This is needed to relate land and sewer invert levels to local authority records.
- *Avoid risks*
  Take great care if inspection demands climbing down into chambers or pits, or climbing up onto roofs. Do not casually place yourself, or others, in a position of danger.

7.8 Surveying and measurement

The main purpose of the initial inspection, or walkover, is to gain an overall impression of the site and its surroundings, noting down the issues which may have a bearing on subsequent design proposals. Unless a full survey has already been carried out at some time in the past, the only accurate definition of the site likely to be available is the Ordnance Survey location plan, ideally at a scale of 1:1250 for land containing buildings. This cannot be guaranteed to be accurate as boundaries and the position of structures may have changed since it was first published. It is a useful guide, but not suitable for design work.

Consequently, the next stage for the building designer or specialist surveyor, is to measure the size, shape and levels of the land, and the dimensions and position of all relevant features, including elements on and immediately around the boundary, and to produce an accurate, scaled drawing. The scale selected is dependent on the size of the site, but is usually to a minimum of 1:500. If existing buildings are involved, they should be measured, and the plans, elevations and sections should be drawn to suitable scales as described in Chapter 3. The relationship between design work and a fully accurate, comprehensive survey drawing cannot be stressed too highly. A good survey drawing is fundamental to the success of design proposals, as errors can be very costly.

There is no doubt that preparing an accurate measured drawing of a site is a skilled task, but in general terms it is one that the building designer should be able to undertake. The process demands a systematic, conscientious approach but is essentially a matter of common sense and attention to detail. It also enables the designer to spend time looking carefully at the site, improving the level of appreciation needed to commence and complete design work. However, as with other issues described elsewhere in this book, the designer must confine his or her work to areas for which they are experienced and qualified, and not produce drawings or offer advice which may be incorrect. The designer must avoid forming unsupportable conclusions about elements of the site and buildings which cannot be seen and proved. Linked with the measured survey, it would be sensible to note the general condition of the site and buildings. There are three different types of survey which can be undertaken for different purposes.

(1) *Condition survey*
A simple condition survey is carried out by walking around the site or through buildings, noting the apparent condition as far as it is practicable to do so. It is inevitably limited because it is not possible to easily look underground, down drain pipes or inside elements of structures without causing damages, an unacceptable proposition anyway for occupied premises. Although damp meters and other equipment may indicate problems, this kind of inspection cannot be fully comprehensive. For example, without lifting floor boards, it is not possible to be certain that there are no damaged floor joists. Consequently, the existence of elements on site or inside buildings which are damaged and require remedial attention may remain unknown until work commences, with a risk of substantial unexpected additional costs.

(2) *Structural survey*
For situations where obtaining comprehensive information is essential, it can necessitate removing coverings or taking samples of materials and causing damages which may be difficult or expensive to repair. This kind of survey, referred to as a structural survey, is a level of investigation which is much more comprehensive, examining elements of construction in greater depth to ascertain the extent of remedial work needed. For example, excavations in the ground may be required to determine ground strength for foundation design, or plaster removed from a wall to check substructure condition. Fitted carpets must be lifted to examine floor boards, floor boards lifted to examine floor joists. In this way it is more likely that all problems are discovered and can be catered for in subsequent design proposals.

(3) *Dilapidations survey*
If a site or building were described as being 'dilapidated', it would suggest that it was in very poor condition. However, in the context of building works, a dilapidations survey, or a *schedule of dilapidations* is a record of the *status quo*, essential prior to commencing work on site so that all parties to the development are aware of existing conditions before changes are made. This is particularly important in relation to existing property around the site boundary, which may be inadvertently damaged during construction. For example, if an adjacent building has visible settlement cracks in its external walls, there is a risk that the owner will associate them with the new construction. This situation can only be avoided if the existing conditions are fully documented, and it can be clearly shown that the settlement cracks were in existence prior to commencement and have not changed as a result of it. It is usual for a contractor to prepare a schedule of dilapidations for agreement with the designer or the client before commencement.

Other areas of investigation, measurement and interpretation which may be carried out by specialists include the following:

- *Structural stability*
  The stability of the ground, buildings, retaining walls and any other existing structures, is likely to require specialist investigation and advice, particularly where additional loading is anticipated.
- *Sub-soil investigation*
  The soil type, bearing capacity, water table level, contamination.
- *Specialist photography*
  Underground photography can be used to investigate hidden voids, chambers and pipework, and telescopic cameras can be used to investigate high level detailing.
- *Damp and infestation*
  Investigating rising damp, dry-rot and woodworm problems are generally specialist matters, particularly where advice is sought about treatment.

7.9 Recording survey information

The technicalities of surveying and levelling can be studied in existing publications for those who are interested. Computerised theodolites and laser tape measures are revolutionising the process, speed and accuracy of collecting and recording dimensions and levels. Specific equipment and techniques will continue to be developed, but the following simple observations are relevant to the application of principles about the way in which measurements should be taken and recorded so as to minimise the risk of errors.

Sketches

Before measuring the site or any buildings on it, the first action is to prepare a rough sketch onto which the dimensions can be added. Sketches of plans, elevations and sections should be separated, clearly distinguishing which is which. Ideally, they should be on different pieces of paper to avoid confusion.

Scale and proportion

When sketching plans and elevations, try to draw elements in the correct proportion to one another, so that for example, the floor plan of a small kitchen is not indicated as being bigger than a large lounge next to it. Although the measurements taken should produce the correct relationship when drawn up, it can be very confusing. When preparing the measured drawing, do not 'cheat' to make something fit if the dimensions do not add up. Indicate openings and fixtures in approximately the correct position. The sketches should be large enough to add notes and dimensions which can be easily read later.

The use of lines

Distinguish between lines which indicate structural elements like walls and those which are fittings and fixtures like worktops and radiators. Make sure that there is no confusion between lines which are part of the building and those used for notes and dimensioning. Use a system of different line thicknesses, pen and pencil or colour coding to clarify each element.

Dimensions

Write dimensions in consistent units, preferably millimetres so that the figures can be read easily at the drawing up stage. Wherever possible, use 'running' dimensions to plot consecutive elements such as the doors and windows in an external wall to reduce mathematical error. If short, individual dimensions are taken, always measure the whole length to confirm the total.

Geometry

Never assume that lines on the site or the walls of buildings are straight, or that apparent right angles are actually 90 degrees. Particularly when measuring buildings, do not assume that floor plans are square or rectangular. Diagonal dimensions from corner to corner will fix the actual shape of the room and free-standing features on the site like trees, can only be positioned by triangulating them against the position of two other elements, with dimensions adjusted for slopes. As the process of preparing the finished drawing, either manually or electronically is based on horizontal and vertical dimensions being at right angles to one another, assumptions of squareness in a distorted building will quickly lead to chaos.

Concentration

Take care to read dimensions correctly, and write down any records taken against the same element on the sketch. Spend extra time recording information where accuracy is essential, such as location of site boundaries, for example, where future setting out may become critical. Make sure that handwriting is legible and that it will make sense on return to the office.

Calculations
Take care if undertaking any calculations on site, or back in the office, that the resulting answer is broadly as anticipated. A calculation that shows the height of a first floor window as 18 m rather than 1800 mm, for example, may be quite difficult to draw up later on.

Photographs
Take photographs to record complicated detail like brickwork coursing or decorative elements around window openings. This will save time in preparing elaborate sketches on site, and is useful to check the accuracy of measurements on the finished drawing.

7.10 Research and enquiry

Useful information can be obtained from external authorities, organisations or companies who may have an interest in the development of the site, or who may possess historical records of its previous use. In the context of site investigation, research is generally taken to be the exploration of possible sources of information which may help to understand the site's characteristics and contribute to the development of design proposals. Enquiries are more specifically directed to bodies known to have interests or powers of control. The principle purpose of research and enquiry is to discover and interpret factors which are not visible and cannot be easily measured.

Possible research sources would include the following:

- *Local or national libraries and museums*
  For reference to historical books, maps, journals and archaeological remains.
- *Universities*
  For historic records, publications, research theses, details of research findings, analysis of current trends and movements and statistical analysis.
- *Local newspapers*
  For old photographs and articles.
- *National organisations*
  National bodies may have detailed records for property, land use or previous activity. The National Trust, Heritage groups, The Victorian Society, the Arts Council and Wild life Trusts for animals and birds.
- *Local societies*
  For personal knowledge, recollections and details of antiquities. Organisations could include local history societies, preservation societies and Village Hall Committees.
- *Churches*
  For details of previous use, ancient burials and protected remains.
- *Ordnance Survey*
  The Ordnance Survey may be able to provide maps prepared at different historical times, showing progressive development in the past. This can be particularly useful if the site has been developed more than once, indicating the possibility of additional forms of disturbance being encountered on top of one another.
- *The Land Registry*
  For up-to-date boundaries and ownership records.

- *Local authorities*
  For records of previous or intended development on site including drawings contained in previous Planning and Building Control applications. Building Control Officers usually have an expert knowledge of the conditions within their own area, and may save considerable time and effort researching possible problems. Technical Services Departments will advise on highways and drainage requirements in relation to existing circumstances. The Environmental Health section will be interested in contamination, pollution and noise.
- *Other authorities*
  Health Authority, Education Authority, Police Authority may all offer useful advice in their own area of specialisation.
- *Governmental bodies*
  The Home Office, Department of the Environment, Defence, etc. should be consulted as may be applicable.
- *Mining companies*
  For information about previous or planned extractions.
- *Technical laboratories*
  For sampling and testing soils in relation to strengths and contamination.
- *Solicitors*
  For possible information about ownerships, covenants, easements and access restrictions.
- *Estate agencies*
  For land and property valuations and availability.
- *Transport providers*
  For details of services to or near the site.
- *Research authorities*
  The Building Research Establishment, for example, undertake research into many aspects of construction, publishing findings in digests as examples of good practice.
- *Trade associations*
  There are a variety of trade associations which produce literature on the correct use of their materials. For example, Timber (TRADA), Cement and Concrete (C&CA), Bricks (BDA), Lead, Clay Pipes, Copper, etc. Manufacturers and suppliers also produce information specifically about their own products.
- *Specialist consultants*
  Specialists can supply important information, including matters to do with Planning, Building Control, property development and marketing.

All of the above can be contacted or visited to see what information exists that may throw any light on the site and its surroundings.

Formal enquiries are made to other agencies who will generally have a direct working involvement with the development of the site. The principle aims are to:

- Discover and locate elements above or below ground for which the agency may have an interest.
- Determine if this interest has any consequences for the project development in terms of how it might be designed or constructed.
- Find out about any legal constraints which the agency may be likely to impose, or if negotiations are required to proceed.

A formal approach is made to the agency as a courtesy notification of intention to develop and to enquire to what extent they have any control over the development, and if their activities would be affected by it. It may be that they have no material interest but it is better that this is confirmed earlier rather than later, when design work is well advanced. A letter of enquiry is sent to the director of the organisation outlining the nature of the project, and asking for information, together with the name of an appropriate contact who will deal with the development. This may yield the desired information, or the name of a representative with whom the designer can discuss proposals to ensure that the interests of the organisation are taken into consideration.

Initial enquiries will yield information which is essential at the outset, followed up later with further detail when proposals are made. For example, the electricity board will supply details of existing high-voltage cables on the site which will affect the overall design layout. Later on, the board will supply details of how their existing services can be modified, or extended to suit the proposal, together with the associated costs. Preliminary enquiries likely to be required at this stage include the following:

- *Planning Authority*
  For general or site-specific planning requirements (Planning Brief), preliminary guidance on standards for density, highways, car parking, suitability for development as intended, etc. Some issues may be directed to other responsible local authority departments for roads, site access, sewers and Building Control implications, etc.
- *Fire Officer*
  The fire officer should be consulted with regards to all the aspects of fire safety outlined in Chapter 4.
- *Health and Safety Executive*
  The Health and Safety Executive must be notified of the intention to develop, and may be able to advise on specific hazards and difficulties.
- *Environment Agency*
  The Environment Agency should be consulted with regard to any development affecting existing water courses, and for development on contaminated sites.
- *Environmental Health*
  Environmental Health concerns the use and occupation of the new building and its effects on adjacent property.
- *Utility Providers*
  Service providers for gas, electricity, water and telecommunications can supply details of existing services and discuss proposals for new ones.
- *Adjacent Land Owners*
  Businesses or individuals who own land or premises adjacent to the site may have considerable interest in new development, whether it directly affects them or not.

7.11 Preliminary design ideas

Analysis of the client, the building type and the site continues throughout the design development period, adding to or refining information about elements in an increasing scale of complexity. The information immediately available will generate ideas and thoughts about a possible design for the new building, mental pictures of conceptual ideas about the shape and size of the building, its form and massing, its location and orientation on the site. This is a very common way that clients brief their advisors, by explaining ideas in their minds. Some of these ideas can be explored with feasibility sketch studies, examining possible arrangements for parts of the building or looking at how the whole building might fit into its environment. Initial ideas based on quite limited information stripped of the complications of detail, help to focus on the important issues which establish the framework. In this way, a layout or form of construction can be tested against the briefing information, checking for strengths and weaknesses to see where further research or investigation might be needed.

It is doubtful whether the exact starting point to the design process can ever be identified or defined. As described in Chapter 2, the design process is not universal and different people will always see things in different ways. The process can start from many positions including definition of size, volume, activity, cost, appearance or impact on neighbours, any of which could suggest an idea. Thoughts may derive from elements of function, seeking to resolve practicalities first, or from aesthetics with a greater emphasis on the appearance of the building, irrespective of practicalities. Concepts which have no regard to cost will not be the same when constrained by the client's budget, and concepts in the client's mind may clash with the needs and demands of neighbours and the local authority. Judgements must be made about the relative importance and priority of these major elements before attempting to account for all the other lesser elements in due course.

Whichever is the starting point, the conclusion will be an idea presented in visual form for scrutiny by the other members of the design team to persuade them that it is worth proceeding with. The form of presentation will vary depending on the means of production and the degree of detail illustrated. Traditionally, feasibility studies would be sketches, possibly drawn quite quickly in freehand to communicate the idea with minimal detail. Today, electronic drawings are by definition more accurate, precise and formal. The operator is obliged to be much more specific in defining the elements of construction and cannot be as vague about detail as was the case with a pencil sketch. Whether this is a good or a bad thing is open to debate. Fixing ideas too quickly can limit options later, but remaining fluid for too long may prevent progress. In any event, conceptual proposals produced in any way at all should be treated with great care. It is all too easy to sketch ideas loosely, cutting corners to make things fit, deceiving both designers and clients into thinking that the scheme is satisfactory. Persuading the client to accept a design on the strength of rough sketches, only to find later on that the proposal cannot be realised can cause serious embarrassment. Equally, fixing any element absolutely may prevent the client from getting a better building in the end.

Of course, the search for a *better building* cannot go on indefinitely. There is some evidence in practice that as much as 80 per cent of design, cost or management decisions are taken or 'locked in' by the end of the first 20 per cent of the design process. The 'Pareto or 80:20 Rule' which readers may encounter in texts about *management* simply confirms the proposition that the fundamental decisions must be, and are made first before dealing with the associated detail. This should not be confused with making decisions in a hurry without sufficient supporting rationale, but rather seen as the need to concentrate on getting the elements which are most important right so that subsequent elements fit in with minimal difficulty.

However, rough the sketches may be, they should be drawn to scale, or at least to the correct proportions, so that if the idea or ideas are adopted, there is a reasonable chance that they can be translated into reality when they are drawn more accurately. There is little point in drawing a workshop roof at a particular height because it looks right on the elevations if in fact it is too low for the equipment and activity that is to take place within it. It would be misleading to describe a 2 meters × 2 meters space on the plan as 'Reception' when that is only large enough for two people at any one time.

A common problem when designing larger buildings is the concentration on the arrangement of the principle spaces, often assuming that incidental spaces or activities can somehow be fitted in later. Toilets are a good example, often represented as a simple block in a left over corner as shown in Figure 7.8. Irrespective of the design criteria for toilets and the eventual full specification, if the block is to have any realistic meaning, even at this preliminary stage, some research must be undertaken into how many toilets will be needed related to anticipated demand, or as a standard number based on experience of the building type. It is not practicable at the beginning to research every area of the new building to a construction level, but if '*x*' number of toilets will be needed, then allowance must be made for a realistic floor area, in the knowledge that the space needed for cubicles, urinals, basins, hand-dryers, circulation space and privacy lobbies can be accommodated in due course.

This may appear to be an obvious observation, but one that many practising designers will surely recognise. Aside from the matter of personal embarrassment, which may be overcome if the deficiency can be corrected, there is the serious risk of negligence if it can not. The programme commits all the members of the design team to working at increasing levels of accuracy as time goes by and the consequences of delays, or redesign because an elementary error can be substantial. To discover that there is insufficient space for the necessary number of toilets, or that there is insufficient head room for a roller shutter door to operate can be a major problem, depending at what stage the discovery is made.

Of more general concern at the early stages to the designer, the consultants, the builders, the authorities and the clients is the degree of self-deception that is incorporated into conceptual ideas. The viability of the whole concept could hinge on a misunderstanding, a misinterpretation or a misrepresentation of any part of the brief leading to the choice of the wrong concept, or cutting short the search for a better one.

7.12 Project File content

The general aims at this stage are as follows:

- Development of the client's brief into the design brief.
- Approval of the client to proceed on the basis of the agreed design brief.
- Communication of accepted information to all members of the design team.
- Completion of site investigations and analysis of advantages and constraints.
- Final details of procurement arrangement documents to be agreed, and exchanged with the client and the other members of the design team.
- A cost review in the light of the design brief where subsequently discovered points of information may have changed initial ideas.

File material could include the following.

Notes, letters and minutes

- Further exchanged information between members of the design team.
- Letter to the client confirming the composition of the design team, design programming intentions and anticipated costs.

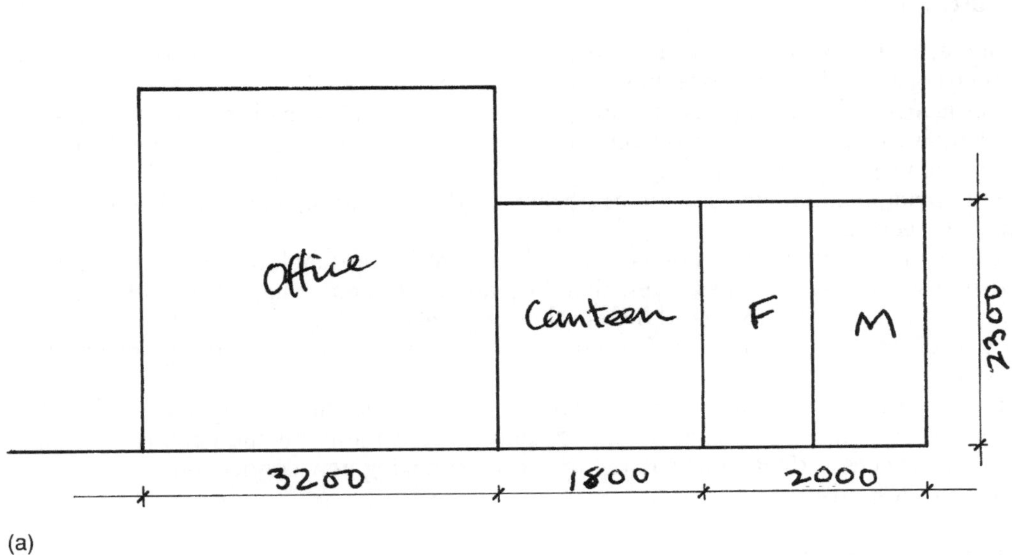

(a)

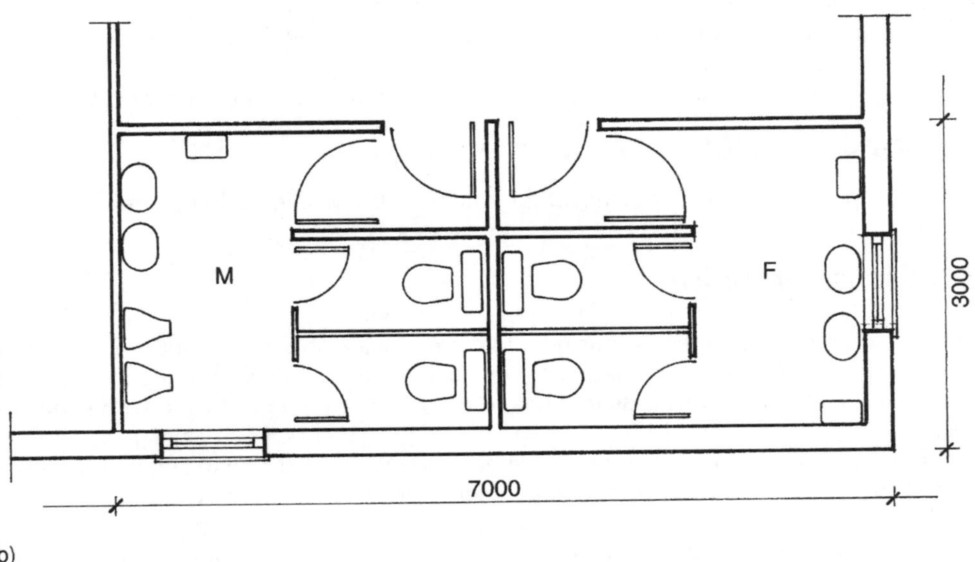

(b)

Figure 7.8 Transition from ideas to reality. (a) The initial sketch plan. (b) The reality of the working drawing. Both are meant to be to the same scale, but the allowance at the sketch stage is shown to be significantly too small.

Site survey notes

- Annotated site plans and records of survey information, including location of site boundaries, dimensions, levels, relevant features and conditions.
- Investigative trial hole reports, including a site plan indicating the location of the test holes.
- Soil analysis reports with engineering constraints or recommendations for foundation design, retaining walls, cut and fill, etc.
- Land contamination analysis, risk assessment and details about any remedial treatment that may be necessary.
- Detailed record photographs of the site and relevant features including boundary conditions, existing structures, entrances, trees, chambers and covers, etc. Dilapidations photographs are particularly important to record the conditions of features which are already in poor condition in order to avoid subsequent accusations of damages caused by design or construction works.
- Formal enquiry letters to external authorities, solicitors, etc. asking for confirmation of boundaries, ownerships, rights of way, etc. and requests for information about the availability of local services and the practicalities and costs of making new connections.
- Copies of responses.

Completed survey drawings

- Layout plans to a scale of 1:500 or 1:200 suitable for overlaying to prepare sketch designs with all relevant features, dimensions and levels.

Checklists

- Information sheets for distribution to the members of the development team about construction, environmental services, areas, heights, finishes and specialist equipment.
- Bubble diagrams and exploratory sketches of activity and flow, links, possible accommodation and staffing numbers, etc.
- The project design brief confirming agreed design criteria with preliminary cost limits.

Sketches of possible feasibility ideas

- Initial concept drawings for elements of the building, the whole building and the site layout.

7.13 Discussion points

(1) Is the location of a commercial business like a car-dealership important to its profitability? Is there a limit to the distance that customers will travel to a new building development? Is it useful to locate a new business in a mixed-use area, or better to locate with similar activities?

(2) Can designers learn from other cultures? What differences are there in similar building types located in different areas; communities, neighbourhoods, districts, cities, counties, countries? How can building designers find out what would be best for any particular location?

(3) What have been the critical influences on the shaping of the UK-built environment? How have car-dealerships changed during the twentieth century? Has the essential business/customer relationship altered? What might happen in the future if and when individual travel becomes much more expensive?

(4) How should building designers respond to commonly understood activity or space definitions? Is there room for individual interpretation? How do designers get to understand the needs of other people?

7.14 Further reading

Baldry B (2003) *Facilities Management.* Oxford: Blackwell Science.
Clancy J (1991) *Site Surveying and Levelling.* London: Edward Arnold.
Cullinane JJ (1993) *Understanding Architectural Drawings: A Guide for Non-architects.* Washington: Preservation Press.
De Chiara J and **Callender** J (1990) *Time-saver Standards for Building Types.* 3rd Edn. London: McGraw-Hill.
Finch E (2000) *Net Gain in Construction: Using the Internet in Construction Management.* Oxford: Butterworth-Heinemann.
Hollis M (2002) *Pocket Surveying Buildings.* Coventry: RICS Business Services.
Hollis M and **Gibson** C (2005) *Surveying Buildings.* Coventry: RICS Books.
Irvine W (1995) *Surveying for Construction.* 4th Edn. London: McGraw-Hill.
Karlen M (2004) *Space Planning Basics.* 2nd Edn. Hoboken: Wiley.
LaGro JA (2001) *Site Analysis: Linking Program and Concept in Land Planning and Design.* Chichester: Wiley.
National House-Building Council (NHBC) (1994) *NHBC Standards* Volume 1.
Pevsner N (1979) *A History of Building Types.* London: Thames and Hudson.
Quinney A (1990) *The Traditional Buildings of England.* London: Thames & Hudson.
Reid E (1988) *Understanding Buildings.* London: Longman.
Rich P and **Dean** Y (1999) *Principles of Element Design.* 3rd Edn. Oxford: Butterworth-Heinemann.

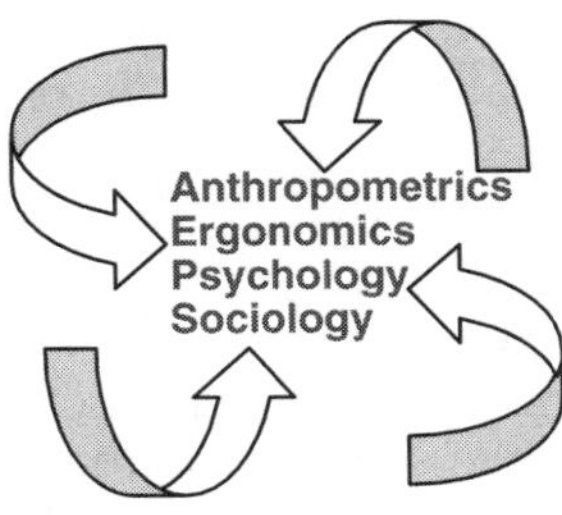

8

The Design: Function, Part 1
How buildings are used

8.1 Introduction

The next two chapters are concerned with the function of buildings, the practicalities of how they operate or 'work', firstly from the point of view of human use, and secondly in terms of how they can be constructed. The term 'work' is not entirely satisfactory as it is almost impossible to define in the context of the performance of a building. There are too many elements, too many perceptions and expectations to reach any universal conclusion about optimum conditions. Almost all buildings are designed to fulfil human needs in some form or other, which can be examined in the context of human dimensions, scale and performance characteristics. Some needs can be accurately defined and precisely satisfied, whilst others are more open ended and uncertain. Some buildings are designed specifically to cater for the activities within them, others are more flexible, permitting or encouraging options. Expectations and management theories change from time to time. For example, the lecturers in a school of the built environment at a university might each have their own private office, or they might share with one or two colleagues, or they might be in a large open-plan space, all working together. These three significantly different alternatives radically affect individual and collective performance and behaviour. For one lecturer, a private office is the only situation which 'works'. For another, sharing with colleagues 'works', as 'I can't bear to sit in silence all day long'. This is a classic design dilemma associated with function which the building designer may have some difficulty in resolving; should the spaces be designed to suit the preferences of the occupants, or should the occupants be encouraged to use the spaces in a different way?

This chapter will look at the way in which activities can be defined, associated or separated and how there is a flow between activities which may eventually define the size and shape of the building. It is important to appreciate the nature of such links, particularly for those building users with special physical needs. The chapter concludes with a review of other issues concerning the use of buildings, including the design of circulation space and the orientation of spaces within a complex.

8.2 Form and function

Formulating the design brief described in the previous chapter is a process of collecting and understanding the project-specific elements related to the client, the building type and the site. It is not an isolated activity to be 'got over and done with' before moving on but is continuously developed through the design stages by adding and refining information about the project-specific elements as they are mixed together with the general elements of design to create the new building, as discussed in Chapter 2. The elements are not mixed in a random fashion in the hope that they will all fit together. The process of mixing is much more sensitive than that requiring careful organization to see that they inter-relate with each other to achieve the desired result. For buildings, at the simplest level, mixing the elements can be focused on the following:

- *Function: The use of the building*
  The practicalities of the way that elements work or perform together and the way in which the building can be constructed and used.
- *Form: The appearance of the building*
  The two-and three-dimensional shape of the building and all the elements within it, as perceived by the eye.

Function and form are inextricably linked and decisions about one will always affect the other. For example, the practical considerations of arranging the penetration of natural sunlight into the interior of the building will dictate the size and location of windows, which affects the internal and external appearance of the wall. Larger or smaller openings in a load-bearing brick wall will alter the solid to void relationships, changing the visual balance. If it were decided that the whole wall should be glazed for aesthetic reasons, then a different structural support system would be required to hold it up. The need for a high, open space with sufficient headroom to accommodate specialist machinery, located next to spaces which could be lower may dictate the form of the shell of the building itself. It is therefore, not possible to dictate the choice of elements by selecting criteria relating to either function or form on their own, without considering the other.

Louis Sullivan, an American architect, promoted a simple axiom with regard to the creation of new buildings; '*form follows function*'. That is to say, the shape and appearance of the building and all its constituent elements are a consequence of intended use. It is a profound philosophy suggesting that there is an inherent beauty in the expression of spaces, structures and materials for what they are, not covering them up or disguising them to look like something else. Recently, some architects have even chosen to display all the building's services, notably for example, the Beaubourg Centre in Paris designed by Richard Rogers.

On the other hand, successful buildings have been designed on the basis of '*function follows form*' where structures and materials have been selected to create buildings that do not derive from their inherent qualities. Consider the Sydney Opera House for example, which is clearly a form into which activity has been arranged.

With respect to the whole building, this offers the two diverse options:

- The spaces for each activity and the choice of structure and materials used to build them are individually expressed, creating the form and appearance of the building as a consequence.

For example, in the car-dealership, the showroom, parts department, workshops and offices could all be treated individually and combined together as units of varying size, shape, height and construction to create the finished building as shown in Figure 8.1(a).

This strategy can also be applied to individual spaces and construction details. For example, a simple pitched roof could be constructed with exposed finished timber purlins, rafters and

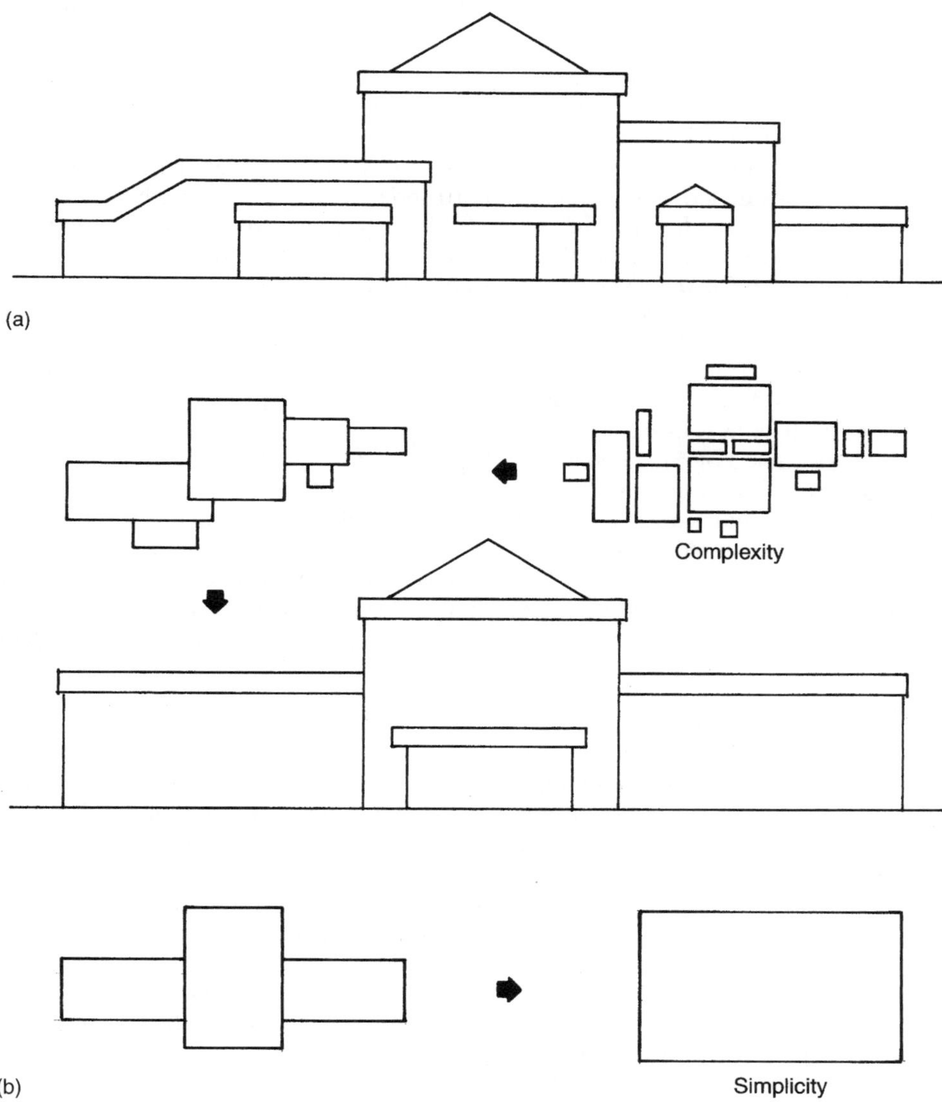

Figure 8.1 Options of form: (a) The individual spaces are all expressed in their own right creating an overall form with each clearly visible. (b) The individual spaces are simplified into clearer forms, each fitted in as necessary.

tie beams with no underlining ceiling. The triangular volume becomes part of the space below and the structural system is visible and comprehensible. Even the fixings of one member to another can be seen and understood as both practical elements of construction, and as interesting details.
Or alternatively,

- The spaces for each activity and the choice of structure and materials used to build them are expressed collectively, incorporated into a simple form or moulded to create a unified appearance.

The form and construction of the whole car-dealership can be predetermined, creating a simpler shell as shown in Figure 8.1(b), with appropriate or necessary activities fitted in as overall space and appearance allows.

For individual spaces or construction details, take the example of the pitched roof once again. The roof could be constructed with prefabricated trussed rafters, with an underlying ceiling of plasterboard so that the space in the roof and the structural system is invisible. To the building's occupants, the roof is effectively a horizontal surface with no indication of the volume above or how it has been constructed. The details and finishes are additional, used to cover up something else.

There are many factors leading to the choice of one option or the other depending amongst other things on the preferences of the client, the demands of the building type and the imagination of the building designer. The new building may utilize both options, collecting some activities together in a simplified form whilst expressing others separately. For example, in the car-dealership the spaces used by the customers could be a single structure, but others for repairs and body work, primarily occupied by members of staff kept distinct. Capital and maintenance costs, flexibility and appearance are important considerations, but for almost all buildings function is the key factor if they are to withstand the critical examination of users and be regarded as successful.

Understanding functional requirements will lead to decisions about the size of floor areas in *two-dimensions* and the volume of spaces in *three-dimensions.* Both must be continuously compared and related always bearing in mind the practicality and economy of construction and its appearance. Good building design is all these things moulded together to create the finished product.

8.3 Activity and flow

All the activities likely to take place in and around the new building can be identified, analysed and understood in isolation. This is a useful exercise both before and during the design process to make sure that important activities are not overlooked, and to attempt to understand what each activity involves. All buildings contain a collection of activities, some of which follow one another in sequence, or come together at fixed points, whilst others are quite separate and incompatible and should be kept apart. For each individual person, there is a *flow* between activities as they move from one situation to another and sometimes back again, and each material flows through manufacture or distribution processes as it is received, worked on and despatched. The whole of the building and the site are criss-crossed with invisible interlinked strands rather like the web of an over-excited spider.

Some activities are 'static' and last quite a long time, like listening to a concert or staying awake during an interminable lecture, but in most cases activity is 'active' involving movement of hands, feet, bodies and materials with the aim of satisfying a need, which is in itself made up of numerous other relatively short individual activities, carried out one after another, which can be identified as separate steps. The steps may make a recognizable pattern or they may be more unpredictable, depending on the needs of the moment. For example, the mechanic replacing a broken exhaust system will repeat exactly the same task over and over again. The necessary steps can be predetermined and arranged, but every customer entering the building will, or can be persuaded to take different steps in the quest for their objectives.

It would not be practicable or interesting to describe all the activities in all building types in any detail, or even for the car-dealership in particular. There are far too many and it would be excruciatingly repetitive. The following rather obvious example, with which the reader will be familiar, shows the extent to which design can influence the general process of activity and is a useful analogy for any activity however complicated it may be.

Consider the number of steps required to make a mug of coffee in a kitchen:

- Unplug the kettle, lift the kettle and move it to the sink.
- Hold the kettle under the tap with one hand and turn the tap on with the other.
- Put the kettle down and wipe splashes off the front of shirt.
- Fill the kettle, straining as it gets heavier.
- Move back to the electric socket, plug the kettle in and switch it on.
- Retune BBC Four to Five Live on the radio.
- Lift the mug off the mug tree.
- Lift the coffee jar out of the cupboard, lift a spoon out of the drawer, measure and place an amount of coffee in the mug and replace the coffee jar in the cupboard.
- Walk to the fridge and get out a milk bottle.
- Walk back to the mug and pour out some milk.
- Walk back to the fridge and replace the milk bottle.
- Walk back to the electric socket and switch the kettle on (the first time was the toaster switch).
- Empty the dishwasher while waiting for the water to boil.
- Lift the kettle, pour hot water into cup, add sugar (optional) and stir.
- Walk to the table, sit down and wait for it to cool down.
- Drink it whilst reading the paper (Guardian).
- (Eventually) walk to the sink, wash and dry the mug, return it to the tree.
- Possibly repeat exercise if sufficient time available.

This is a simple process which most people do all the time without a moments thought. They will not think about the process of making and drinking their mug of coffee or generally appreciate the part played by both product and building designers, providing everything goes to plan. However, a failure at any single point in the process can render the objective, the hot mug of coffee, difficult or impossible to achieve. Short-term problems such as a broken mug, an empty coffee jar or a temporary power failure can be overcome, but slipping on a polished floor, tripping down a step carrying hot water or straining to reach a shelf which is too high for comfort are much more significant and not so easy to live with or correct. A simple activity becomes irritating, difficult or dangerous causing inconvenience or creating risks of injury, further compounded by the fact that constant repetition, which increases familiarity actually diminishes care and attention.

Responsibility for the performance of elements within the process varies. The loose equipment like the mug, the spoon and the kettle are the responsibilities of the client who will choose them to suit their own preferences and any failure in the performance of these elements cannot be attributed to the building design team. The fixtures and fittings like the sink and the taps are likely to be standard products, selected and purchased from a manufacturer's range. The manufacturer's responsibility relates to the supply of products which perform in accordance with expected standards, but this can be negated by incorrect specification or by proposing the use of a product in a situation for which it is not suitable. In this case, the specifier would be at fault for choosing the wrong component or material, either in error or in direct contravention with the manufacturer's recommendations.

The design and installation of the kitchen layout itself however, including the height of work-tops and cupboards, the specification of the floor finishes and the positioning of artificial lighting are all elements that can be varied to suit the specific needs of the client or the kitchen, ensuring that the simple activity of making a mug of coffee is as convenient, efficient and as safe as possible. The aim in this small example can be summarized as creating an environment where making a mug of coffee is taken for granted and drinking it as a pleasure.

Every activity in the building can be seen in this way, as a series of steps leading to an end. Making a mug of coffee is a simple process and relatively easy to understand, and there can be little excuse for it to be incorrectly designed. Others are much more complex either because they involve specialist machinery or processes or because it is not certain what the activity taking place will be. For example, one activity within the car-dealership is processing cars through a Ministry of Transport (MOT) test. This involves examining a range of features and functions on existing cars and through testing, checking that they are performing satisfactorily. It may take the form of an examination leading to 'pass' or 'fail' or it may include resetting or replacing parts of the car to guarantee success. In either case, the process is a methodical one, working through stage by stage to test all the relevant items, with the aid of elaborate machinery and tools. In principle, it is the same as making a mug of coffee, proceeding step by step to a conclusion. The building designer may not appreciate the content of all the steps, or how they are linked one after another, so this activity must be carefully researched taking advice from experts in the subject, perhaps in this case the client themselves who have the necessary previous experience to understand the requirements and who know exactly what they want. In this way a layout of space and equipment with a specification for all the elements associated with this activity can be prepared in the knowledge that it will work as intended.

Elsewhere in the car-dealership, consider the customer who enters the building to arrange for an MOT test at some time in the future. In this case, it would be possible to organize their flow through the building just for that activity alone, but as there are so many other things that the customer might wish to do, or which the client would like to encourage them to do that it would not be a sensible way to plan the building. It is not certain what the customer will choose to do or how they might respond to other things if they were available. As the elements of the activity of booking an MOT test are a small part of the overall activity of a customer moving through the building, the layout should be flexible enough to accommodate all the options to suit the customer's convenience. Flexibility is also advantageous to the client, who may wish to alter the layout of elements within the building from time to time to incorporate technological advance or changes in sales techniques. If the building is designed too rigidly to suit specific activities, flexibility will be unnecessarily limited. The position of fixed structural walls and columns, location of windows and doors, and associated sterile corridors will limit the client's ability to change the layout to suit new needs which may not yet be possible to anticipate.

Every activity within the car-dealership is a process, including the customer looking at new cars, the mechanic fitting a new 'big end' and the parts receptionist locating a replacement light bulb. The designed solution must cater for them all. Examining each activity and flow at its fundamental level of detail is dependent on an understanding of the capability of people to grasp, reach, lift and see what they are doing, together with an appreciation of the arrangement of equipment and surfaces needed for them to do it. The two principal sources of data available concern ability and work.

8.4 Anthropometrics

Anthropometrics is the scientific study of the measurements of the human body. The human sciences of physiology, anatomy and psychology provide essential information for designers of all products intended for human use. Data concerning 'typical bodies' includes physical dimensions and performance such as head height, foot size, arm reach, strength and mobility, sensory acuity in areas of eye sight, hearing, cognition and breathing, and experiential information about behaviour, culture and emotions. Statistical average figures are used as guidance for the design of many elements within the building, which will suit most people, but thought should also be given to the fact that maximum and minimum ranges may not be so convenient to some special user groups. In the example in the previous section, making the mug of coffee depends on human dimensions and faculties including the following:

- *The height and depth of the worktop*
  The height of the worktop is selected to provide comfortable working conditions when standing in front of it, and the depth restricted by the ability to reach to the back against the wall.
- *The height of cupboard door handles and the depth of high-level shelves*
  The ability to reach up to open doors and lift out stored items, and the ability to see into the back of the wall cupboard.
- *The shape of the tap handle*
  How the handle is gripped by the fingers and the amount of force needed to turn it on and off.
- *The position of the socket outlet*
  So that it can be reached comfortably to insert the plug top and switch on.
- *The level of illumination*
  Which should be at least the minimum required to be able to pour out the boiling water in safety.

At the risk of being 'politically incorrect', it should be mentioned that for some of these issues there is a statistical variation between men and women which should not be ignored. On average, men are taller than women, can reach higher and further, and can lift heavier weights. Obviously, wall cupboards and shelving in the kitchen designed for a six foot tall young man would not suit a five foot tall elderly woman. There is a need for sensitivity and consideration for the needs of likely occupants if the design is to be successful.

Elsewhere in the building there are other elements dependent on human dimensions such as the size, shape and strength of furniture, the dimensions of stair treads and risers, the height of door openings and the positioning of light switches and hand rails. The shape and depth of baths and shower trays are governed by ability to step up or climb in comfortably and safely. The height of window cills or the position of transomes may be related to eye level so as not to interfere with views, and the position of handles and locks are placed to be within easy reach of adults, but out of the reach of small children.

Some of these issues are subject to mandatory controls. For example, all the dimensions associated with stairs are defined in the current Building Regulations including widths, head heights, tread and riser dimensions, and handrail and balustrading requirements. Others are matters of good practice for which there are many printed reference sources of data already available.

8.5 Ergonomics

Ergonomics is the study of the relationship between work and the environment in which it takes place. Sometimes referred to as 'human engineering', ergonomics concerns the relationship between people, activities, tools and the interior or exterior environment that they find themselves in. Individual or organizational 'work' in the context of building design can be defined as any activity involving people, materials or goods. The individuals' physical and psychological characteristics referred to in the section about anthropometrics help to determine workplace arrangements, particularly important for comfort, health and safety, attitudes and motivation. The car-dealership as an organization will have a management structure with procedures, systems and policies designed to promote teamwork and co-operation towards defined ends.

An ergonomic process can be static like reading a book, almost static like tapping away at a keyboard or involve various degrees of movement and strenuous effort. The example of making a mug of coffee is 'work'; a task made up of a series of movements which can be made easy or difficult depending on the arrangement of all the elements needed to undertake it. The distance between the kettle and the tap, the relative positioning of the cup, spoon, coffee and milk and the distance between the dripping spoon and the sink can be organized by design to make the task simple and uncomplicated, or difficult and messy, assessed by the phrase 'user friendly'.

Making a mug of coffee is just one task in the kitchen. Bringing in raw materials, storage and preparation, cooking, serving food, washing up and putting everything away are all tasks involving work, and a good example of an ergonomic process, requiring easy access to components in the right order so that tasks are completed conveniently and safely.

As with anthropometrics, ergonomics is a combination of physical abilities and the performance of faculties. Walking, climbing, bending, reaching and lifting are obvious physical activities relating to carrying out tasks which can be made easy by design, or hazardous by neglect. For the car-dealership, the mechanics working on a damaged car are surrounded by equipment, tools and materials and need practicable access to parts, oils and waste bins. They also need good levels of natural and artificial lighting to see what they are doing as they perform their different tasks, environmental controls to remove harmful fumes and dust and an efficient heating system to keep warm when the external doors are opened in winter. In a commercial situation, ergonomics is very important to avoid injury, tiredness and wasted time. Information about working processes is often gained through 'time and motion' studies, seeing how tasks are being undertaken and how improvements could be implemented to make them more comfortable, safe and efficient.

Tasks need not be confined to manufacture and production. The health risks associated with looking at a computer monitor for long periods or prolonged operation of a keyboard are now well recognized. Repetitive strain injury in the fingers and wrists of keyboard operators could become a major problem of our time. Simply opening a door has its dangers, walking into and breaking glass, banging into someone on the other side and trapping fingers in the jambs are all elements of the design that must be considered if the task is to be made safe.

It is not necessary to examine all ergonomic processes from first principles as standard data is available on minimum space and arrangements for undertaking common tasks. For the car-dealership, many of the arrangements for working are well established, and the client will also be able to advise about their preferred requirements.

8.6 The needs of special groups

Standard anthropometric and ergonomic data is generally related to people of 'average' height, reach, strength and ability, and guidance is statistically based on anticipated ranges. If this data was used to design all the elements of the building, some people would be excluded or their use of the building would be made unnecessarily difficult. For those people who cannot operate at 'standard' levels, such as the disabled, infirm, elderly or small children, some elements of the building must be adjusted to maximize their comfort and safety. Those unable to walk or who have difficulty walking or climbing, who have poor eye sight or are short of hearing, who cannot bend or reach will be unable to use some elements of the building unless their needs are recognized. Standard design solutions will prevent them from entering and moving around the building, and even if they can do so, they may be at risk of injury.

This is one area of design that had been neglected until relatively recently. Special design considerations for people with particular needs had been regarded as optional, and the additional costs involved deemed to be uneconomic because it was thought that it concerned very few people who could make their own arrangements if they wanted to use the building. Fortunately, this attitude has changed and all new buildings which are available for public use are now designed to accommodate people with special needs as far as is practicable so that they can use the building with a minimum of inconvenience. The current Building Regulations (Part M) impose requirements for some issues which are periodically revised and improved in the light of experience. The Disability Discrimination Act imposes conditions on the design of all buildings which are accessible to the public. Design guidance is also available in the form of good practice publications for the elderly, the infirm and for children, where their needs are not presently covered by specific building regulations. In fact, many design issues are not cost related but are more a matter of consideration and helpful thought. Installing a lift to floors above ground level is obviously very expensive but ensuring that doors are wide enough to accommodate wheelchairs, or using bright colours to mark hazards like steps, add little or no extra cost to construction, but make a significant difference to many people. As with the analysis of any activity within the new building, due consideration should be given as to how those with special needs are likely to be able to undertake them.

The general design issues to consider include the following:

Access onto the site

- In order to avoid possible accidents, it is best to separate vehicle and pedestrian routes, which should both be clearly identifiable, well lit at night time, and have appropriate direction and warning notices.
- Traffic calming measures help to reduce speeds, and clear visibility lines are essential at bends or junction. Protective barriers (walls, fences, railings and planting strips) help to prevent pedestrians inadvertently stepping into traffic flows.
- Extra large car parking spaces should be provided close to the entrance to the building, ideally under cover to provide shelter from the rain.
- For people walking or pushing wheelchairs and prams, kerbs and steps can be difficult to negotiate. Single steps or unexpected changes in ground levels can cause discomfort and be

hazardous. Dropped kerbs and ramps will help, but all changes of level should be clearly identified (textured surfacing) and include suitable handrailing and balustrading.
- Steep gradients should be avoided wherever possible (maximum 1:12, ideally 1:20).
- Surfaces should be designed to minimize the risks of slipping or tripping.

Access into the building

- The main entrance to the building should be ramped at a comfortable gradient with handrails and there should be a flat landing area in front of the door so that wheelchairs do not roll away while the door is being opened.
- The door leaf should swing inwards in the direction of movement, and preferably outwards as well to assist easy exit. Automatically controlled doors should be considered to assist those whose movement is slow or restricted in any way.
- The space between draught lobby doors should be sufficient to allow manoeuvrability.
- Handles and push plates should be positioned at a reachable level for wheelchair users and areas of glazing marked to be clearly visible.
- Lighting and direction signs.
- Disabled toilet facilities.

Access around the building

- The width of all door openings should be designed to accommodate wheelchairs.
- Lifts may be considered for access to floors above ground level, and stairs designed to suit the needs of ambulant disabled and those with poor eyesight.
- Fixtures and fittings should be designed and positioned to prevent children from climbing up them or getting stuck in them. Open rise steps and horizontal balustrading should be avoided together with low-level fittings in front of windows.
- Colour-coded handrails, doors, ironmongery and notices can help those with poor eyesight to negotiate their way around the building.
- Textured surfaces can be used to identify areas of risk such as the top and bottom of flights of stairs.
- Single steps and raised threshold strips should be avoided as they are easy to trip over.

Visibility

- The height of window cills and transomes determines the view available to those in wheelchairs.
- People with poor eyesight are more likely to walk into full height glazed panels if they are not clearly marked, to fall up or down poorly lit stairs or trip over thresholds.
- Bright colours can help to identify critical elements such as stair tread nosings.
- Lighting levels generally.

Equipment controls

- Consider the strength and reach of those likely to use equipment, the operation of locks and handles, taps and switches.

- Automatic door closers should not be too strong and restrictor stays will prevent people falling out of windows.
- The temperature of heating devices and hot water from taps should be controlled to prevent accidental injury.
- Be mindful of the positioning of dangerous features like door swings, radiators and glass for those who might fall onto them.
- Think about the position of socket outlets in relation to the equipment that they may serve to avoid risks of falling over trailing flexes.

These are just some of the issues which require careful consideration. In some cases, specific installations are needed for the known occupants of the building. For schools, hospitals and homes for the elderly, for example, other specialist equipment or arrangements will be required which must be fully researched before making design proposals.

8.7 Understanding individual movements

Both the building and the site layout are designed to accommodate many individual movements, some running together or overlapping; others kept apart to minimize conflicts. People and materials are organized so that they move easily and efficiently through the building avoiding frustration of the user's needs. A movement can be simple and focused, such as opening the door and entering the building. It is a movement which the user will generally concentrate on as a means to an end, and providing that the door has been designed with the users health and safety and convenience in mind, is one which is quickly completed. Other movements offer many choices and options as they take place like walking through the showroom to a reception point. This is a much more flexible movement with no specific fixed points along the route. It is a movement that may invite deviation to some other point of interest depending on what is encountered. The original target objective may be replaced by another one which had not been thought about until it presented itself.

To understand how these movements work, what is required or what could happen, it is useful to play the role of a person or a material coming into the building and identify the possible sequence of activities associated with their journey through all the spaces before they go out again. In the process of making this imaginary journey, it may be possible to establish which other issues are associated and which are not.

For example, the people using the car-dealership could include the following:

Private customers

- There are many different customers who come into the building for specific reasons including looking at something, buying something, collecting something, making an appointment or paying an account.
- They may come in to look at a new or second-hand car, to find out the cost and availability of a part for their own car or simply to browse through the items on display.
- They may be looking for a valuation for their own car with a view to part exchange or come in to have a new ignition key cut.
- They may purchase petrol or anti-freeze, cigarettes, sandwiches or a token for the car wash. They may be determined to concentrate on any one single activity because they are in a hurry,

or alternatively, they may have sufficient time to spare to take an interest in something that they had not thought about before arriving.
- They may know exactly what they want, or may have no idea and want to talk to someone who can show them the options.
- Customers may not enter the building at all, but look for incidental services on the site like fuel, water and air for tyres or a car wash facility. Some car-dealerships now even have cash withdrawal machines as an extra service for their customers.

Trade customers

- Regular customers can be expected from other dealerships, and include traders in spare parts and repairs who are more specific in their needs and requirements.
- They may have special arrangements for access and be permitted into parts of building to which other customers are excluded.
- The client may employ staff specifically to deal with this aspect of trade and may prefer to separate them from other enquiries.
- The nature of display and information may be different for these people, who generally know exactly what they want and do not have to be presented with a range of options to consider before making a purchase.
- Special facilities may be available for ordering in advance and collecting pre-assembled selections of parts without waiting.

Visitors

Visitors to the building could include friends and relations of members of staff, health and safety inspectors, the fire officer, insurance valuers or parties of university students who want to know how a car-dealership works. Reception and security arrangements may be specially geared up for casual visitors or organized groups. Somewhere in the building there could be a space set aside for formal or informal meetings.

Business representatives and suppliers

Many external companies will be involved with buying and selling cars, parts and materials to and from the car-dealership. They need access to the management or particular members of staff to discuss their products, make presentations and negotiate sales. This may essentially be a private activity, separated from customers so that staff attention is not diverted from service and sales.

Management and employees

There are many different members of staff including the managing director, departmental managers, sales, parts and servicing people, receptionists, drivers, valets and cleaners. Some will work in a private, self-contained manner whilst others will be located in noisy, communal spaces. The operations of some staff will be mostly clean and quiet, intended to create an attractive, safe environment for customers, whereas others will work in messy or dangerous areas, specifically excluding customers who could be put at risk of injury. The layout of the accommodation within the building must be considered carefully if staff and customers are required to pass through these diverse environments.

The requirements of all these people will vary as they enter, pass through and leave the building. They all need access, car parking, reception, direction and satisfaction of the objective of their

visit. The environmental conditions and ambience of spaces are different for each activity within the building. The quiet, peaceful atmosphere needed to conclude sales does not mix easily with the noise and fumes of the workshop. The managing director would be reluctant to discipline a member of staff in front of customers, who in turn would be seriously irritated to have to walk through an oily puddle to get into their new car.
The materials within the car-dealership could include the following:

- *New and second-hand cars*
  These can include deliveries for immediate sale to customers who have ordered them, for display to attract potential customers or for attention and repair before being offered for sale. They may all require servicing or cleaning before being driven away. Space will be needed to store them in such a way that they are accessible and not at risk of accidental damage or theft.
- *Parts*
  Parts will arrive for specific repairs in the workshop or for general stock available for potential customers. Some parts are small and can be easily stored, others are large and only obtained as and when needed. Some parts are required in relatively large numbers and purchased in bulk, others are wrapped for customer display.
- *Fuels, oils, paint, etc.*
  Consumable materials are needed for general workshop use related to car repair and servicing, and for sale to customers. The storage and presentation arrangements may be separated.
- *Paper, brochures and samples*
  The general administration of the new building will involve the use of consumable paperwork together with handouts for potential clients illustrating available products and maintenance manuals for goods purchased.
- *Catering supplies and incidental goods for sale*
  Many car-dealerships will offer their customers a range of food and drink to be consumed on the premises as a courtesy facility whilst waiting, or as an addition to their other incidental sales. Self-service machines may be used or fridges, cupboards, cabinets and shelves. Products must be handled in accordance with current legislation with regard to storage and display. A range of other items could be made available to passing customers such as newspapers and magazines, toys, CDs, flowers and barbecue charcoal.

All materials must be delivered, received and recorded, stored, distributed, worked on and despatched or presented for sale. Some of them will generate waste which must be collected, stored and taken away. Some activities produce by-products which can be recycled, others are scrap. There is liquid waste from washings and servicing requiring special considerations for health and safety, and disposal. Each material must be managed by someone as it comes onto the site and goes off again.

As well as the principle activity associated with any flow, there can be numerous associated considerations or possibilities that the design can accommodate. For example, take the flow of the customer entering the building to arrange for an MOT test described earlier:

- Driving onto the site
- Parking and walking to the entrance to the building
- Arrival at reception
- Being directed to the workshop reception
- Waiting to be served
- Inspecting a fault on the car with the receptionist or a mechanic
- Leaving the keys and arranging for collection when work is completed

- Checking the account and paying the bill
- Collecting the car and driving away

As they move through the building, they might also want, or be encouraged to do other things such as:

- Looking at the display of new cars, inside or outside
- Looking at second-hand cars
- Taking a test drive
- Looking at parts, accessories and incidental goods
- Arranging for a future service
- Buying something
- Waiting for work to be carried out
- Watching a video or studying a display
- Making a phone call
- Going to the toilet
- Having a drink
- Resting and reading a newspaper
- Working while waiting
- Occupying small children
- Going somewhere else while work is completed, returning when it is finished
- Arranging to use a courtesy car if the work will take longer than you are prepared to wait

Bubble diagrams can be used to look at each area and the incidental issues associated with it. The bubble diagrams can relate activities and help to decide where in the overall layout they should be. This exercise can be done without regard to size or shape, positioning complementary activities together or separating conflicting activities from one another. Colour coding is a useful aid highlighting activities that are associated or perhaps capital letters for words of greater importance. Arrows can mark the direction of movements, plotting the working routes through the building. Any system of graphic identification is usually easier to read and appreciate rather than a jumble of words. It is unusual for this kind of analysis to be presented as a finished document but it can be very helpful in discussing and explaining issues to help to justify a particular arrangement as being better or poorer than another. As mentioned elsewhere, it is always useful to record and keep any form of analysis for future reference as a check against design proposals to see that decisions already made have been correctly incorporated.

It would not be practicable to show bubble diagrams for all the eventualities in the new car-dealership in this book. The illustration for the customer arranging to replace a warn tyre shown in Figure 8.2 as an indicative analysis of a particular movement, can be systematically applied to them all. Exactly the same system will reveal the movements in any building type related to their specific demands. It may not be practicable to examine all the movements and flows in the new building from first principles. It is perfectly reasonable to refer to historical precedent to see what can be learned from existing buildings of the same type. There may well be satisfactory arrangements which have been established through use and experience which do not need to be reconsidered or methods of good commercial practice which dictate certain solutions. For example, the layout of shops and supermarkets is very specifically arranged for commercial reasons, leading customers around in a particular way. Even the order in which goods are stacked on shelves is significant in achieving the best sales. Houses, churches, cinemas and restaurants all have their own flow patterns, satisfying customers, users, staff and processing materials and goods in their own way.

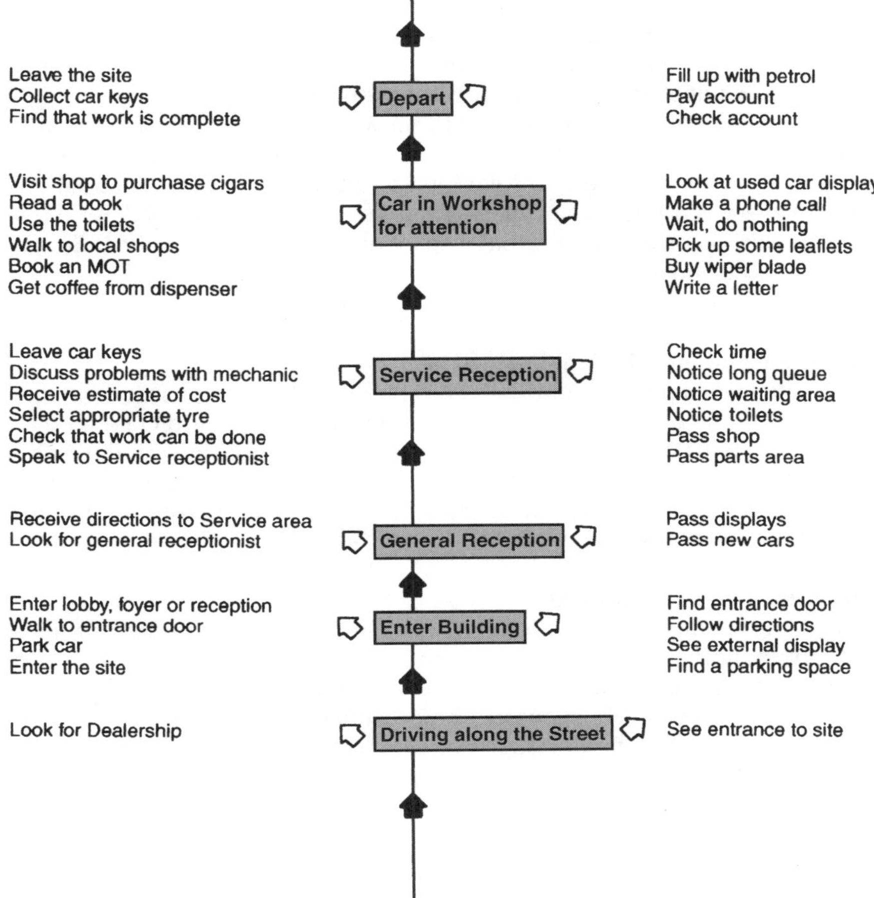

Figure 8.2 A structured analysis of customer movement.

In many cases though, existing precedent should only be regarded as indicative and should be used to examine the possibilities. 'If it's not broken, don't fix it' is a rather glib saying, but has an element of truth about it. On the other hand, in design terms, only if 'it' is examined carefully, can any judgement be made as to whether 'it' is broken or not. The fact that any element within the building has existed in any form for a long time does not mean that it is correct or cannot be seen in a different way.

8.8 Orientation

When exploring the plan arrangements with the bubble diagrams, it is important to be aware of the orientation of existing site constraints, which could be at any point on the circumference of a circle, facing north, south, east or west. The location of some activities on the site and within the building will be affected by these constraints, particularly with respect to access, visibility and prevailing environmental conditions which vary around the circle and which can be exploited or minimized depending on user requirements or preferences.

For example, the windows in the car-dealership showroom might be best positioned in a wall facing passing, potential customers, but if that means that the showroom windows face due south, there is a likelihood of excessive solar gain and glare making the internal environment uncomfortable. If there is a busy, noisy railway line running along one side of the site, this is unlikely to be the best place for the external display of cars. However, perfect design solutions are not always possible and will inevitably contain compromises.

Orientation can affect design and use including the following factors.

Access onto the site and into the building

Access onto the site from a public highway is commonly only available from one direction and the entrance to the site is predetermined. This single factor may well dictate the layout of the whole site and the building itself, positioning activities accordingly. The visibility of display, customer convenience, security and control begin at the entrance, and unsightly areas of storage and waste disposal are generally kept away from it. Car parking, deliveries and turning space will be positioned in relation to the entrance.

Privacy for internal and external spaces

Some activities can be shielded from external view, and equally some be positioned so that looking out of the building does not infringe the privacy of adjacent occupants. For the car-dealership, some activities are unattractive to look at and can be hidden from customers view whilst others need to be positively exposed. The potential for privacy into and away from the site is often already established by development around the site over which there is little control, but the layout of the site and the new building can make matters better or worse.

Views

The appearance of the building from a distance is dependent on the view points available to onlookers. A prominent site is permanently on view to neighbours and passers by and the design must acknowledge the publicly visible elements of the development, not only for the 'common good' referred to elsewhere but to protect and enhance the client's image. From the site and the new building, attractive views may or may not be available. For the car-dealership, this is unlikely to be a significant factor in the designed layout, but for other building types, notable residential development, an attractive outlook can be a significant bonus to the future purchasers and users, which should not be wasted. Any building in which the occupants will spend time such as houses, restaurants or rest homes can be designed to maximize good views and minimize the effect of poor ones.

Exposure

For some sites, the direction of the prevailing wind can make the use of external spaces uncomfortable unless they are appropriately sheltered. Even the design of structures and the fixing of roofing materials must take into account the strength of gusts of wind.

Sunlight for all internal and external spaces

The position of the sun in the sky varies throughout the day, rising in the east in the morning and setting in the west in the evening. It also changes its angle of elevation in relation to the seasons during the year rising and falling either side of mid summer. These variations affect the penetration of the suns rays into the building and onto spaces around the building. At different times of the year and throughout each day, the warmth, brightness, shade and glare in each space changes, adding to or detracting from their usefulness to the occupants. For example, as the sun rises in the east it will shine into spaces during the morning. Setting in the west, the sun offers afternoon and evening light to spaces at the end of the day. The highest and brightest sunlight of the day is approximately at mid-day, when solar gain and glare is at its greatest. Direct sunlight penetration into spaces at this time of the day may be uncomfortable for the working occupants.

Over-shadowing

Considering the effects of the sun extends to the position of adjacent development, trees and planting, including the implications of future unrestrained growth, which can lead to parts of the site or the building itself being in the shade for lengthy periods of the day. Similarly, the new development itself may cause overshadowing of adjacent property, denying them the benefit of the sun's warmth. In a climate such as the UK, losing sunlight unnecessarily is disappointing to most occupants.

Noise

Existing development around the site will generate a variety of unwelcome noise including road traffic, trains, industry, sports and leisure facilities, schools and deliveries. Careful design of the building and construction detailing can minimize problems. Noise will be generated within the new development affecting the occupants and those in neighbouring property. For the car-dealership noise sources could include panel beating, drilling and engine revving interfering with staff and customer privacy and thinking, and upsetting the amenity value of local residents.

Pollution

Existing activity around the site may generate dust, fumes and smells drifting in from one particular direction. Equally, the car-dealership may create pollution drifting out onto adjacent property. Both eventualities may require design considerations if the nuisance is to be minimized or eliminated.

8.9 Circulation

For any activity in the new building, or within the site around it, the space provided must be at least large enough to accommodate it, but some additional space is needed to enable people and materials to move or be moved from place to place. The activity may be completed at one point, but it is not possible to start the next one without moving to the space where it will take place. Space between activities is called circulation space, and in large buildings it will occupy a considerable floor area, representing a significant construction cost. Some forms of circulation space are essential to separate activities so that they do not interfere with one another, but in many cases, if handled sensibly can be incorporated within the plan as an attractive feature offering flexibility for additional uses. The extent of floor area needed for circulation depends on how activities and spaces are related to each other in the plan layout. In most cases, unusable, 'dead' circulation space, which can be used for nothing other than movement should be avoided.

Circulation space also has other meanings apart from the purely functional one of moving about. It can be public or private and often marks the boundaries of transition between one and the other. For example, in a house the hall is a relatively public circulation space where visitors can enter and be received. The staircase, although visible and accessible is private in the sense that visitors would not be expected to go upstairs on their own, and the landing is usually invisible from the hall and a completely private area for the use of the occupants only.

In other buildings, the circulation space is used to reflect prestige and importance. In the concert hall, the foyer is a large entrance space used in many different ways, but is a 'grand' environment, often of theatrical proportions where people can meet and display their evening dress to one another. In an office or bank, the foyer expresses the style of the company, its wealth, power and the substance of its 'professional' activity.

Circulation space is also used to control the flow of people in different ways. In the cinema, the foyer is exiting and lively, brightly lit, stimulating anticipation of the forthcoming entertainment. It is where future events are advertized, tickets are bought, and sweets and popcorn purchased. The passage leading to the auditorium narrows and the lighting dims as people are calmed before entering. The aisles and gaps between rows become smaller and smaller, obliging visitors to slow down even more and take greater care in reaching their seats, so that they are quiet and relaxed ready to enjoy the film.

Other forms of circulation space facilitate easy and comfortable exit or escape in an emergency to enable occupants to reach points of safety without sustaining injury.

Circulation space can take the form of the following.

External circulation, paths, roads, drives and hardstandings

Circulation space allows vehicles to drive onto the site, turn around and go out again. Large vehicles need considerable areas of hard surfacing to manoeuvre into position to be unloaded or loaded. Parking a car requires extra space over and above the size of the car itself to edge backwards and forwards into line. Parked cars need space to walk around them and so that doors can be opened without banging in to the next one. People need space in which to walk, and push trolleys and wheelchairs.

Lobbies, foyers, halls and landings

Some spaces in buildings are defined circulation spaces where the principle reason for their existence is to permit occupants to pass through them to get somewhere else. In a small house, these spaces would be the minimum for human movement and contain little else. In a larger, public building, they are often oversized so that above and beyond the needs of specific movements they can also be used to accommodate other activities such as reception, meetings, waiting areas and coffee lounges.

Passages

Passages are narrow routes connecting spaces with no other function than providing the ability to move. They are very wasteful and costly and should be avoided if possible. They often result in a planned layout because the subdivision of accommodation into rooms placed one after another permits no alternative. Grouping rooms around a larger hall is much more efficient and attractive.

Undefined pathways through spaces

Within every activity space there must be sufficient room to enter and leave and to move around within it. This circulation space is not defined but it amounts to a large floor area. Consider the circulation space within a typical room like an office, illustrated in Figure 8.3 and look at how the circulation space is useful or wasteful, depending on the location of governing elements.

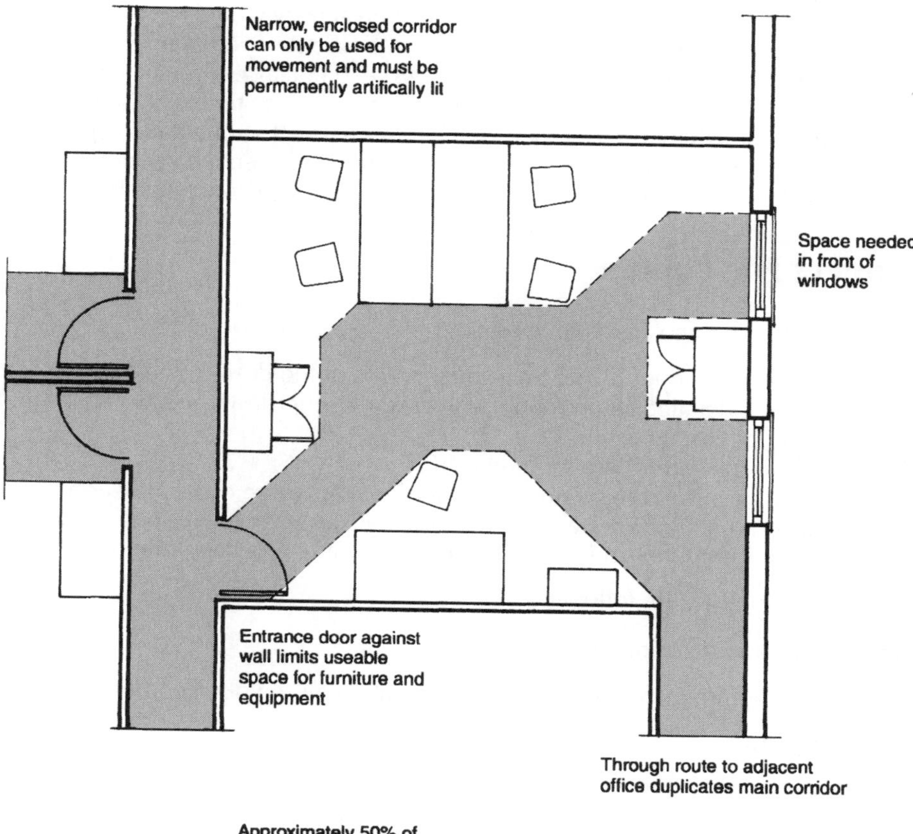

Figure 8.3 An example of wasteful circulation space.

Stairs, ramps and lifts

These are required to move vertically between the floors of a building, for people, goods and materials. They are legitimate 'single function' circulation spaces, deliberately intended to focus users on the task of moving in order to minimize the risks of injury.

Some buildings are designed as 'open plan' giving the occupants the flexibility to determine their own internal arrangements and circulation between activities. This reduces the likelihood of imposing prejudices on the layout and can also be a matter of sensible economy, particularly in changing times when future use may be uncertain. This is a common situation for spaces used for display in art galleries or shops, or for group working in offices. The layout may regularly change to suit current demands. Perhaps the most obvious example is the main hall in a sports centre, which caters for a wide range of different activities at different times, all within the same volume of space. Not long ago, each individual sports and recreation activity would have had its own special facilities, designed and constructed for the purpose, but a single open plan hall now accommodates basketball, tennis, badminton, judo, cricket nets, keep fit classes and anything else that may be in demand, sometimes with more than one going on at the same time. The hall can be used for play and recreation, for exhibition, and viewing and spectating. The circulation space that would have been necessary to get from one activity to another has now been incorporated into the layout without specific definition creating a more useful and much more economical building.

8.10 Project File content

The general aims at this stage are as follows:

- Collation of consultants input as per the requirements of the design brief.
- Cost checks against detailed information, ensuring that the design proposals do not appreciably change initial cost constraints.
- Detailed consultations with interested authorities.
- Assessment of the new buildings value in use and market value on completion.
- Preparation of presentation material summarizing design conclusions including rough sketches of feasibility studies and possible solutions to any aspects of the development.

File material could include the following:

Letters, meeting notes and minutes

- Further details of communication between all members of the development team.

Bubble flow diagrams

- Consideration of relationships of activities and spaces for the whole complex or any component parts such as reception, showroom, parts and workshop.
- Analysis of customer and staff flow patterns.
- Analysis of deliveries, storage and despatch of materials and cars.
- Assessment of anthropometric data about the sizes of spaces and equipment in relation to the human body.
- Investigation of ergonomic data about human working performance.
- Research into processes data about working patterns and necessary equipment and space to undertake specific operations such as customer sales, car repair and maintenance, body panel beating and spray painting, car washing and refuelling, etc.
- Consideration of the implications of providing access and facilities for disabled workers and visitors.
- A revue of health and safety issues for customers and staff.

Further briefing information

- Catalogues and details from manufacturers and suppliers about materials, systems and equipment that may be under consideration for use in the development.
- Additional information clarifying design criteria improving the understanding of requirements, developing and refining the design brief.
- Detailed cost analysis of possible design proposals including considerations of flexibility and adaptability.

- Cost assessments leading to design decisions such as floor areas, space heights and style and quality of potential finishes.
- Analysis of any likely effects on the client's existing business operations during construction.
- Considerations of phasing development, if appropriate, including the client's cash flow during the construction period.

8.11 Discussion points

(1) Should form follow function? How is practicality reconciled with decoration? Should decoration be funded at the expense of practicality or vice versa?

(2) Some UK buildings are being designed to respond to environmental conditions in an 'intelligent' manner. Is this likely to be negated by 'unintelligent' people? Does quality of life depend on labour saving devices? How can designers be expected to understand process which they may never have experienced before?

(3) Can buildings be designed to accommodate the needs of everyone, including the very young, very old and those with disabilities? What are the options for maximizing flexibility for building users to suit there own needs at any time? How has building design responded throughout the twentieth century to changes in human needs, and what changes might be required in the future?

(4) To what extent are/should building designers and developers be concerned with the health and welfare of the future occupants of their buildings? Who is to blame if someone trips and falls down a simple straight flight of steps?

8.12 Further reading

Adler D (1999) *Metric Handbook: Planning & Design Data.* 2nd Edn. Oxford: Architectural Press.

Bone S and **Bright** K (2004) *Buildings for All to Use 2.* London: CIRIA.

British Council of Disabled People: www.bcodp.org.uk

Burden E (2000) *Elements of Architectural Design.* Chichester: Wiley.

Clements-Croome D (ed.) (2000) *Creating the Productive Workplace.* London: Spon.

Dul J and **Weerdmeester** B (1998) *Ergonomics for Beginners.* 9th Edn. London: Taylor and Francis.

Foster L (1997) *Access to the Historic Environment: Meeting the Needs of Disabled People.* Shaftesbury: Donhead.

Goldsmith S (1997) *Designing for the Disabled: The New Paradigm.* Oxford: Architectural Press.

Lambert S (1993) *Form Follows Function? Design in the 20th Century.* London: Victoria and Albert Museum.

Levy M and **Salvadori** M (1994) *Why Buildings Fall Down.* US: WW Norton & Co.

Neufert E (1988) *Architects Data.* 2nd Edn. Oxford: BSP.

Newman O (1973) *Defensible Space: People and Design in the Violent City.* London: Architectural Press.

Pheasant S (1999) *Bodyspace: Anthropometry, Ergonomics and the Design of Work.* 2nd Edn. London: Taylor and Francis.

Royal Society for Prevention of Accidents: www.rospa.com

Salvadori M (1990) *Why Buildings Stand Up.* US: WW Norton & Co.

Syms P (2002) *Land, Development & Design.* Oxford: Blackwell.

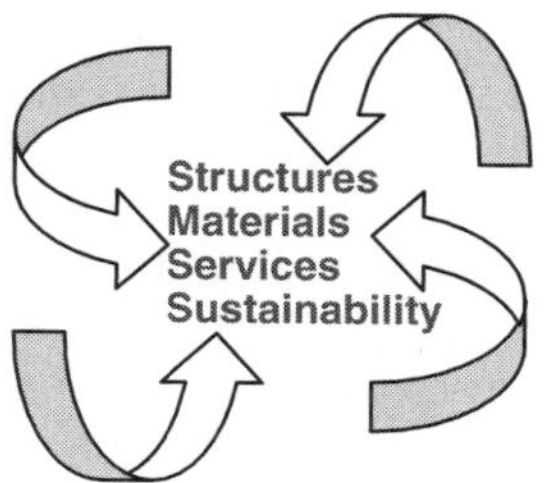

9

The Design: Function, Part 2 Design and construction constraints

9.1 Introduction

The generic consideration of functional requirements in the previous chapter was concerned with how buildings are, or can be used by people. This chapter considers some of the practical constraints inherent in the use of materials, the possibilities of structures and construction, and the way in which environmental controls and energy conservation are influencing the design of buildings. The next stage of the design development process attempts to apply an understanding of human and process needs with the potential means of satisfying them in the context of the practical demands of the project. The previous chapter explored the way in which the car-dealership might be used by staff, customers and visitors. Analysis of flow patterns began to indicate the relationships of activities to one another in principle only, unhindered by the need to limit or shape actual spaces. There are well-understood characteristics associated with the *building type* which may determine the sizes of some floor layouts and enclosed volumes, but in many cases, these will be determined uniquely in relation to the specific project, incorporating briefing from the client and others in the design development team. Various technical constraints will influence the choice of structures, materials and environmental controls, and political initiatives concerning sustainability and energy conservation may dictate other aspects of the design. The nature of the site itself can suggest or require certain forms of construction and any decisions will inevitably be governed by financial control. The statutory controls referred to in Chapter 4 begin to dictate design decisions.

9.2 Size and dimensions

The identification of activity and their flow patterns has so far been considered in isolation, determining the content of possible activities and the broad relationships between them.

Establishing the main flows or movements will begin to indicate where activities should be located on the site and within the envelope of the building. The bubble diagrams used for this preliminary analysis can be drawn without regard to actual size or shape, but once relationships have been found which may be practicable, the diagrams can be developed further by introducing the element of scale, adding dimensions to the bubbles either in terms of overall size or to proportion with specific lengths and widths. For example, the showroom may be shown as a 25 m × 25 m square or a 40 m × 15 m rectangle. Factors governing shape will be discussed later in this section but determining size has two important implications.

Firstly, the anticipated movements, processes, work, storage, etc. must be given realistic sizes so that they can actually be accommodated. This is particularly important in the development of a 'real' plan from the schematic bubbles when the relative size of spaces to one another becomes significant. For example, a rough sketch drawn without regard to real size might show the toilets as being the same size as the showroom, or the reception and customer display area drawn the same size as a single toilet cubicle. Such discrepancies are of no value in conceptual design and cannot lead to any sort of sensible floor planning or consideration of what the building might look like.

Secondly, establishing the 'real' size of all the spaces will determine how big the building is going to be once all the elements of activity have been added together, generally referred to as 'accommodation'. This is a critical point in the design process because the result will decide if the site is large enough for the client's requirements and, even more significantly, if the client's budget is large enough for the preferred building. Consequently, there are two possibilities when it comes to establishing 'real' sizes:

- *Add up all the defined space sizes to arrive at an overall size and cost*
  This method is based on the dispassionate assessment of the client's needs following research and analysis of their requirements, activities and processes. In terms of cost it is 'open-ended' and may lead to the design of a building which is either too big or simply too expensive.
- *Generate floor areas to fit into a predetermined overall size or to match a maximum allowable budget cost*
  It is more common to find that reductions, restrictions or compromises in 'real' sizes must be accommodated whilst still maintaining the integrity of the design. In practical terms, the total size of spaces *is* limited by space available on the site and *must* match the client's budget.

With regard to economy, the quantity surveyor (QS) can equate the client's budget to a possible '*x*' square metres of floor area or '*y*' cubic metres of volume. If this is taken as the projected total cost of the building, then clearly it must equal the sum of all the separate parts. However, if the list of accommodation adds up to x or y + 25%, then either some elements must be omitted, reduced in size, or combined with other elements so that the overlap space is only counted once. For example, to define a showroom space and sales office spaces as separate areas might add up to x + 25% floor area, but a showroom with sales office space as open plan within one single space still adds up to '*x*' floor area, meeting the desired target requirement.

Assessing potential floor area on the basis of cost requires some determination of relative value, in the sense of needing more or less expenditure in relation to the nature of the accommodation. For example, the costs associated with the quality of space and finishes required in the showroom area are likely to be greater than that in the workshop area. Notwithstanding the cost of equipment which must be added to the cost of any construction, cost planning may dictate floor areas in line with the available budget, apportioning capital investment between public and private spaces in a pre-determined ratio. A 50/50 cost split may equal a 30/70 floor area split, which is a useful indicator of how large spaces can be.

In any event, the space provided must be at least the minimum for the desired activity. Returning to toilets once again, there are recognised minimum standards for each type of publicly accessible building for male, female and disabled use, including the number and size of cubicles, the space needed for urinals and washing basins, and for entrance lobbies and circulation. The minimum floor area for this element of accommodation is easily determined.

Other activities are more flexible and floor areas can only be established in relation to the use of the space. For example, consider the internal showroom. Analysis of its function will begin to add up to an actual size. The sizes given in this example are entirely fictitious, but not totally unreasonable as an indication of the process.

How many cars will be displayed?	
9 @ 15.0 m² each	135 m²
How much space is needed to walk around them?	
1-metre pathway around each car	
9 @ 18 m² each	162 m²
How will they be brought in and taken out?	
Additional space equal to three cars	
3 @ 15.0 m²	45 m²
How many sales staff are employed?	
5 @ 6 m² each	30 m²
Customer reception space?	
2 Receptionists @ 6 m² each	12 m²
4 Customers @ 2 m² each	8 m²
How do they relate to each other?	
A desk area of 12 m²	12 m²
How is the sale concluded?	
2 Private areas of 9 m² each	18 m²
What do customers do while they are waiting to speak to a member of the sales staff?	
Waiting area of 45 m²	45 m²
Are refreshments available?	
Drink machine area	5 m²
How do incidental sales feature?	
Display/information space	20 m²
Initial total:	**492 m²**

Say for design purposes: 500 m². This figure is a net internal floor area, exclusive of space needed for structural elements, like walls. It is also net of circulation space needed to get from one activity to another. One way of accounting for this space is to add a small percentage to the above figure, say 10 per cent. Therefore, total desired floor area is now approximately 550 m² for the showroom activity. Does this match the budget allowance or must elements be revised? If a showroom of 550 m² is too large or too expensive, which of the activities described above can be reduced in size, combined with others or eliminated?

Additionally, economy of design may be related to the shape of the space. Is it to be square, round or rectangular? If it is rectangular, what proportion should it be? Are any of the activities likely to take place within the space easier or more difficult to undertake because of the chosen shape? This may not become apparent until a shape is suggested, which could, for the sake of argument be a 2:1 rectangle. The elements of accommodation defined above can be arranged within it to see if they all fit, bearing in mind at the same time the possible appearance of a structure which is a 2:1 rectangle on plan. Maybe 3:1 or 1:1.75 would be better starting points or maybe a 1:1.5 rectangle is the actual result of adding all the elements together.

This process can be repeated for each of the defined activity areas: office, administration, management, parts, workshop, parking, external display, etc. Each can be developed into an actual size, based on the original bubble diagram, related to each other as determined by the earlier analysis. A possible sketch layout for the showroom is illustrated in Figure 9.1. At this early stage of producing sketch possibilities the point is now reached when it is possible to see if all the elements of the new building fit together in a meaningful way as shown in Figure 9.2. This is the point when reference to three dimensional design and appearance becomes fully apparent. Does the chosen combination of spaces offer the possibility of a simple, recognisable three dimensional form, or is it better to be a collection of individual, identifiable forms brought together? Are there any odd elements on the plan which do not fit? How is one element which is left sticking out at the side to be viewed? Is the building a collection of parts, self-generating,

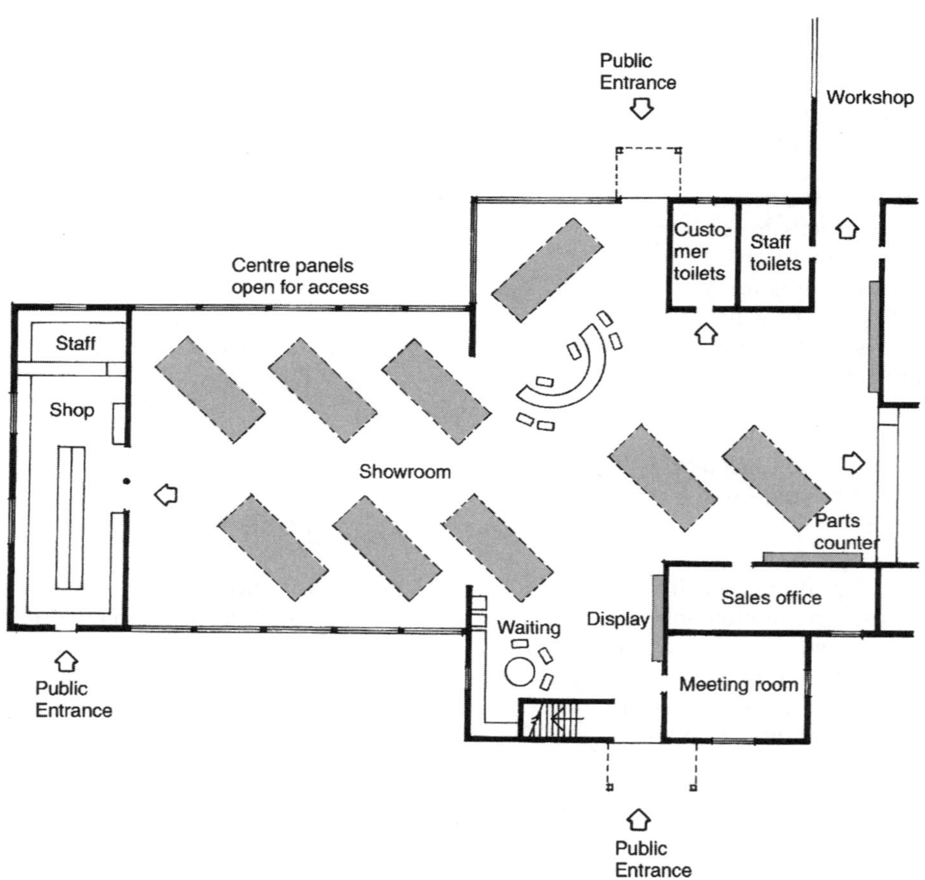

Figure 9.1 A sketch layout for the showroom.

or is it a desired form into which all the bits should fit? Is the odd space to be like a small shed attached to the side of a larger one, with its own walls and roof? If so will it look satisfactory, or will it look like an afterthought? Will it introduce an element of interest or contrast that is actually valuable, or will it conflict with the desired form and spoil the appearance? Consider also the economy of construction, are the additional walls and roofs unnecessary and extravagant? Are the construction details with valley gutters and flashings creating added risk of building failure through rain penetration?

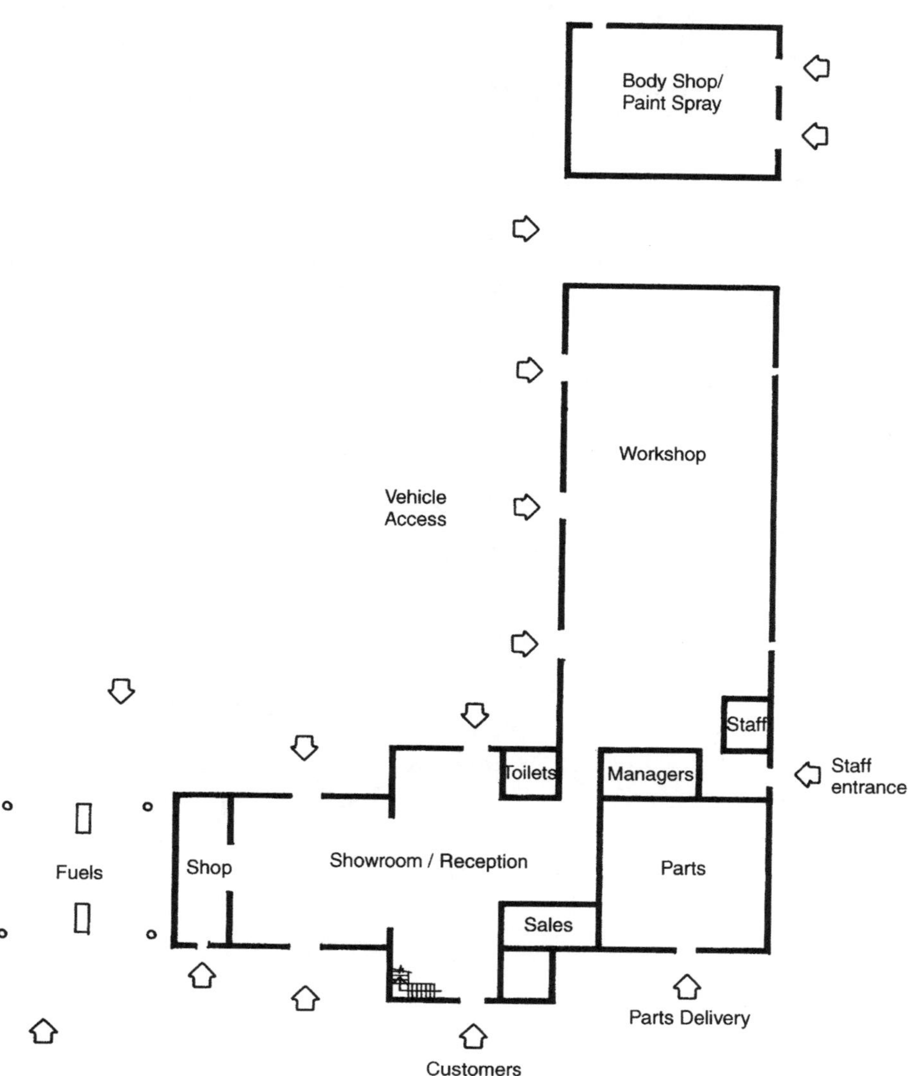

Figure 9.2 A sketch layout for the whole building.

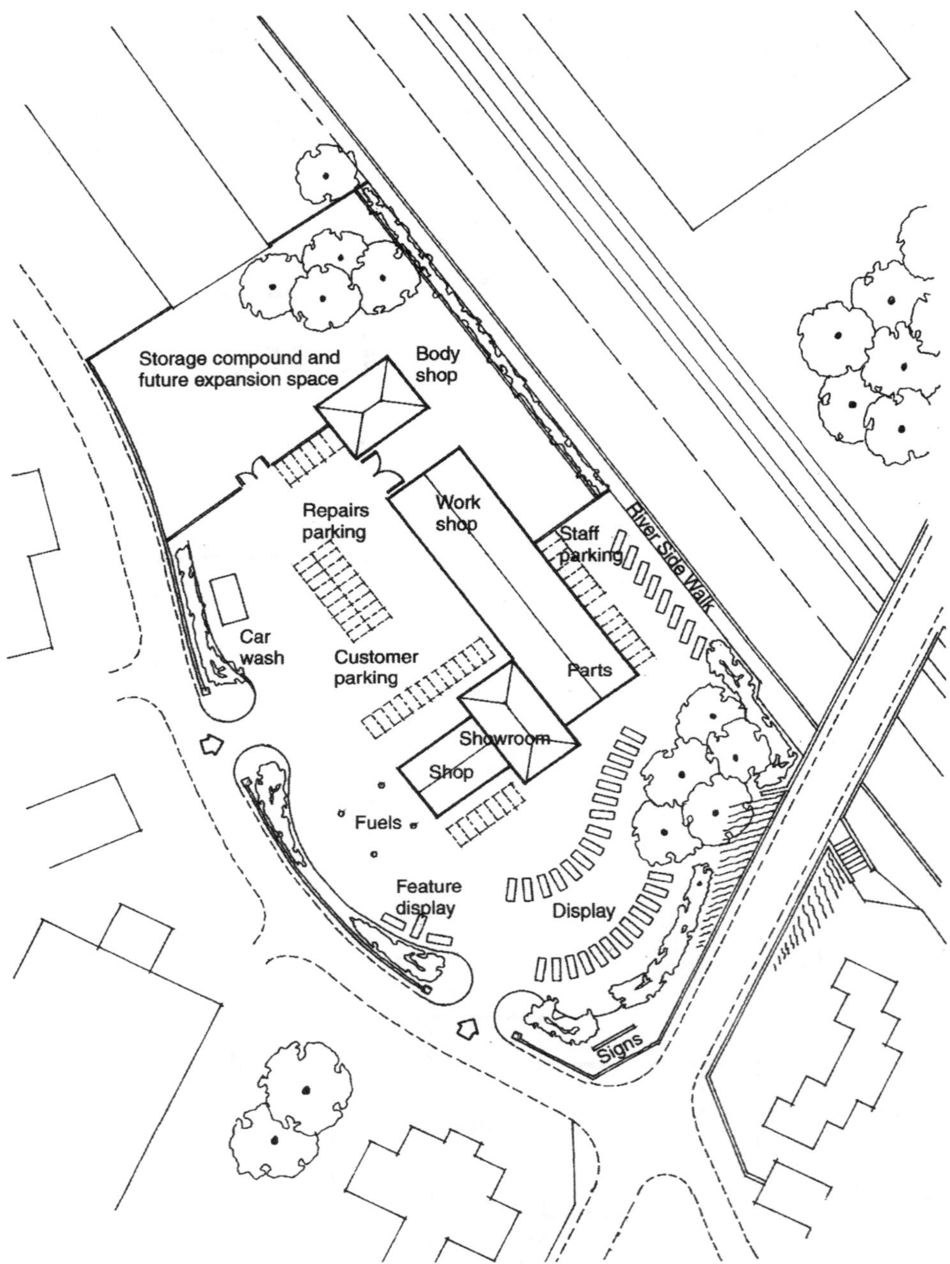

Figure 9.3 A sketch layout for the site.

The design evolves from a consideration of all the elements of the car-dealership. The early bubble diagrams lead to preliminary layouts for individual spaces, all the spaces collected together, and the development of the whole site. A possible sketch of the site layout is shown in Figure 9.3. The analysis described above can lead to a number of different solutions, with or without practical merit. Each one can be checked to see that issues which have already been stated as requirements or established as priorities have actually been incorporated or achieved.

9.3 Construction constraints

Exploring possibilities at the beginning of the design process cannot be divorced from the realities and consequences of construction. Ideas about forms, structures, materials and details can be imaginative and appear quite convincing on sketch design proposals, but in reality may be impracticable and uneconomical. Proceeding with such ideas may depend on the views of other members of the development team with regard to the viability and/or practicality of developing them. The client in particular, must be convinced that the ideas are appropriate for them, and that their development is likely to be worthwhile. For the majority of design projects, there are four important considerations as ideas emerge:

(1) Ideas should be capable of being designed (eventually) in detail.
(2) Ideas should be capable of being built (eventually).
(3) Ideas should not involve wasteful or unnecessary expense.
(4) Ideas should not carry a disproportionate risk of failure.

For example, the use of parapet walls to obscure rainwater guttering for visual effect can be a poor detail when compared to a traditional overhanging eaves. The construction of a parapet wall with damp proof courses (dpc), drips, cavity trays and flashings is complex and costly. The probability of subsequent weathering damage and water penetration is a significant future problem which the client should not be asked to deal with without very good reasons. The decision to use parapet walls is therefore dependent on the certainty that they can be constructed so that there is no risk of failure and that the additional expense is merited for aesthetic reasons.

The theory and practice of building construction influences all design decisions at each of the stages of building work, from excavations to decorations. Detailed factors and methodology can be studied in the many excellent publications already available, but the following general observations about design considerations for elements of construction should be borne in mind.

Materials, components and assemblies

Building construction combines basic materials with manufactured components. Sand, cement and aggregates are used in a 'raw' state whereas stone, slate and timber are 'worked' creating components with attractive, durable finishes. Clay, plaster, glass and plastics are moulded or cut into units, creating bricks, blocks, tiles, sheets and boards or into special shapes for pipes, extrusions and sanitary ware. Various fluids are used for gluing, jointing, protecting and decorating. A typical building will commonly contain the following range of materials:

- *Cementitious*: cements, lime, aggregates for concrete and blocks, sands for mortar, screeds and renders.
- *Burned clay*: bricks, roof tiles, ceramic and terra cotta floor and wall tiles.
- *Stone*: aggregates, hardcore, random rubble, ashlar, slate.
- *Gypsum products*: plaster, plasterboards and coving.

- *Glass*: windows, doors and mirrors.
- *Timber*: softwood, hardwood, processed boards, tongued and grooved floorboards, chipboard and hardboard.
- *Metals*: *steel and alloys*; rolled, stainless, galvanised, reinforcement, structural members, nails, screws and sundry fittings.
- *Aluminium and alloys*: door and window frames, ironmongery, rainwater goods.
- *Copper and alloys*: brass, bronze and cupronickel.
- *Lead*: flashings, dpc.
- *Plastics*: unplasticised polyvinyl chloride (UPVC), polythene, sheet vinyl, rubber and laminates.
- *Glass fibres*: insulation and canopies.
- *Porcelain*: enamelled sanitary goods.
- *Asphalt and bitumen*: dpc and roofing materials.
- *Joints and seals*: mastics and neoprene gaskets.
- *Paints and coatings*: glues, paints and stains.
- *Paper*: linings and decorative wall coverings.

Materials, components and assemblies can be defined or assessed at a number of stages in the lifetime of the building: design specification, manufacture, delivery, handling and manipulation, assembly, fixing, finishing and in use. Factors which shape performance levels include the location and degree of visibility, and the extent of exposure to aggressive forces within the completed construction. Traditional practice is a guide to performance, but 'innovative' or novel solutions to building construction cannot alter basic characteristics of materials. In many cases, performance is related to the intrinsic features, composition, method of manufacture, size, shape and cross-section. Cohesiveness, response to compression, tension, bending, sheer and impact forces, resistance to permeability, absorption, expansion and contraction, and the smoothness or abrasiveness of finished surfaces are all matters of fact which influence the selection of materials as well as determining how they may successfully be combined.

As well as intrinsic qualities, design requirements present a range of choices, which can be seen as essential, desirable or preferable, and which demand assessment in advance of deciding how they may be achieved. For example, the need for finished construction to be strong and stable, able to withstand imposed loads without moving and cracking is an essential aim, which can be achieved in a variety of alternative ways. Other aims include resistance to the hostile effects of the climate, water, frost, wind, sunlight and atmospheric pollution so that deterioration, weathering, and normal wear and tear can be reasonably controlled by occasional or regular cleaning and/or economical maintenance over a planned lifetime. Construction may be permanent or may offer flexibility permitting partial or complete dismantling and reassembly.

Construction must be safe and comfortable, minimising the risk of injury through fire damage, structural collapse and spread of dangerous fumes. Damages may be caused too by abrasive chemical and biological attack, and by unwanted fungi and insects which can be extremely destructive if undeterred or undetected. The control of excessive heat loss, humidity, solar gain and sound transmission have a bearing on acceptable comfort levels, and construction should generally be seen as attractive and aesthetically pleasing. All of the above are further influenced by availability, time programming, skill levels, economical capital cost against maintenance costs and depreciation, and personal perceptions of quality. See the product design wheel in Chapter 2.

On site, consideration must be given to the implications for labour and plant, wastage, theft and vandalism, testing, inspecting, delivery and storage. Supply issues such as sourcing, packaging, conditions of delivery, continuity of supply, solvency, factory testing and guarantees, technical support and quality assurance are all pertinent to the selection and specification of materials, components and assemblies. For some materials and components, the process of manufacture causes unavoidable variations in dimensions and colouring, which may be regarded as inherently attractive if handled correctly by careful selection, mixing and anticipation of tolerances. Tolerances at junctions between materials and components are affected by ease of cut and fit, fragility, protection of exposed surfaces, simplicity of shape, repetition of patterns, ease of handling, laying, fixing and fitting, time and skill.

The potential for failures is a significant issue in the construction industry, for the professionals involved, and for clients, customers and users. The possibility for failure ranges from an unexpected breakdown of the basic material through to an inability of a whole system to perform as anticipated. Materials may be unable to withstand local conditions, breakdown, lose cohesion and develop unplanned voids. Chemical reactions may cause efflorescence and corrosion, differential movement cause fractures at jointing boundaries and loss of structural integrity cause settlement cracking. Systems may come to the end of their natural, useful life and be replaced through planned maintenance. Quality assurance at the design and construction stages is essential to maximise lifetime and minimise potential for defects. See also Chapter 11, which considers how construction information describes materials, components and assemblies. The following sections consider design constraints at the various stages of construction from demolitions through to completion of decorations.

Demolitions and clearance

It may sound odd to start a section on construction constraints with 'demolitions', but the increasing reuse of 'brownfield' land, development sites which have been used before, and which already contain buildings, means that the building designer may be involved with renovations, alterations and extensions, or even with demolition and total site clearance. Considerations include:

- Potential 'useful life' of structures and infrastructure on site in terms of financial viability.
- The extent to which historical or architectural merit may influence retention.
- Concern for effects on wildlife and local ecology.
- Protection for retained items of value on site and on adjacent property.
- Health and safety issues for anyone involved on site.
- Possibility of recycling useable demolition waste such as crushed concrete or steelwork.

Excavations and earthworks

Construction below ground is an expensive element of the new building, often not fully appreciated because once completed, it is invisible. Considerations include:

- Protection for mature trees and retention of reuseable top soil.
- Unless the site is absolutely flat, it is likely that the existing ground surface levels will be unsuitable for the building and the site works around it. Initial excavation requires a careful balance between required internal and external floor levels in relation to the natural levels of the site. As large buildings may require a continuous, level internal floor, changes in the natural profile of the land to accommodate slopes and depressions may be unavoidable, either cutting into the ground or building it up. The way that the building is placed on the site can significantly affect costs of construction depending on the extent of proposed reshaping.
- Cut and fill is generally acceptable for landscaped areas of the site, but excavated fill cannot be used under buildings or areas accessible to vehicles where the quality of compaction is important.

- Problems of excavations adjacent to site boundary (see section in Chapter 3 about undertaking a dilapidations survey). Retaining walls may be needed to hold up parts of the site or neighbouring land and buildings or to contain basements.
- The extent of excavations needed may also depend on any contamination of the sub-soils and the level of the underlying water table.
- Poor ground and deep excavations, supports, shoring, drainage/flooding/pumping.
- Waste disposal/storage/reuse.
- Soil conditions, etc. for foundation design (leads to next section) impurity removal, contaminants, carboniferous material.
- No works to take place beyond site boundary or to interfere with adjacent property. See the Party Wall Act on the web pages of the Office of the Deputy Prime Minister.

Foundations

The purpose of foundations is to transfer the weight or load of the building and everything inside it to the ground, in such a way that the completed building remains stable. Foundation design must balance the forces imposed on the ground by walls and columns against the strength of the sub-soils below ground so that there is a minimal subsequent movement. For large buildings like the car-dealership, this is a specialist task undertaken by the structural engineer, who will liase with the Building Control Officer of the Local Authority to agree the design of foundations and supervise their construction. Foundation design is based on the knowledge gained from trial holes or boreholes. However, these are not always totally conclusive, and it is common to find in due course that excavations will reveal local problems such as soft or hard spots, rock, areas of contamination, sulphates or coal. In this event, the design of the foundations must be modified as the problems are discovered. This is one element of the new building that cannot always be accurately costed in advance of construction, and it is usual to include a provisional sum to cover any unforeseen additional work. Considerations include:

- The system adopted for the superstructure will influence foundation design. Traditional small-scale construction using brick/block cavity walls and partitions for small rooms, will generally have simple concrete strip footings, or a reinforced concrete raft.
- The need for large open internal spaces with wide floor and roof spans associated with a car-dealership, will require the use of steel or reinforced concrete frames. The point loads on each column will be supported on reinforced concrete pads, or piles for sites with particularly weak sub-soils and ground beams.
- Internal load-bearing columns or walls will have their own foundations or be supported off a thickened reinforced concrete raft.

Substructure

The substructure is the construction above foundations up to the finished ground-floor level. Considerations include:

- For large buildings, construction below ground offers significant economy of land use, for basements, cellars, car parking or services, and for the car-dealership, pits in the workshop.
- Spaces below ground need special attention to detailing to avoid problems with damp proofing, ventilation and natural lighting.
- Space will be needed at first-floor level for access downwards.

Superstructure

The superstructure is the whole building above finished ground level: the structure, the walls and cladding, the floors and roof, and the doors and windows.

External and internal walls

The outside of the building is often referred to as the external envelope, protecting the inside spaces from the elements. Considerations include:

- Traditional masonry construction has limited load-bearing capacity for multistorey buildings and cannot cope with large open spaces.
- Under such conditions, the most common solution is to use load-bearing framed construction in reinforced concrete or steel, with non-load-bearing infill panels or cladding.
- The structural elements may be hidden, or exposed for aesthetic reasons or the need for fire protection.
- For multistorey buildings, the continuity of elements of structure will affect floor-to-floor relationships, circulation, lifts and staircase positioning.
- Spaces must be made available for distribution of services: pipes, ducts, drainage, etc.
- The positioning of openings such as windows and doors in walls may be determined to some extent by the structure of the building. For example, a roller shutter door cannot be located in the wall between columns where there is a diagonal brace and large areas of glazing must be designed with regard to the supporting structure either behind or in front of them.
- All the elements of the superstructure must be designed to prevent rainwater, damp and uncontrolled cold air entering the building, and excessive amounts of warm air from leaving it.
- Internal spaces must be ventilated and surfaces kept warm to prevent condensation.
- Framework close to the site boundary requires special detailing to avoid damaging adjacent property in the event of collapse.

Floors

Floors are deigned to withstand anticipated superimposed loadings, with in-built factors of safety. Structurally, they are either in situ or suspended. Considerations include:

- Loading requirements
- In situ or suspended
- Self-finished/applied finish
- Wearing capability

Roofs

The principle function of any roof is to keep rainwater out of the building and to collect it, so that it can be transferred to drains at ground level. Considerations include:

- Construction must also minimise heat loss which can be considerable as warm air rises to the highest level.
- The appearance of the roof is a major aesthetic consideration.
- The roof can be flat with a minimal fall for rainwater run-off, or can be sloping from a minimum 5 to 45 degrees + pitch depending on the structural design. Buildings with large roof spans will have a shallower pitch to reduce the volume within the roof space, which may not be useable.
- Large roofs can also incorporate translucent panels to let natural light into the interior of the building.

Openings: external and internal doors and windows

The size and positioning of openings is related to the functional requirements of the plan, but is an important aesthetic consideration as well. The proportions of the openings and their relation to the whole elevation will be considered in the next chapter. Design decisions are needed on:

- Views in and out, vision and privacy.
- Opening size, fixed and opening lights.
- Materials for frames and glazing.
- Ironmongery.
- Secured by Design issues.
- Heat loss, solar glare, ventilation and passive solar gain.
- Fire protection and compartmentation.

Fixtures and fittings

Fixtures and fittings are elements within the building which are permanently built-in and are not intended to be periodically moved or repositioned including the following:

- Staircases, ramps, lifts and hoists.
- Benches and worktops.
- Storage spaces designed as units fitted into the structure of the building.
- For the car-dealership, hydraulic ramps and specialist machinery for servicing will require purpose designed housings, affecting the structure of the walls and floors.

Ironmongery

Ironmongery is the general term used to describe all the fittings needed to operate doors and windows. In almost all cases, they will be selected from a manufacturer's standard range, which will offer a multitude of options for size and style. The choice of ironmongery will depend on the requirements of each door or window for issues such as:

- The degree of security needed.
- Ability to withstand the wear and tear of normal usage.
- Hinges, handles, locks, bolts and automatic closers.
- Choice of materials including naturally finished steel, brass, aluminium or coloured plastic.
- The quality of selected fittings can significantly enhance doors and windows, and colour coding can help to identify spaces. For example, in the car-dealership, sales, parts and repair spaces could be defined by different coloured door ironmongery including room or activity titles on push plates.

Sanitary goods, plumbing and internal drainage

Sanitary goods include all the fittings which are connected to water supplies and drains, and will generally be selected from manufacturer's standard ranges. There are many choices for size, shape, style and quality of finish depending on requirements for:

- Wc's, urinals, baths, showers, sinks and basins.
- Materials include vitreous china, steel, stainless steel or fibreglass.
- Connections to hot and cold water supplies and drainage outlets.

- For economy these are best kept as close together as is practicable to minimise pipe runs and heat loss.
- For multi-storey buildings spaces containing sanitary goods are best kept above one another to simplify drainage arrangements.
- Bathrooms, kitchens, utility rooms and toilets require ventilation to remove steam, water vapour and odours. Mechanical extract fans, required under current Building Regulations will achieve this, but it is useful if these spaces are located on external walls, so that natural ventilation can be obtained as well by opening a window.

Mechanical and electrical services

The provision of environmental services in a large building such as the car-dealership will be relatively complex and expensive, designed and installed by specialists based on current standards and practices. Systems and equipment must be co-ordinated in relation to spaces and surfaces with regard to their effectiveness and appearance:

- Power and lighting (normally electrical).
- Heating and hot water (gas, electricity, oil or solid fuel).
- Ventilation.
- Telecommunications systems.
- CCTV and security issues.

Detailed design work is often based on performance specifications, definitions of the maximum, minimum or general standards required for particular circumstances. For example, the required capacity and location of power outlets, the level of artificial lighting, the desired internal working temperature and the number of air changes per hour needed for the activity being undertaken can be determined as necessary so that the specialists can decide how they will be achieved.

Finishes, decorations and furnishings

As previously discussed in Chapter 2, the choice of materials for each element of construction can depend on whether or not the material is visible on completion. For example, carcassing timber used for a framed wall can be rough and relatively crudely fixed if it is to be covered with finished match boarding or plasterboard. A concrete floor under a sand cement screed need not be laid perfectly level or smooth. However, if the timber framework is to be visible then it must be made of finished timbers expertly jointed, and in the absence of a screed, the concrete floor must be power float finished ready to receive a chemical sealant.

Apart from considerations of the 'honesty' of using materials for their own inherent qualities, decisions for selecting them depend on a variety of factors including cost, practicality of construction, tolerances, the level of acceptable deterioration from wear and tear and weathering, and how easy it is to maintain them. Some elements such as facing bricks, floors and ceramic tiles, for example, are generally regarded as permanent and not expected to be decorated or changed for some time. They must withstand attack from anticipated sources if they are to remain in a reasonable condition. On the other hand, walls and ceilings might be periodically refreshed or changed to create a new internal environment. The most common surface finish for these two elements is plaster, which is easily repaired, accommodates an uneven rough masonry substructure leaving a smooth hard-wearing finish which can be painted or papered as required.

For the car-dealership, all these possibilities could be incorporated. For example, the showroom could have high-quality pre-finished materials such as glass, steel, aluminium, brick or stone requiring little, or no decoration or future maintenance. The workshops might be constructed using more economical fair-faced blockwork, perhaps sealed or painted so that they can be re-decorated if and when they get dirty. The offices could be plastered so that unsightly electric conduits can be concealed and so that colour schemes can be easily changed by future occupants.

Achieving high-quality finishes in new buildings can be a difficult proposition. Conditions on site are not always conducive to obtaining the degree of accuracy and tolerance that had been envisaged. For example, the demand for fair-faced blockwork with perfect blocks and consistent mortar jointing, may be an unreasonable expectation of the quality of the supplied materials and the ability of the block layers. It is much more likely that a plastered wall will result in a finish with which both specifier, contractor and client will be satisfied. The building designer should always beware of alienating builders by asking for the impossible when it is their own detailing that is responsible for introducing difficulties.

The selection of incidental furnishings such as carpets, curtains and furniture is usually a matter for the client or building users. For large buildings the services of an interior designer may be used to select finishes co-ordinate colour schemes. It goes without saying, of course, that this is the element which may be most apparent and all the good design work so far could be spoiled by inappropriate finishes.

External works

The external works are usually the last part of construction, completed just before handover of the building to the client. The desolation and waste, the rubbish and mud will be cleared away, and the spaces around the building transformed with a mixture of hard and soft landscaping:

- Large areas of hard surfaces, particularly concrete and tarmac can look very unattractive, especially if laid out in regular geometric patterns associated with vehicles and car parking.
- Harshness can be reduced by introducing curves, changes of materials and by interspersing planting beds and trees. Sufficient space must be given to accommodate plants as they mature.
- Small areas should be avoided, particularly tapering strips, which are not big enough to support vegetation and difficult to maintain.
- Species should be selected which are suitable for their location and which can be reasonably looked after.
- The position of inspection chambers must be considered for convenient access.
- Drop kerbs installed for disabled access.
- External lighting contributes to safety and helps to reduce crime.
- Boundary walls and fences, gates, security gates.

9.4 Environmental and communication services

One of the basic aims of building design, described briefly in Chapter 2 is to create an internal environment to meet the needs of the building's occupiers. Generally, the internal spaces must be designed to exclude wind, rain and damp and to keep them warm, ventilated and well lit so that occupation is comfortable, enjoyable and efficient. Specific environmental needs vary between individuals, activities, building types and geographical location. For example, one person will need or prefer a higher temperature in their house than another, and a comfortable working temperature for sitting still in front of a computer would be too high for someone doing a physical task, moving around regularly. The heating levels in a hospital must be maintained at much higher levels than those in a warehouse, and buildings in the UK generally, will need heating for longer periods in the year than those in the south of France.

Consequently, the building must be designed to provide a basic level of heating but include a flexible control system to accommodate desired variations. This is generally true of all the environmental services installed in buildings, with considerable attention being paid to minimising the consumption of energy and costly resources.

The principal environmental services in buildings are as follows:

- *Space heating and hot water supply*
  Traditionally, heat has been generated by burning gas, solid fuels and oil or by various electrically powered systems. These radiant energy sources have traditionally had little effect on the plan form of buildings, where heating systems have been regarded as add-ons, subsidiary to the spatial design itself. Continued political and financial pressures to conserve energy and minimise wasteful heat loss are leading to improvements in thermal insulation techniques, but the development of equipment to harness the natural renewable resources of the sea, wind and sun is likely to lead to changes in the way that buildings are designed and managed in the future.

  With respect to the individual building, the most significant developments are taking place with solar power, either with direct, passive systems, where the orientation and form of the building is designed to take advantage of penetration of sunlight to generate warmth, or with indirect systems using attached or free-standing mechanical equipment such as solar panels fixed to the roof.
- *Power*
  Machinery and equipment is powered by electricity designed and installed to suit the loads expected for the use required. This can range from a single-socket outlet to run a kettle to a complex three-phase installation to feed powerful machines.
- *Lighting*
  Artificial lighting is usually powered by electricity and will be selected to provide a variety of lighting levels as appropriate to undertake tasks.
- *Ventilation*
  Mechanical ventilation, powered by electricity ranges from extract fans in bathrooms and kitchens to remove the build up of water vapour in the air, to full air-conditioning systems to introduce fresh air or re-circulate treated air. Passive systems of air circulation encourage the movement of air through buildings as a result of internal and external temperature differences and wind-induced pressure differences, improving air quality at no cost in an environmentally friendly way.
- *Cold water supply*
  Cold water is supplied directly from main feeds or from storage tanks to individual outlets like toilet cisterns, sinks, showers and baths, or to some industrial and manufacturing processes.
- *Telecommunication links*
  Communications today are through the telephone system, computers, fax machines and the Internet. At present these operate via fixed lines or cable connections brought into the building and distributed to individual outlets. A separate connection is normally required to receive television transmissions.

The design of complex installations and control systems is undertaken by specialist consultants or subcontractors who can ensure that proposals are correctly designed in accordance with current legislation and standards, and who are familiar with current theory and technological advance. The design of some installations is closely related to the design of the building itself, often requiring joint decisions affecting both. The heating system, for example, is related to the size of volumes to be heated, potential heat loss through the structure, the number and size of doors and windows, and their orientation to the sun and wind. It is affected by the number of people working in the space and the heat produced by equipment. The lighting systems may provide a basic minimum level of illumination for set tasks, but the location and style of fittings and additional light output are matters which may need further thought in order to achieve certain desired effects. The appearance of equipment and the means of feeding it are also important considerations. Essential ductwork, pipes and cabling can be very unsightly. Means of easy maintenance and convenient replacement are needed together with the possibility of relocation without causing major disturbance.

Electricity, gas, cold water and telecommunications links are supplied to buildings by service providers, who are currently all independent private companies. Generally, underground services are brought onto site from local mains, sometimes via new substations and terminated at stop taps or meter boxes. The service providers will modify their existing infrastructure where practicable to supply the new feeds. The cost of this work depends on the availability of existing services adjacent to the site and the extent of works required. The installation and connections beyond the stop tap or meter box are the responsibility of the client, designed and installed by the specialists. The service providers are able to offer advice about installations.

The principal services consultants were listed in Chapter 2. The main implications concern the design and location of the following:

- Plant and equipment, the floor area needed to accommodate it and how access is to be made available.
- Ductwork, pipework and cabling.
- Power outlets.
- Ventilation outlets, grilles, flues and chimney stacks.
- Fixtures and fittings.
- Storage tanks for water, chemicals and gases.
- Specialist supplies such as compressed air.
- Integration into design and construction programmes.
- Cost provision (see Chapter 11 for details of Prime Cost Provisional Sums).

Although the technical design content and the physical constraints of installing systems are a matter for the specialists, the building designer can play an important role as co-ordinator, ensuring as far as is practicable that the elements of function and appearance are compatible with one another. Positioning and feeding equipment and outlets is a complicated business and sometimes very difficult to visualise until the installations have been completed. It must be decided in advance if functional elements are to be visible or 'hidden'. Visible service installations can be attractive or interesting, adding to the appearance of the building, like the exposed overhead pipes and ductwork in fashion in many modern student union bars. Alternatively, they can spoil the appearance and look like an afterthought. If they are to be hidden, space must be made available in the design for suspended ceilings or ducts often seen in large open-plan offices or public toilets to conceal the plumbing arrangements. The choices may be matters of economy, client preference or design intent, but as mentioned elsewhere, is rarely successful if it occurs by default. The development of the design for spaces and surfaces should acknowledge the constraints of environmental services at the earliest possible stage.

9.5 Sustainability

The concept of sustainability is becoming increasingly significant as the relationship of the natural and built environment at both local and global levels becomes better understood. Until relatively recently, concern for 'the planet' was a matter for fringe protesters, responding to the escalating late twentieth century destruction of the environment and dwindling energy resources. Now there is genuine political concern, although not agreement, about many issues extending into areas of life that are being seen to be diminished by overexploitation, waste or simply social breakdown. Matters of concern include the following:

- Greenhouse gases and global warming.
- Fossil fuel energy consumption.
- Renewable natural materials.
- Recycling waste.
- Community behaviour.

Serious thinking about sustainability emerged in 1983, when the United Nations (UN) recognised the need to address the worldwide conflict between the interests of economic development and its apparent harmful effects on the environment. A UN commissioned report, chaired by Norwegian Prime Minister Gro Harlem Brundtland set off a chain of events. 'Earth summits' were held in Rio de Janeiro (1992), also known as the International Climate Change Convention; Kyoto (1997), which generated the Kyoto Protocol, and Johannesburg (2002) on sustainable development. The UN's worldwide programme is known as 'Agenda 21', involving governments and groups in every area of human activity which has an impact on the environment.

In the UK, energy consumption had been regarded as a natural right, as if resources were limitless and pollution an irrelevance. According to the Building Research Establishment, 'buildings account for approximately half of the UK carbon emissions' and yet until very recently, the commissioners and designers of new buildings have generally settled for transferring sustainability problems onto future occupants as a consequential cost of occupation. Little attention was paid to procurement of building materials from renewable sources, disregarding the destruction of rainforests for hardwood as 'being nothing to do with us!' Similarly, waste in the construction industry has always been seen as a 'fact of life'; an inevitable consequence of the way things have always been done.

Add to this the location of new development encouraging the use of the private car, particularly to 'out-of-town' shopping centres and for new residential development. In the UK, congestion on roads is exacerbated by the morning and evening rush hour traffic flows. Despite the urban regeneration taking place in most towns and cities in the UK, there remains great pressure on green belt land, constantly leading to the widening of the boundaries of the built environment, which is not sustainable in the long term. The politics of land use is somewhat beyond the scope of this book, and of course, beyond the control of the building designer, who inherits the site from his or her client. With regard to energy conservation, building design and internal environmental controls for an individual project, there are two conflicting issues which are the subject of much debate and concern.

The increase in user demand and expectation

The standards of environmental comfort achieved in buildings today bears no comparison to those constructed even a relatively short time ago. Few houses in the 1950s would have contained central heating, tumble dryers, dishwashers, microwave ovens, freezers or instantaneous showers. Few would have had more than one bathroom or conservatory. Yet today, all these features are commonplace, almost standard. They are taken for granted along with the massive increase in energy consumption that goes with them. The rapid advance in the design and manufacture of equipment is adding more and more complex installations and controls to further fuel user expectation. Closed circuit video surveillance cameras, electronic gates and garage doors which open automatically on arrival are becoming ever more familiar, and without doubt, even more innovations are on their way.

The increase in capital, maintenance and running costs

All these features, together with all the other materials, fixtures and fittings in any building are not only relatively expensive to purchase, but their manufacture consumes scarce resources. There is an energy cost to making and distributing them, to maintaining and disposing of them when they become obsolete. There is an 'embedded energy' in the process of transforming raw materials to finished, installed products. There are also 'running' costs, consuming energy generated by burning non-renewable fossil fuels, which is polluting the atmosphere to an unacceptable level. The Building Research Establishment's Environmental Assessment Method (BREEAM) has been used since the mid-1990s as a way of assessing the environmental performance of both new and existing buildings, including offices (eco), homes, industrial units, retail units, schools and others. It focuses on a wide range of issues including management, energy use, health and well-being, pollution, transport, land use, ecology, materials and water consumption. Each area is scored to give an accumulated rating of pass, good, very good or excellent.

Of course, an individual can choose at present to determine the energy consumption in their own building, regulated only by their ability to pay for it. But when this situation is multiplied to include all the buildings in a town, city, country and the world, it is easy to see that it is unsustainable and rightly the subject of considerable political intervention. Building design can make a contribution to the energy costs of both construction and occupation, which in many cases also offers opportunities for marketing design proposals to clients and users.

The reasons for energy conservation include the following:

- Counteracting rising fuel costs.
- Reducing commercial overheads and improving competitiveness.
- Contributing to the reduction of 'greenhouse' gases.
- Complying with legislation.

The consequences for building design and designers could include the following aims:

- Choosing the elements of the building's fabric so that their performance lasts for the lifetime of the building.
- Specifying materials that are obtained from renewable sources.
- Reducing uncontrolled air movements and heat loss through openings.
- Considering the location and orientation of openings with respect to heat loss.
- Relating buildability to the difficulty of site conditions.
- To ensure specification complies with recognised standards.
- Ensuring construction follows specification.
- Keeping up to date with technological advance and ensuring that the client is aware of the cost advantages of particular forms of construction and usage.

- Treating the design of buildings in an holistic manner, responding to their environment in ways which allow modern technology to control it.
- Seeking specialist advice about appropriate energy conservation measures.

Other areas of interest pertinent to building design include the following:

- *Green building technology*
  The term 'Green' is used widely in the UK to describe a political view of managing the environment. The use of materials and power systems which limit the use of finite resources is generally described today as being 'green'. Examples include solar and wind power, heat pumps, improved thermal insulation, sewage composting, recycling waste water, grass roofs, using materials from renewable sources and those which could be reused for other buildings at the end of their useful life.
- *The Climate Change Levy*
  To help the UK meet its targets for reducing greenhouse gas emissions, the Climate Change Levy (2001) taxes industry, commerce and the public sector (not domestic property) on its energy use with offsetting cuts in employers' National Insurance Contributions and additional support for energy efficiency schemes and renewable sources of energy.
- *Sick building syndrome (SBS)*
  SBS describes symptoms of ill health and discomfort linked with spending time in a building. Headaches, eye, nose or throat irritation, dry coughs, itchy skin, dizziness and nausea, difficulty in concentrating, fatigue and sensitivity to odours are thought to be caused by chemical pollutants and vehicle exhausts drawn into the building, or chemical pollutants given off by equipment, carpets and furnishings, or bacteria, pollen and mould breeding in drains or air-conditioning systems, or high, low or changing humidity and temperature levels linked with poor ventilation.

The complaints may be localised in a particular room or zone, or may be widespread throughout the building. In contrast, the term 'building related illness' (BRI) is used when symptoms of diagnosable illness are identified and can be attributed directly to airborne building contaminants.

Perhaps the most well-known building illness is Legionnaires' disease, a type of pneumonia caught by inhaling small droplets of water suspended in the air which contain the Legionella bacterium. Outbreaks occur from purpose-built water systems such as cooling towers, evaporative condensers, whirlpool spas, air-conditioning and industrial cooling systems, and from water used for domestic purposes in buildings such as hotels, where temperatures are warm enough to encourage growth of the bacteria.

- *Sustainable development*
  This is a concept of managing an economy to meet the needs of its people whilst ensuring that natural resources are maintained and enhanced for their benefit and for future generations. There is a concern that costs associated with construction and use are assessed at a community level:
- *Remediation of contaminated land*
  The redevelopment of previously occupied sites is becoming increasingly common. Defunct commercial undertakings can leave the site and its underlying sub-soils polluted with a variety of obnoxious chemicals, which can be difficult and very expensive to remove or neutralise. The remediation of contaminated land is the subject of much academic study particularly in

the assessment of risks associated with the nature of the treatment needed to ensure safe future occupation and use.

- *Gaia*
  Gaia is the Greek word meaning Earth Mother Goddess. Promoted by James Lovelock, a British biologist, Gaia is a theory that the Earth is a self-regulating organism where all its components, systems and creatures live together balanced in a homeostasis. They are all dependent on one another for their survival.
- *Feng shui*
  A Chinese principle that location in the universe affects our destiny. When buildings or interiors are 'harmoniously positioned in accordance with feng shui practice, life patterns are in balance and harmony with nature and the universe. Prosperity, health and equanimity follow.'

Make what you like of that!

9.6 Landscaping and planting

Landscaping is the shaping and forming of the spaces and 'hard surfaces' around the building, and may include the siting of the superstructure of the building itself. The main elements of landscaping are levels, shapes, materials, views, vistas, steps, walls, fences and surfaces including features like pergolas, arched openings and sculptures and occasionally, water in ponds, pools, fountains and waterfalls. Landscaping is the creation of the spaces around the building which for some building types is as important as the spaces within the building.

Planting is the treatment of 'soft surfaces', the choice and disposition of grass, shrubs and trees. Planting can have a major effect on the appearance of the building, enhancing some features and obscuring others, and it changes through the seasons and throughout the lifetime of the selected plants as they grow and increase in size.

The design of the external spaces around the new building should not be forgotten or left to the end of the design process as if it were a separate, decorative add-on, included as an afterthought to use up odd left over spaces. The analysis of activity and flow within the building, applies equally to the movements around the building as people come and go, work, sit and relax. Spaces can be created which are attractive and interesting, taking advantage of good views or concealing poor ones. To the passer-by and visitor, the visual appearance of the landscaping and planting can be every bit as powerful as the building itself, and in many cases may become the dominant feature.

Landscaping and planting can define movement around the site, limiting access to certain areas, and can influence security. A dense border of prickly shrubs deters access, but is equally a shield for illicit activity. Similarly, a high wall or fence is a good security barrier, but easy to hide behind.

Landscaping can be regarded as a specialism and there are designers who are expert in this area. For many building projects, however, landscaping is a part of the building designer's work, detailed and specified like any other part of the new building. Planting is an important element in landscaping as expanding trees and shrubs become significant 3D, sculptural features which together with areas of grass define shapes and volumes around the building. The building designer need not be a specialist in horticulture, but should have a concept of planting in terms of shapes and colours, and be able to appreciate the long-term effects of plant growth. The understanding of appropriate species, and the mechanics of planting and growing can be obtained from specialists.

The Planning Authority will take a keen interest in both landscaping and planting, particularly where they can make a contribution to the local environment around the site and for the benefit of the community. The formal Planning Approval will usually contain the condition that consent is subject to satisfactory planting being carried out on completion of the building and that it must be maintained until well established.

9.7 Cost control

The analysis of flow patterns and dimensional requirements for spaces to accommodate activity leads to conclusions about floor areas and volumes appropriate for the new building. Estimates of the cost of materials and construction can be prepared by the QS based on rates available in published books like Spons, or in relation to experience of local current trends. At this stage, the focus is on the capital cost of the general superstructure of the building, with internal floors and walls plus a notional cost for drainage, services and landscaping. As well as these costs, however, the client's budget must include for the costs of purchasing the site, specialist fixtures, fittings and machinery, decorations, furnishings and all professional and statutory fees. It is worth bearing in mind, for example, that the cost of purchasing a site for a single house can be the same as the cost of building the basic superstructure, and that the client's expenditure on furnishings, fittings and decorations could be as much as a further 20–40 per cent on top of the cost of the superstructure. A prospective purchaser could add even more to that by increasing the specification of standard items like baths, taps and kitchen fitments.

The client will also be faced with the continuous costs of running the business and maintaining the new building once occupied. Design decisions influence the capital cost of materials and labours but can be equally important with regard to the implications for future use. Considered selection of materials and details of construction, along with many other factors can increase or reduce the client's future expenditure on energy consumption, maintenance and replacement.

Detailed cost planning and cost control are major subjects in their own right and are beyond the scope of this book. The design team must, however, endeavour to ensure that the elements are selected with regard to the total budget available. This can be a delicate business as each person will have different priorities which may lead to decisions which are detrimental to eventual use. In many cases there is a need to consider the balance between capital expenditure and future running costs. A better, but more expensive material or finish will last longer than a cheap one. Investment in thermal insulation to reduce energy consumption represents a higher capital cost, but lower fuel costs, and may be self-financing within a relatively short period of time. In areas subject to a high level of usage, hard-wearing, long-lasting materials may be advisable, but conversely it may be more cost efficient to use cheaper finishes in the knowledge that they will be replaced or redecorated on a regular basis.

Significant considerations include the following:

- *Designing broadly in compliance with the client's budget*
  The client's budget is made up of the elements already listed above, but consider with a building like a car-dealership that the cost of specialist fixtures and fittings and 'prestige' materials for finishes could add a further 30–50 per cent onto the cost of the building itself (maybe even more for some specialist activities). It is quite conceivable that the cost of the building is less than half of the client's total outlay. Hence, if the cost of a feasibility study proposal for the

building was estimated to be 75 per cent of the client's budget, then clearly this is a profound constraint to proceeding any further with that idea.

- *Thinking about economy rather than wasteful extravagance*
  Careful consideration should be given as to whether or not design features have been introduced for 'effect' without proper reason. Complexity can involve considerable expense which may not be justifiable.
- *Considering the sensible use of space*
  Spaces should be arranged so that they are not too large or too small for the possible activities which they may contain, and avoid wasted space which cannot be put to good use at all. Every square metre of floor area costs money and wasted circulation space or 'strange' shapes on plan which cannot accommodate 'standard' equipment or furniture can quickly amount to an unreasonable percentage of the cost of the building.
- *Selecting economical structural systems*
  The enclosure of spaces and means of structural support have cost implications. For example, an expensive steel framework providing a large clear span could be made much more economical through the introduction of intermediate supports, providing that they can be fitted conveniently into the plan. Identical repetitive elements will be more economical than a mixture of special shapes and sizes. For example, regular column spacings and consistent portal frame spans will be cheaper to design, manufacture and erect.
- *Designing details and choosing materials which perform as required and do not deteriorate too rapidly in time*
  Specified materials must be capable of withstanding the effects of normal use, wear and tear, and the effects of weathering without deteriorating unreasonably quickly. Of course, the word 'unreasonable' is subject to interpretation, depending on what is expected. This matter is discussed further in the chapter on construction information and specification writing.
- *Creating environments that minimise energy waste and reduce the costs of maintaining internal environmental conditions*
  The design of spaces and the selection of materials and construction systems should contribute to limiting the waste of energy. Power consumption to maintain comfortable lighting and heating levels represents a significant running cost once the building is occupied.
- *Aiming to minimise the costs of future maintenance*
  The selection of finishes and design of installations should be such as to limit the need for redecoration, cleaning, repair and replacement to the minimum. This should include a consideration of the costs and difficulties of gaining access to high-level elements and the location of equipment like heaters which may need regular inspection and adjustment.
- *Carefully considering potential health and safety risks and aiming to minimise future problems as required by the Construction Design and Management Regulations (CDM)*
  All the element of the building should be designed in such a way as to minimise hazards during construction and at any time during subsequent use. This matter is discussed further in the section on CDM Legislation in Chapter 4.
- *Working towards the appointment of contractors, subcontractors and suppliers in such a manner that their costs are reasonably competitive*
  Depending on the client's chosen procurement method, the preparation of information about the requirements for the new building should ensure that it can be obtained at a competitive but realistic price. Documents should contain sufficient information so that the client is fully aware of cost commitments, ideally before construction begins on site.

- *Supervising construction so that additional costs are minimised*
  Contract administration must be targeted towards achieving the requirements agreed with the client before construction commences, so that extra costs do not escalate to a final account which may be beyond the client's means.

9.8 Project File content

The general aims at this stage are as follows:

- Translation of analysis bubble diagrams into relatively accurate, scaled design proposals.
- Development of proposals incorporating all necessary professional advice and input, in the knowledge that they can be achieved when examined in detail at later stages without destroying initial concepts.
- Offering design solutions which are broadly in line with the client's initial budget.
- Obtaining and incorporating all necessary statutory requirements.

File material could include the following:

Notes, letters and minutes

- Further details of communication with all members of the development team.

Briefing development

- Additional design criteria information improving the understanding of requirements.

Completed checklists

- Checklists of activities and spaces with complete design intentions, including environmental figures and calculations for power and heating, drainage and structural calculations. Standard Assessment Procedure (SAP) evaluation of energy performance.

Samples

- Samples of materials and assemblies obtained for discussion or presentation to the client for approval.

Details of consultations

- Local Authority Planning Officer for information about development possibilities and constraints including Planning Briefs and information about Planning Appeals for difficult sites.
- Building Control including information about ground conditions and drainage.
- Fire Officer including a fire safety plan showing how the requirements of the Fire Officer will be addressed.
- Highways Authority including details about access to the site, alterations to adjacent highways and highways drainage.
- Environment Agency including details of affected watercourses and contamination.
- Conservation or Civic Society, statutory undertakers, archaeologists, Parish Council, neighbours or local residents.

Applications to Statutory Authorities

- Completed application forms for Planning, Building Control and other interested authorities including appropriate fees.
- Copies of the necessary accompanying documents, such as drawings and supporting letters.

Legal agreements

- Relating to any aspect of the development including adoption of sewers (Section 104: Water Industry Act 1991), adoption of highways (Section 38: Highways Act 1980), improvements to local infrastructure outside the site boundary (Section 106: Town and Country Planning Act 1991), provision of services for water, building water, electricity, gas, telephone and cable communications.

Presentation material

- Drawings, including site plans, building plans, elevations and sections showing what the proposed new building will look like and illustrating significant elements of construction.
- Specially prepared illustrations such as perspectives, isometric and axonometric projections.
- Photographs of the locality and the site with superimposed images of the new building.
- CAD generated models of the exterior and interior of the new building.
- 3D models of the whole complex or selected parts.

9.9 Discussion points

(1) How can minimum and maximum floor areas be determined for particular activities? Are there optimum sizes catering for all possibilities and can flexibility always be offered to potential users? What are the implications for building users if designers 'get it wrong'?
(2) The useful lifespan of structures and materials is of considerable interest to designers, developers and building owners. What factors influence initial choices? Is it possible to design a maintenance free structure? Where is the boundary between acceptable weathering and unacceptable deterioration?
(3) What are spaces around and between buildings for? How do land values influence the provision of public open space in urban and rural areas? How is the use of space around buildings affected by the UK climate?
(4) Energy conservation is accepted as being essential, yet most of the UK's older buildings are very inefficient. How should this situation be addressed? Should owner-occupiers be encouraged or obliged to save energy? How can buildings be designed to take maximum advantage of passive solar gain?

9.10 Further reading

Anink D, **Boonstra** C and **Mak** J (1998) *Handbook of Sustainable Building.* London: James and James.
Baggs S and **Baggs** J (1996) *Healthy House.* London: Thames and Hudson.
BEDZED: Beddington Zero Energy Development. www.bedzed.org.uk

BREEAM: Building Research Establishment Environmental Assessment Method. www.breeam.org
Brundtland Report. www.brundtlandnet.com
Burberry P (1997) *Mitchell's Environment and Services.* 8th Edn. Harlow: Longman.
Building Research Establishment. *BREEAM Environmental Assessment Method.* www.breeam.org.uk
Building Research Establishment. *The Environmental Building.* http://projects.bre.co.uk/envbuild/envirbui.pdf
Centre for Alternative Technology. www.cat.org.uk
Crowe S (1994) *Garden Design.* 3rd Edn. Woodbridge: Garden Art.
Dean Y (1994) *Materials Technology. Mitchell's Building Series.* Harlow: Longman.
Earthtrends environmental information portal. www.earthtrends.wri.org
Edwards B (2003) *Green Buildings Pay.* 2nd Edn. London: Spon Press.
Edwards B and **Hyett** P (2001) *Rough Guide to Sustainability.* London: RIBA Publications.
Eurocodes Expert. www.eurocodes.co.uk
Everett A (1994) *Part 1: Mitchell's Materials.* 5th Edn. Harlow: Longman.
Glasson J, **Therivel** R **and Chadwick** A (2005) *Introduction to Environmental Impact Assessment.* London: Routledge.
Harris C and **Borer** P (2005) *The Whole House Book: Ecological Building Design and Materials.* 2nd Edn. Machynlleth: Centre for Alternative Technology.
Hockerton Housing Project. www.hockerton.demon.co.uk
Houses of the Future. www.housesofthefuture.com
International Institute for Sustainable Development. www.iisd.ca
Keeping M and **Shiers** D (2004) *Sustainable Property Development.* Oxford: Blackwell.
Lyons AR (1997) *Materials for Architects and Builders: An Introduction.* London: Arnold.
McMullan R (1998) *Environmental Science in Building.* Basingstoke: Macmillan.
NGS: National Green Specification. www.greenspec.co.uk
Poyner B and **Fawcett** W (1995) *Design for Inherent Security: Guidance for Non-Residential Building.* London: CIRIA.
RIBA Bookshop. www.ribabookshops.com
RIBA Library. www.architecture.com
Riley M and **Cotgrave** A (2004) *Construction Technology 2: Industrial and Commercial Building.* Basingstoke: Palgrave Macmillan.
Rogers R and **Power** A (2000) *Cities for a Small Country.* London: Faber.
Sattersthwaite D (1999) *Sustainable Cities.* London: Earthscan.
Secured by Design. www.securedbydesign.com
Seeley IH (1996) *Building Economics.* Basingstoke: Macmillan Press.
Sherwood Energy Village. www.sherwoodenergyvillage.co.uk
Stroud Foster J (2000) *Mitchell's Structure and Fabric.* 5th Edn. Harlow: Pearson Educational.
Ubbink (UK) Ltd. www.ubbink.co.uk
UK Government Approach to Sustainability. www.sustainable-development.gov.uk
UK Sustainable Development Commission. www.sd-commission.gov.uk
Vale B and **Vale** R (1991) *Green Architecture; Design for a Sustainable Future.* London: Thames and Hudson.
Wolley T, *et al.* (1997) *Green Building Handbook.* London: Spon Press.
Yeang K (1999) *The Green Skyscraper.* London: Prestel.

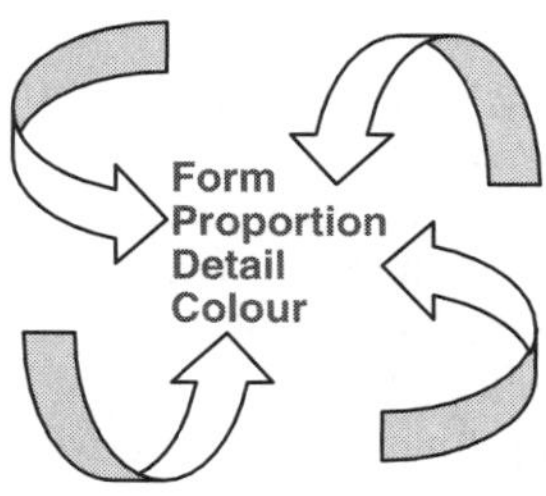

10

The design: aesthetics

10.1 Introduction

The visual appearance of any designed product is clearly of fundamental significance because it is the first, and lasting impression of the product, often establishing and sustaining its perceived merits. Buildings are no exception, exaggerated by the permanence of long life expectation for most structures, and whether a building functions adequately or not, it is its appearance which usually provokes most comment, especially for example as with many concrete buildings from the latter part of the twentieth century. Although deciding on the merits of an image may be seen as subjective, this chapter considers how the appearance of buildings is influenced by individual perception and interpretation. The building designer should be aware of the psychology of visual perception and understand the nature of the composition of a designed image. There is a visual hierarchy in building design from the large scale of the street scene to the small scale detailed design of a door handle, each affected by decisions about form, proportion, detail and colouring. This chapter concludes with brief consideration of the links between appearance and construction constraints, and how the appearance of buildings changes over time through the effects of weathering and unavoidable wear and tear.

10.2 The nature of aesthetics

Our awareness of the characteristics of the natural or man-made environment and all the features within it depends on the information received by our senses; a combination of touch, smell, tastes, sound and vision. The degree to which different individuals are able to use their senses to collect information is not identical and can vary from a level of basic recognition to perception of extremely complex and subtle nuances, depending on their ability to recognise and understand the information available to them.

The most perceptive of the human senses is sight, which is how most people are able to comprehend their situation and all that is going on around them. For much of the time, visual information helps with activities associated with selection, movement and avoidance of risks. For example, it is useful to be able to see and recognise a door, to understand how to open it and to be able to see through and beyond it in order to avoid walking into an obstacle or tripping down a step. For those with poor eyesight, elements of buildings can be designed to enable them to use other senses like a change in the surface texture of the floor before reaching a step, which they can sense through touch, or the voice recording informing them that a lift door is closing, which they can hear.

Basic recognition of features within the environment is generally based on the information available to the senses, but the individuals understanding or interpretation is influenced by memory gained through learning or experience. Consequently, as well as the straightforward practicalities of life, the information received by the senses is used to perceive complexity and will determine individual 'likes' and 'dislikes', assessed objectively or enjoyed subjectively, creating feelings about what is attractive or unattractive, ugly or beautiful. 'Seeing is believing', 'seeing eye to eye' and 'see what I mean' are common phrases linking mind and eye.

Everyone can recognise a piece of music, a glass of wine and a building, but reaching any universal consensus about the merits of the composition of the music, the taste of the wine or the appearance of the building is unlikely, because for every person praising their qualities there will be numerous others who for many reasons are unable to agree with them. In fact, it could reasonably be argued that it is this very diversification of opinion that has developed and continues to maintain the variety and richness of the content of the environment. For some people, enjoyment comes from simplicity, harmony and balance creating a state of tranquillity and peacefulness, whereas others prefer complexity, discord and vitality stimulating them with contrasting dynamism and excitement.

The factors concerning visual perception are often referred to collectively as 'aesthetics', a word which has its origins in the Greek language, formally defined in the dictionary as: 'the theory or philosophy of the perception of the beautiful in nature and art in accordance with the laws of the beautiful or with principles of taste'.

'The laws of the beautiful' and 'principles of taste' are difficult concepts. Even assuming that they actually exist and designers were able to fully implement them, the results of their work largely reflect the values and attitudes prevalent in their own time, and enjoyment and criticism are clearly subject to personal interpretation, about which there is little common agreement. Consider for example the highly decorative and ornate buildings created in the Victorian era, followed by the simple, austere concrete buildings of the mid-twentieth century, followed by a return to elaboration and complexity evident in many new buildings today. Each generation of designers adhered to their own set of principles to create buildings with a completely different appearance. 'Beauty is in the eye of the beholder' is an overused cliché defending individual taste, and it may be impossible to explain every visual image in an entirely rational manner in accordance with 'laws' and 'principles'. However, to say that a flower or a building are *aesthetically pleasing* implies a favourable response to the visual qualities of the images received by the eye which either exist in nature, or have been deliberately composed by the designer.

It is interesting that the dictionary definition of an 'aesthete' is given as 'one who *professes* a special appreciation of the beautiful and *endeavours* to carry his or her ideas into practice', which could be interpreted as either suggesting a little scepticism on the part of the compiler or that the subject of beauty is surrounded with mystique which only the aesthete can understand. It is regrettable that the latter proposition is evident in the writings of some architects and the critics of their work which could reasonably be described as the expression of a superiority complex. For example, in his book 'An outline of European Architecture', Nikolaus Pevsner starts the introduction as follows:

"A bicycle shed is a building; Lincoln Cathedral is a piece of architecture. Nearly everything that encloses space on a scale sufficient for a human being to move in is a building; the term architecture applies only to buildings designed with a view to aesthetic appeal."

As a definition of architecture, this is surely fatally flawed because if a 'bicycle shed' is 'designed with a view to aesthetic appeal', then by Pevsner's own definition, it must be architecture. Perhaps in Pevsner's time, bicycle sheds really had no merit worth discussing, but the use of the term 'bicycle shed' could be read as an analogy for the more ordinary, practical every day buildings which make up the majority of the built environment, like for example a car-dealership. Lincoln Cathedral is undoubtedly a magnificent building, specifically designed to be visually impressive and constructed to last for a long time. The car-dealership may not last quite so long, but its visual appearance is no less important in its own context, within the built environment around it. The fact that the appearance of some buildings is more attractive and interesting than others is surely reflected in the quality of design demanded or offered at the time, creating buildings which are brilliant, good, average or poor. Consequently, those involved with the design of a 'bicycle shed' or a car-dealership should not feel that their work is in principle any less significant than the work of the designer of Lincoln Cathedral, or that they are therefore, allowed to pay any less attention to their work because it may be regarded as being of little importance.

The functional elements described in the previous chapters are the basis for creating the layout of the building and for determining its relationship to the site in 2D. Decisions about function are dependent on the examination of activity and flow, establishing how the building is going to be used, but as a building is a 3D object, decisions about the layout cannot be separated from its appearance. Both the insides and outsides of the building can be seen from a variety of different positions and it is their appearance which defines the spaces, enabling the building's users to recognise and understand them, to a much greater extent than the simple 2D plans. Although the building may function admirably, an unattractive appearance can detract from the pleasure of using it. Conversely, a visually 'beautiful' building which does not provide for the physical, practical needs of its users will rapidly engender feelings of dissatisfaction and displeasure.

10.3 Visual perception

Before looking at the elements of aesthetics in building design, it is important to appreciate how images are perceived. Seeing and understanding is a two-part process as light waves reflected off the image enter the eye, followed rapidly by the transmission of messages to the brain. The image is created mechanically but understanding what the light waves mean is dependent on the way that the brain interprets the information that it receives. The process of receiving and interpreting takes place in different ways from a fully conscious, careful analysis of the image, absorbing and recording all the detail before deciding how to react, to a subconscious, almost automatic response requiring little or no mental effort at all. On many occasions however, understanding occurs at some point between these two extremes when just enough detail confirms recognition.

The first time that images are encountered such as new faces, products and buildings, they will be examined consciously, registering all the detail available in order to understand them. The image is stored away in the memory so that when it is encountered again later on, the brain is able to recognise the image from relatively few clues seen by the eye. The brain does not require a visual re-examination of all the details every time the image appears before deciding what it means. This is a kind of in-built safety mechanism because if every single image had to be examined closely before recognition, the brain would be fully occupied just with interpreting images and have little or no time for anything else.

Explanations of the mechanics of visual perception can be found in a number of psychological study areas but the theory of Gestalt psychology, founded in 1910 by the German psychologists Wertheimer, Koffka and Kohler suggest how complex images are examined to determine their simplest form to permit recognition and how small elements of detail are used as the basis for subconsciously 'adding' the remainder. The theory suggests an ability to read a situation from memory, or from experience to create order. In this way, any image, however minimalist or however complex can be enjoyed for its totality, as it is not necessary to sense or understand every detail or element in order to appreciate the composition. The brain is able to respond to images and sounds without stopping to examine exactly how they have been achieved. Consider the work of poets, artists and composers which can be enjoyed without a full conscious understanding of all the ingredients or how they have been mixed together. Understanding the construction of the composition may of course heighten the pleasure, but it is not essential.

Providing the image being viewed and the picture held in the memory are broadly in agreement with one another, the image can be described as 'passive' in as much as it can be understood without too much effort, but memory is not always foolproof and can on occasions be deceptive. Memory influences vision and through familiarity fosters expectation by anticipating from previous experience. With some images, the lack of completeness can cause subconscious difficulties as the brain tries to be helpful, attempting to make sense of what it is being told. Sometimes it will make the wrong interpretation, be confused or unsettled so that no amount of conscious effort can lead to a resolution. The brain will either conclude its own meaning or be unable to decide what the image is.

There are many examples of this phenomena illustrated by 'optical illusions' with which the reader may already be familiar. The interpretation of line length, solid/void transition and background/foreground domination can be demonstrated with simple diagrams as shown in Figure 10.1 which are 'tricks' well known in the art world which have been exploited for visual effect to unsettle or stimulate viewers. Images of this type can be described as 'aggressive' in as far as the eye is disturbed and unable to settle on any particular point within the image because the brain is unable to make sense of the information it is receiving. There can be a subconscious feeling that something is wrong leading to a conscious struggle to keep on looking for meaning in the image. However, no matter how much effort is applied a resolution cannot be obtained, maybe leading to frustration and annoyance.

Images that come somewhere between the two extremes, being generally passive and familiar but with controlled, aggressive, dynamic elements adding a touch of vitality and excitement will create a sense of interest. The amount of 'interest' that the viewer can tolerate can be a significant factor in the determination of personal 'likes' and 'dislikes' depending on how stimulated the viewer is prepared to be, or how much effort must be put into understanding the composition.

10.4 Visual hierarchy

Aesthetic analysis of buildings is generally done at a 2D level, viewing the sides or elevations of the building as if they were pictures or photographs. This is a convenient, although simplistic method because in reality the appearance of the building alters as the viewer's position changes. Elevations drawn at right angles to the viewpoint can never actually be seen in their totality and in most cases buildings are viewed obliquely from different directions, with more than one elevation visible at any

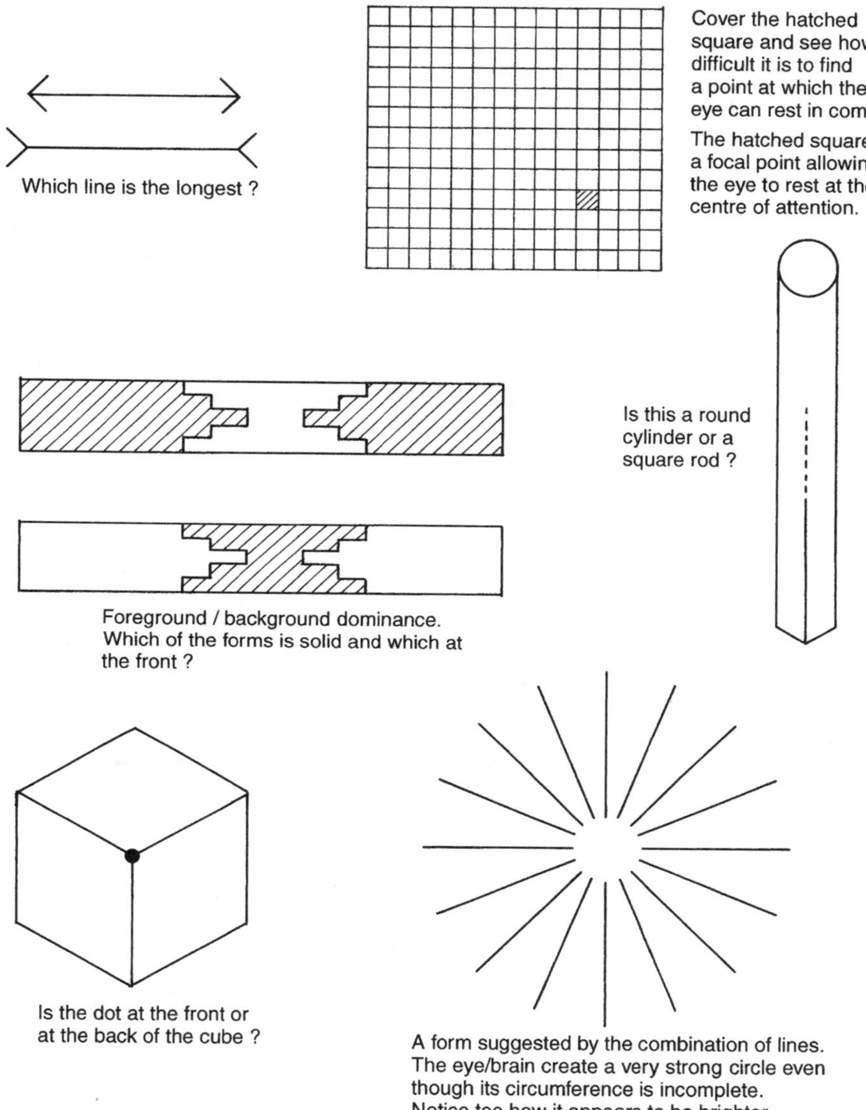

Figure 10.1 Optical effects.

one time. Even a single elevation is distorted by the effects of perspective as the elements decrease in size the further away they are from the viewer. This means that the appearance of the finished building presented to the eye is much more complex than the designer's drawings indicate. The use of computer-aided drawing to create 3D models helps to resolve this difficulty offering the opportunity for a much more comprehensive assessment of the building's appearance at the design stage.

There is a descending hierarchy in the visual perception of buildings from the large scale (*macro*) to the small scale (*micro*). Figure 10.2 shows how the first view of the building, set in its environmental context is of its overall shape and general colouring viewed against adjacent buildings and features. On approach, all the elements within the overall shape of the building; the walls, roof,

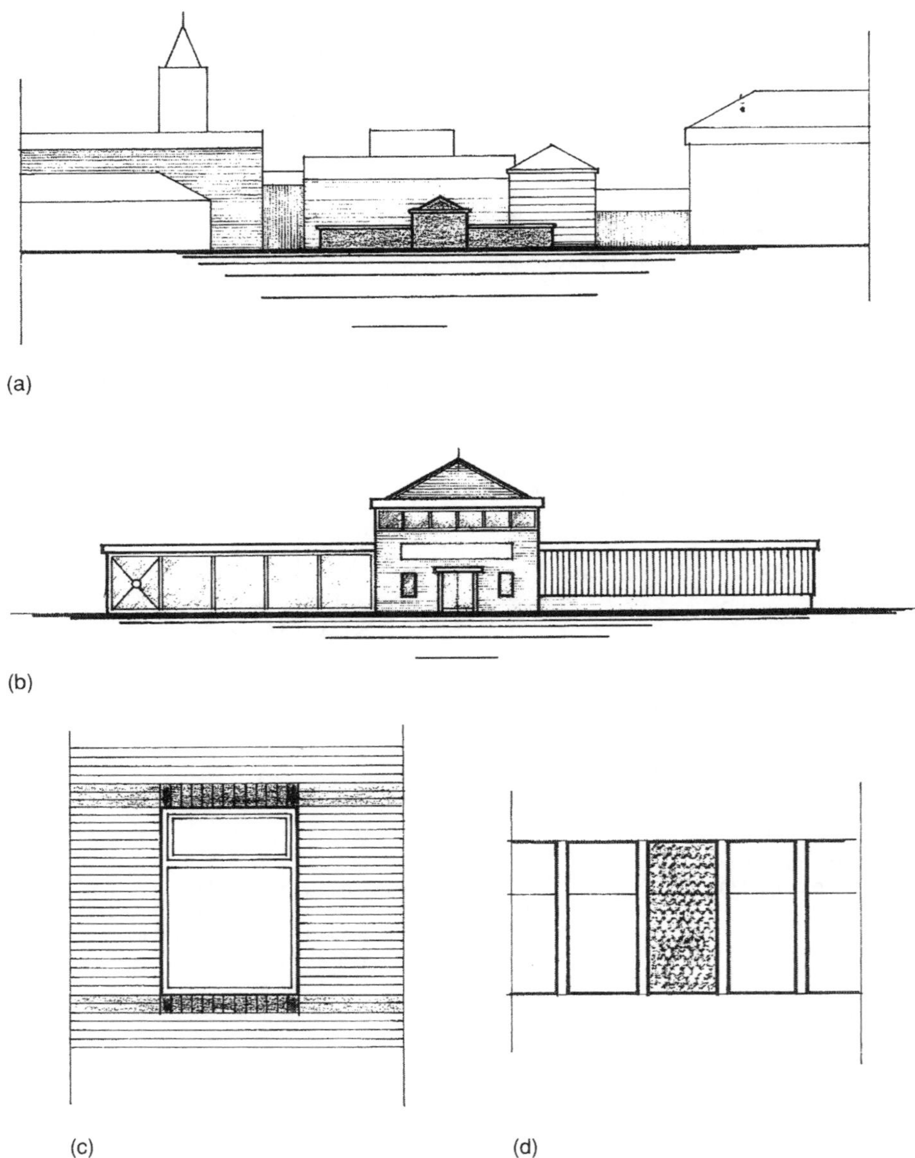

Figure 10.2 Visual hierarchy: (a) the building in the context of its surroundings; (b) the collection of forms; (c) the form and proportion of the individual element; (d) the detail, texture and color of the basic unit.

windows and doors become clearer until finally, in close up, the design of the details, materials, texture, colour and quality of manufacture is fully visible. The appearance of the individual volumes and planes of the building and its surfaces are referred to as forms, defined by lines around them or visible when viewed against their surroundings. Form is perhaps the simplest element of aesthetics, the one most easily seen and understood. Forms can be changed by altering their proportions and visual balance depending on their relationship to each other and they can be enriched, or impoverished through the complexity of detail, texture and colour.

At each of these levels, from the view of the building in its setting to the appearance of any of its component details, images can be regarded as compositions. They have a visual structure, achieved either by default, by accident, or by deliberate intent making them comprehensible or incomprehensible, attractive, stimulating, tedious or ugly.

10.5 Elements of composition

The fundamental structural elements of any 2D composition are points, lines, planes and volumes (Figure 10.3), which can be expressed *positively* as visible objects or which can occur *negatively* in between and around other objects. A 2D drawing of an elevation of a building contains forms which are defined by these elements. The simple rectangular shapes of the roof and walls, together with the windows and doors within them contain visible points, lines and planes in the form of corners, edges and faces but there are others created because of the spaces left between them. They are the result of the composition of the positive forms and can have significant visual force in their own right. For example, a symmetrical elevation has a strong point at its centre although the actual point is not marked, and the row of windows is a strong visual line leading the eye from side to side or in a particular direction towards a point. The spaces between the windows are seen as planes even though the top and bottom lines are missing.

A 3D composition is made up of the same elements, points, lines and planes but with the addition of the extra dimension creating volumes, either as positive enclosures of space, or once again, as negative spaces between other volumes. When looking at any image, the viewer's eye will automatically locate points, move along lines between points and move along lines around planes, either recognising each in turn or attempting to make sense of the way in which they have all been combined together. The way that these elements are viewed or read within the composition can be influenced by the following factors.

Focal points

There is a tendency for the elements in any composition to compete with one another for visual attention and the eye will inevitably attempt to seek out the one which is most important or significant. The main source of interest is referred to as the focal point, the location within the composition that can be said to be the centre of attention. Other elements in the composition can be organised to lead the eye to this position so that the overall image is comfortable and clearly understood. The focal point may simply be the largest or brightest object in the composition, but it can be much more subtle than that as a result of the way that lines and planes direct the eye towards it or as a consequence of the *visual weight* of each of the elements within the composition. In a composition of many similar, repetitive elements, a small variation in visual weight immediately attracts attention enabling the eye to rest at this position. Without this device, the eye will continue to flit from one element to another and back again, making the composition unsettling.

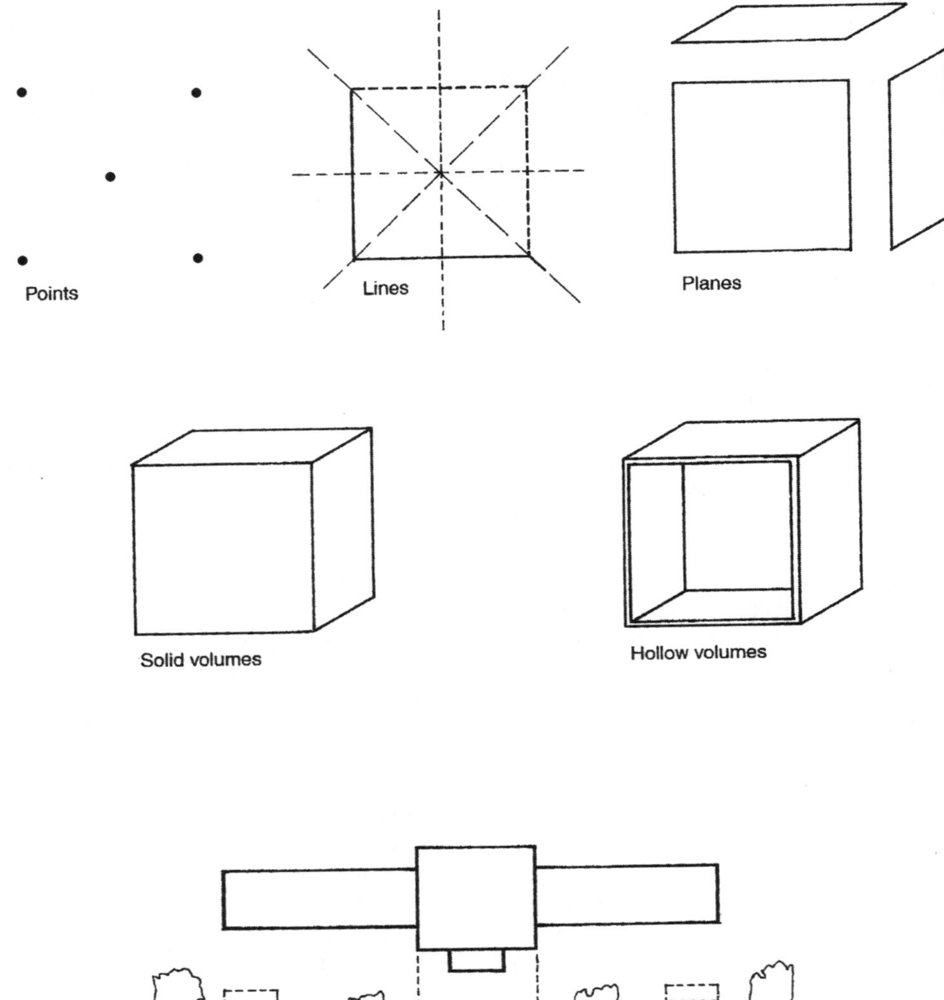

Figure 10.3 Points, lines, planes and volumes.

Duality

A problem can arise in a complex composition containing two or more equally dominant features, each of which is laying claim to be the focal point. In this situation the eye is unable to rest on the most important part of the image and constantly switches from one to another, causing unrest and agitation. Images with equally competing focal points are said to contain a *duality* which splits the centre of attention in such a way that neither position takes precedence over the other. A composition can contain many separate points of interest which can be studied in due course but at first glance, the dominant feature should be the most obvious if the total image is to be understood. The 'additive' and 'subtractive' form of building design illustrated in Figure 10.4 shows how a composition can retain or lose its focal point.

Unity

Compositions made up of various elements, which each have different shapes, sizes, textures and colours, can lead to clashes which are uncomfortable or unattractive. Disparate elements can be *unified* by creating a common link between them which visually holds the composition together. There are a number of ways that this can be achieved including the use of a datum base line to which the elements belong; an straight axis holding elements together on either side; a clearly understood relationship in arrangements of order and hierarchy for repeated elements and a consistency in elements of varying shape, size, texture or colour so that each element has something similar in its makeup. All these features can be said to bring *harmony* to the composition, in a sense making them appear to belong to the same *family* with similar features. They can be recognised in some respect as belonging together rather than be totally separate. Some examples are shown in Figure 10.5.

Vitality

Complete harmony or balance sometimes leads to monotony or dullness. Repetition of the same shape, size, texture or colour can be tedious, lacking any clear points of detail to stimulate the viewer. Variations and contrasts can be introduced to deliberately create points of interest leading to a feeling of *vitality* in the composition. Repeated elements can be arranged to create a rhythm or by treating the odd element in a different way creating a focal point subtly changing the balance, making the eye and brain work a little harder to extract the meaning. Interest can be created by introducing something which is unexpected such as changing the angle of elements from the normally anticipated horizontals and verticals or a bright splash of colour to attract the eye. Consider the A's in Figure 10.5, repetitive elements, but designed to create interest.

All of the elements of composition described above can be applied to the macro and micro scales when viewing a building. Determination of form, proportion, detail and colouring for the whole building down to the smallest element of construction within it can be manipulated by the designer to create harmony and balance, contrast and vitality to make the appearance more or less interesting and stimulating to viewers. In principle, the appearance of each of the building's elevations must be designed as a composition but always bearing in mind that the elevation itself is a contrived view; the true appearance is 3D and from any given position is actually the combination of several elevations seen at the same time. In this respect, the merits of the composition may be judged from any point on the circumference of a 360-degree circle. The following sections are a brief introduction to the visual composition of buildings and the factors which affect both 3D and 3D appearance.

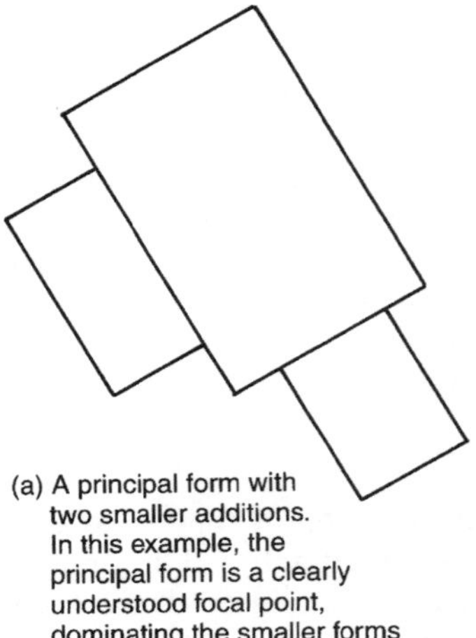

(a) A principal form with two smaller additions. In this example, the principal form is a clearly understood focal point, dominating the smaller forms around it

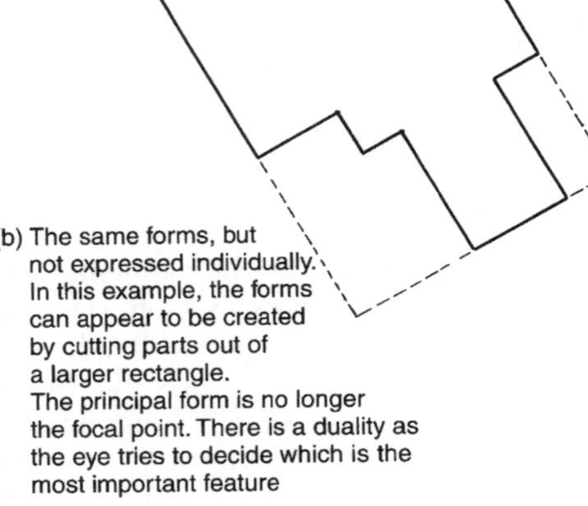

(b) The same forms, but not expressed individually. In this example, the forms can appear to be created by cutting parts out of a larger rectangle. The principal form is no longer the focal point. There is a duality as the eye tries to decide which is the most important feature

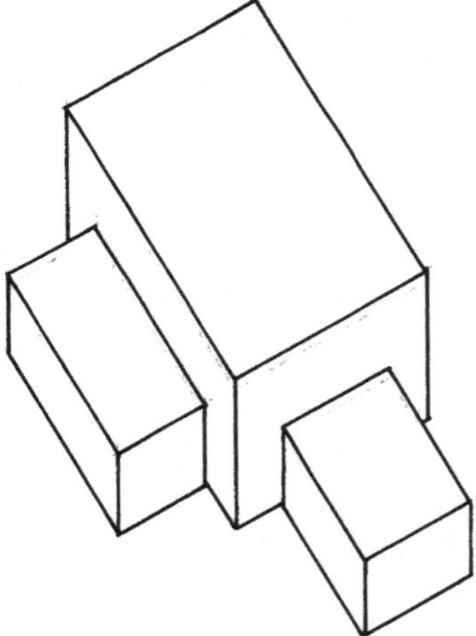

(c) Each form is expressed, retaining the integrity of the original largest form. It is easy to understand what the composition is and how it has been achieved. These forms are said to be articulated

(d) Expressed as a single form it is difficult to tell what all the shapes mean. The eye/brain imagine it to be a large volume with sections missing

Figure 10.4 Transformation; additive and subtractive form.

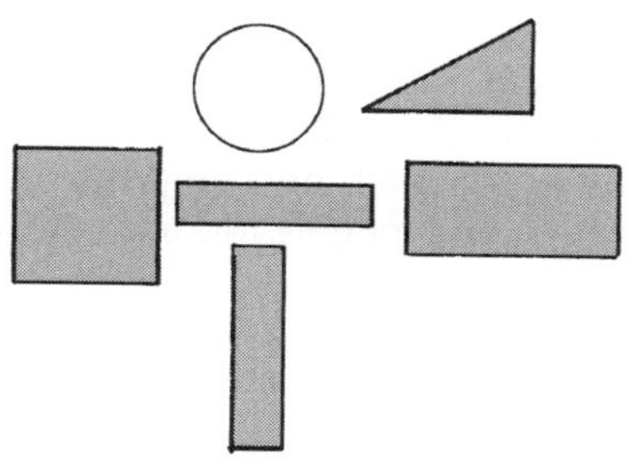

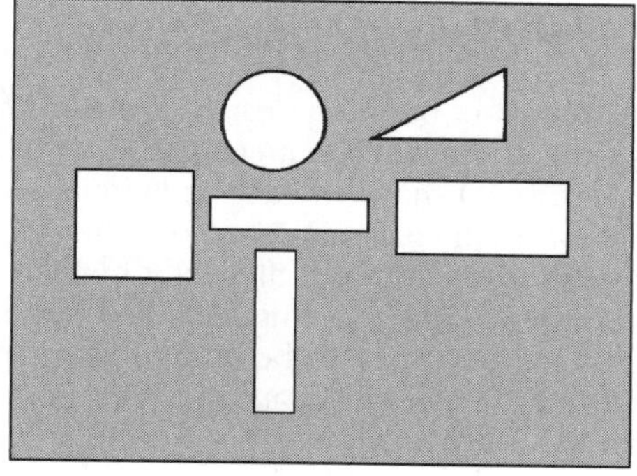

Irregular forms unified by similar colour or texture, or by a common background

Similar forms, but of different size are unified by being in the same proportions to one another

Note the element of vitality introduced by placing one form at an unexpected angle or emphasizing one form to create a focal point

A A A A A A **A** A A A A A A

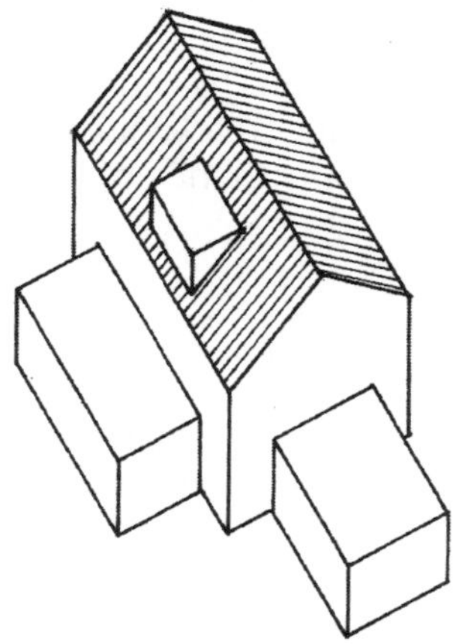

The smaller forms do not respect the qualities of the principal form. The mixture of forms of construction can destroy the unity of the elements of the building

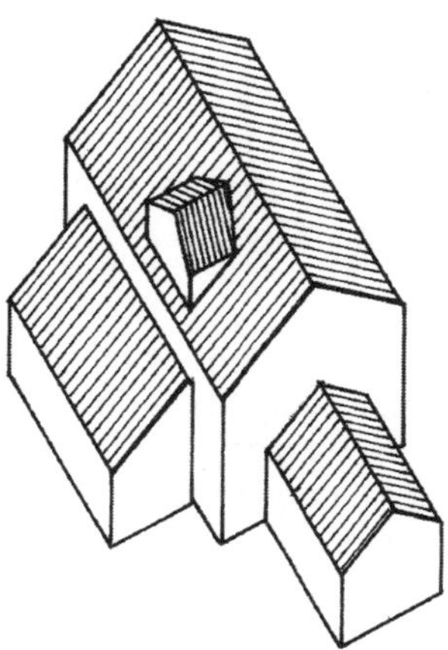

Each form is unified by a similar treatment of the roof construction

Figure 10.5 Examples of unity.

10.6 Form

In the context of building design, the word *form* is generally used in relation to the 3D outline or massing of the whole superstructure or of its constituent structures, but it can also be used to describe the 2D shape of each individual element within them. The walls and roof as a totality have a form which is made up of other elements such as columns, panels, doors, windows, chimneys, bricks and roof tiles, all of which have a form of their own. An object is said to possess *a form* if it has a distinguishable boundary defining its shape, which may be a 'hard' line like a frame, but can also be an arrangement of reference points, which if sufficient in number will permit the eye and brain to fill in the missing gaps, to create a visual boundary line even though it is only imaginary. For example, a tree in winter will still appear to be round or pear shaped, even though the edges are made up of many small, irregular twigs, and a courtyard will be seen to be square if the four corners are marked with some positive features like posts or buildings. Figure 10.1 shows how a strong circle is generated by radiating lines. The most common way in which forms are recognised is by changes in texture, tone or colour between adjacent objects or when an object is seen in front of a contrasting background. The edges of the forms may not exist but the transition point is still clear. For example, a circular form like a tree trunk does not have an edge down either side and its width from any viewpoint can only be seen against contrasting surfaces behind it. This is particularly important as a means of delineating planes at different distances away from the viewer which would otherwise blend into one another and lose their individual clarity.

Forms can be seen as simple geometric shapes or as the combination of different shapes placed together, or depending on the viewpoint, the simple form can be distorted by perspective, creating other forms which may or may not be recognisable. The basic 2D forms are the *square* or *rectangle, triangle* and *circle*, which in 3D can become *cubes, pyramids, cones, spheres and cylinders*. These are all expressions of pure mathematical forms which, without modification are harmonious to the eye and easy to understand. Most shapes and volumes visible in buildings are made up of these simple forms abutting, overlapping or surrounding each other, sometimes resulting in quite complex and apparently new forms, but providing that individually they remain *visually determinate*, they will still be comprehensible. Changing simple forms into more complex ones, as shown on Figure 10.4 is often referred to as *transformation*, which can be achieved by either adding something to the original form, or taking something away from it.

Additive form

Any 2D or 3D form can be extended by adding other forms to it, creating a new form. Highlighting the new form or shape by a strong boundary line or border, or using a texture or colour which crosses the boundaries between the original forms can lead to confusion and a lack of clarity if the new image is not recognisable as the sum of each of the constituent parts. Additive form is generally most successful when the original form retains sufficient of its characteristics to still be discernible in its own right. In compositional terms, the original form retains its position as the focal point, surrounded by smaller *articulated* extensions retaining their own characteristics. Increasing the proportions of the additions, or overlapping the forms so that they are no longer *visually determinate*, will lead the eye to believe that it is actually looking at a form which is incomprehensible, or a larger form from which parts have been removed, leading to uncertainty. At some point, the elements of the composition may begin to compete, creating a duality described in the previous section.

Subtractive form

Conversely, the original form can have parts removed or cut away from it, leaving out elements. In this case, the eye and brain may try to replace the missing pieces as an attempt to return the image to its original, complete form. Once again, the extent to which this is done will either permit recognition of the original form, or lead to a conclusion that the resulting image is a new, unfamiliar form. At some point the image becomes a combination of other smaller forms and the original form vanishes altogether. Subtractive form taken to extremes can result in ambiguity or visual instability which can be quite unsettling to the viewer leading to misunderstanding and sometimes creating a feeling that the building might actually fall over.

There are some historical examples of buildings designed and constructed in simple, pure geometrical forms, but most buildings are mixtures of additions and subtractions. The pyramids in ancient Egypt, the tepee, the igloo and the rectangular form of many early skyscrapers were more or less mathematical in their construction based on a single, recognisable shape or volume. For others, their complexity demands close attention to the way that the volumes and shapes are put together to create a coherent whole. Factors influencing interpretation of volumes and shapes include the following.

Size and scale

The size of any form or plane is a matter of fact. For example, a particular wall may be 10-m long by 4-m high, and a window in the wall may be 1800 mm wide by 1050 mm high. However the perceived scale of these dimensions depends on the distance that the viewer is away from them and their relationship to other features with which the viewer is familiar. The simple example in Figure 10.6 shows how the scale of the wall and window appears to be quite different when seen by the side of a human figure. Cross referencing in this way takes place all the time as the eye and brain attempt to rationalise images based on previous experience. For the building designer, scale is significant because some elements of the building designed to be in scale with others when seen from a distance could be quite disproportionate when seen from close up. For example, a substantial civic building or a church will probably have a relatively large feature entrance doorway in scale with the remainder of the elevation, but the actual entrance door itself is often reduced in size within the larger feature to be more appropriate to regular human use. Similarly, the window openings may be much bigger than is actually needed in order to retain the balance of scale within the overall facade.

Scale is also important when designing buildings adjacent to or close to other buildings. The precedents already established for the strong horizontal lines of doors, windows, roof eaves and ridges can be reflected in the new building so that there is a similarity in scale, a recognition of the visual strengths of existing features which will be viewed together with the new ones. Differences in relative scale between adjacent buildings can ruin the qualities of both.

Unity

Forms of different sizes combined together can be unified by being the same shape or by constructing them to similar proportions. Additions to the original form can be designed in the same style or the selection of materials can be on the basis of using sympathetic, complementary colour schemes. Unity creates harmony and avoids unsatisfactory clashes. For example, extensions to a building with a pitched, tiled roof will look poor if they are constructed with flat, felted roofs and the use of steeper or shallower pitched roofs than the existing building will also look out of place. Examples of unity are shown in Figure 10.5.

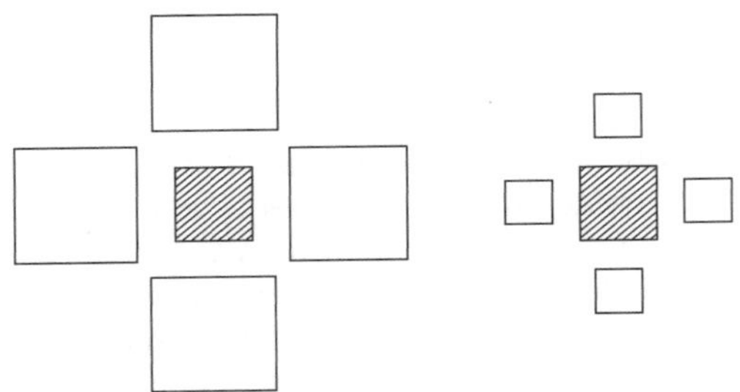

Visual scale is the size of any element when seen against some form of reference standard. The black square on the left appears to be smaller than the one on the right because of the differences in scale with their surroundings

(a)

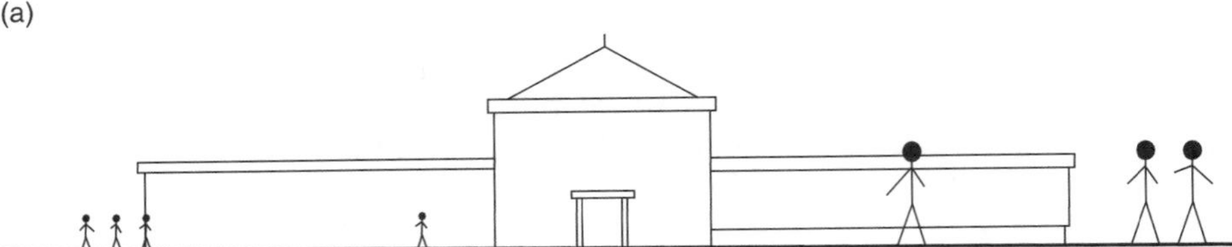

The appearence of the scale of the buiiding and elements within the composition is affected principally by the comparision with the understood dimensions and propotions of the average human body

(b)

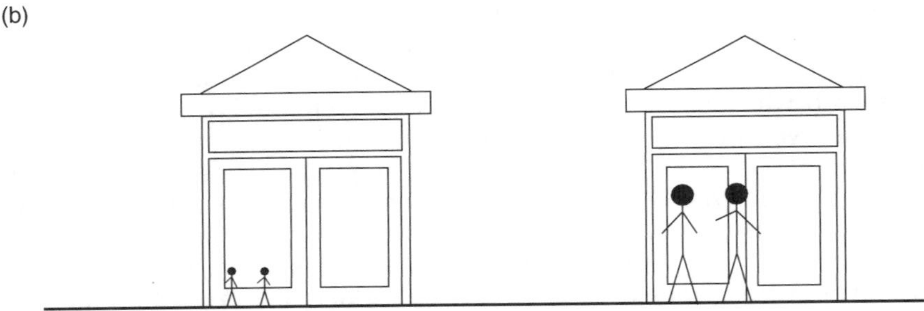

Is this a grand, imposing entrance, or really quite a small, ordinary one?

(c)

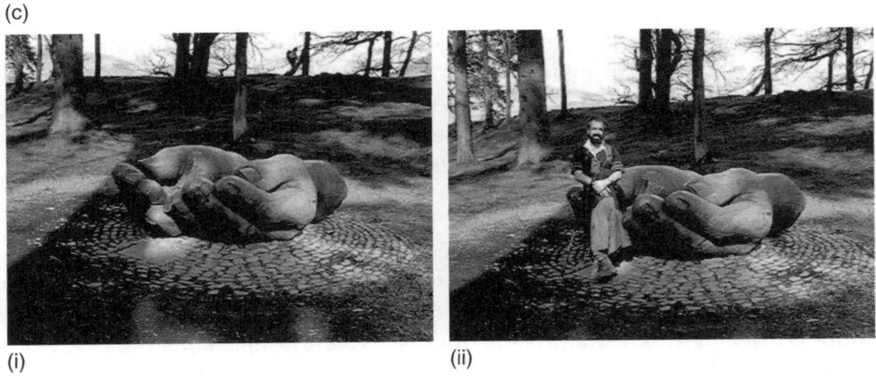

(i) The size of the wooden hands is not possible to understand until compared with a reference such as a human body, which gives the image scale. (ii) The author in the lake district.

(d)

Figure 10.6 Implications of scale.

Repetition and rhythm

Many forms within the building are repetitive including, bricks and blocks, columns, panels, windows and doors. When elements are repetitive, there can be a simple elegance in keeping them exactly the same, avoiding unnecessary, perhaps arbitrary changes. Exact repetition, rather like a heartbeat is reassuring, comfortable and harmonious. Sometimes, however repetition which maintains exactly equal spacing can become predictable and monotonous. The composition can be made more interesting and stimulating by introducing subtle rhythms, grouping repetitive forms together in clusters. Arranging the windows in pairs for example with more and then less space along the elevation creates an additional pattern adding a richness to the appearance of the wall.

Logic and ambiguity

The eye and brain's anticipation of some elements of construction is based on logic. Visual perception of roof structures and the location of entrance doorways for example are often predetermined in the mind, so that where they are constructed or located in ways which are unexpected, the brain will attempt to correct the situation by trying to imagine a different building. For example, the entrance is expected to be at the front of the building rather than at the side. The simple rectangular plane of the roof tiles is much more clearly the front of the building than the pointed gable end, which is automatically perceived as being the side. The span of the roof is expected to be across the shortest dimension of the building, not the longest. Buildings with these features located out of place can lead to uncertainty.

Symmetry

Forms will be perceived as balanced if they are copied either side of an imaginary axis, either vertically or horizontally. The simple mathematical forms described earlier all have inherent symmetrical lines running through them, and the most familiar form, the human body itself is made up of matching parts either side of a centre line running from the head to the ground. Lopsided forms which deny the axis or create another, off-centre axis can lead to visual instability and unease. Mixtures of varying forms and shapes can be held together if balanced on either side of an axis which in itself acts as a strong unifying factor as illustrated in Figure 10.7.

Static and dynamic forms

Forms can be seen as absolutely fixed in their location, stable and unmoveable. Most buildings take this form, sat on a datum of firm ground. An element of vitality can be created by changing the position of the form so it is no longer balanced or inert, but has an appearance of instability and potential movement. In fact the appearance of most buildings in reality is often like this because of the angle of view and the effects of perspective. The familiar rectangle when seen from the front is inert, but seen obliquely from the side, is tapered and becomes more dynamic. Ribbed steel cladding on the external walls is generally fixed in regular rectangular panels with simple horizontal or vertical joints. Changing a horizontal joint to one with an angle creates movement and vitality in the elevation, challenging the viewer by offering something unexpected. The same may be applied to glazing, which is usually rectangular with horizontal and vertical edges; static openings in the structure of the building. Angled jambs or cills change the visual balance leading the eye in a different direction.

Shading, the effects of sunlight

The perception of the forms of the building in reality is strongly influenced by the quality of natural or artificial light illuminating them. Externally, buildings are rarely viewed as bland

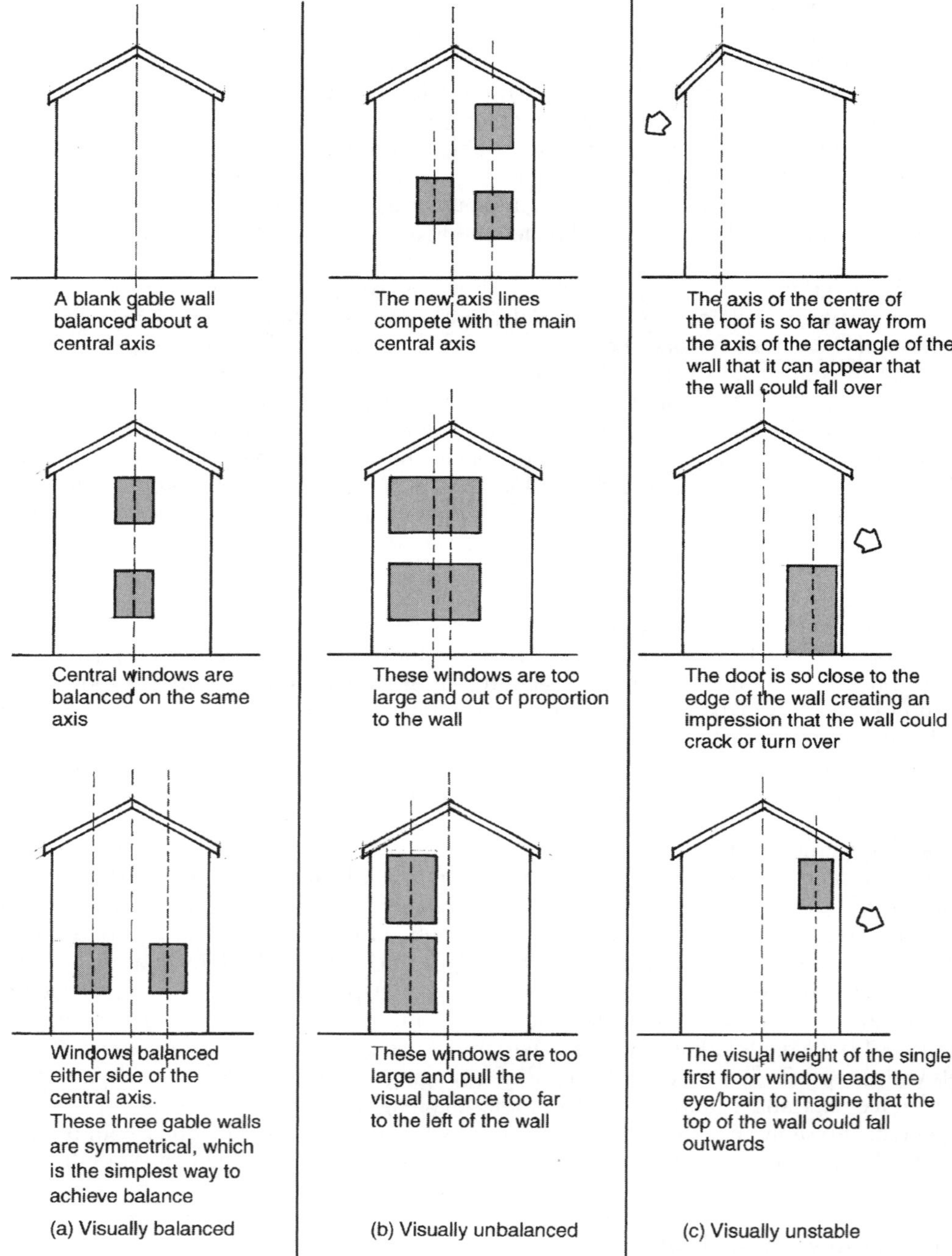

Figure 10.7 Balance and visual weight.

elevations because the sunlight shines onto and into them, changing its angle of penetration throughout the day and the year. The texture of the surface, the depth of projections and reveals will change the appearance of the image as areas become darker or lighter with more or less shadowing, changing the visible form itself.

Visual weight

As well as the size and shape of the forms, perception of the composition can be dependent on the *visual weight* of the constituent elements. Small, dark shapes are visually strong and dominant and can visually overwhelm much larger lighter shapes. For example, consider the windows in the front elevation of any building, which from the inside looking out are clear and light, but which appear almost black when seen from a distance outside. In the composition of the elevation the visual weight of glazing can be greater than the visual weight of a brick wall surrounding it. Determination of the proportion of forms and their relative visual balance will be considered in more detail in the next section.

10.7 Proportion

Any form can exist at almost any size. The regular cube can be a whole building, a single room, a shoe box or a dice. Although there is a substantial difference in the actual size of each of them one when compared to another, the relative size of one to another can be transformed by the effects of scale and under certain circumstances it can appear that they are the same size dependent on the closeness of the viewer and the juxtaposition with some other feature acting as a reference point. Forms can also be transformed by altering their horizontal and vertical dimensions, consequently changing their shapes. The regular cube can be stretched into a low, horizontal block or a high, vertical tower. The roof triangle can be flat or steep and the windows can be wide or tall. In principle, the forms are the same, but they have been modified or transformed into new shapes.

To create the composition, decisions must now be taken about the relative scale of the elements, the choice of the size or shape of forms in relation to one another and with regard to the whole composition. The visual relationship between the dimensions of height and width, the degree of steepness of the roof pitch and the area of the wall compared to the area of the openings in it for example, can all be adjusted so that the proportions of each element are balanced and ordered.

There are technicalities which determine the proportions of some elements in buildings associated with materials, manufacture and construction. However in many cases, the design team have sufficient discretion to manipulate the proportion of elements as they think fit. In this respect there are no absolute rules leading to conclusions that something works or does not work. There are mathematical systems which have been used periodically by successive civilisations, which they have employed to create their buildings. The Greeks and the Romans strongly believed in attempting to relate themselves to their gods through the purity of mathematics and both the Renaissance designers and more recently, Le Corbusier in the 1950s with his Modular system (Le Modulor) harmonising the dimensions of architectural elements in proportion to human stature, a system based on the Golden Section and the mathematics of Fibonacci numbers, first put forward by Pythagoras, subdividing and proportioning forms based on the dimensions of the human body.

Their interest in the purity and 'beauty' of mathematics through the implementation of the Golden Section, and other forms of numerical progression such as the Fibonacci Series, which has been demonstrated to exist in the natural world, created buildings which were designed in accordance with strict rules generating spaces, volumes and features which were dimensioned to suit a formula. The theory was that mathematical progression and logical relationships imposed an order which was more satisfying to the eye than arbitrary or random selections. Their objective was to achieve proportions for the elements of their buildings which were in perfect harmony with one another and collectively, with the whole. An explanation of the workings of these systems is an interesting academic study and there are many examples of buildings from the Greek, Roman and Renaissance periods illustrating the results. On the whole, these buildings were very formal, designed as images of power or domination, or as expressions of wealth and importance. In many respects they were contrived as works of art or as sculptures intended to create respect, wonder, awe or fear. The reasons for their creation came from the perceptions of their own time. Successive designers have drawn on these systems as inspiration for their work, improvising and re-interpreting in order to create their own styles, but they have also had to account for new building types and activities, advances in forms of construction, different materials and skills all contributing to changes in appearance.

Setting aside the merits of a purely mathematical approach to proportioning the elements of the building, perhaps the most interesting concept to consider is the relationship of the building and its parts to the dimensions of the average human body. Designing with regard to anthropometric data was discussed in Chapter 8, but there are also aspects of visual perception which relate to the human form. For example, low, flat horizontal forms are seen as tranquil and passive, associated with lying down, whereas tall, upright vertical forms indicate activity and dynamism, associated with standing up. The form of the building and the form of any other visible elements in it can be predominantly one or the other, or a mixture of both.

Consider the composition of a simple elevation shown in Figure 10.7 and how the location and proportion of the forms within it affect its appearance.

Balance

Each of the forms in the elevation is a simple shape, which in a balanced state has an invisible axis running down or across its centre point. In the gable end example, the axis of the whole elevation is central and if it is to remain balanced, other elements in the elevation should respect this axis. Creating other off-centre axis lines changes the balance of the composition and creates other forms within the elevation which are no longer symmetrical. Altering the shape of any of the forms or their relative alignment with one another adds further confusion, or interest depending on the design intentions. The opportunity for controlling this element of the composition may of course, be limited by the need to locate a window or windows in particular locations in relationship to the floor layout. The design of the window itself can be made balanced or unbalanced. The position of opening lights either retains the central axis, or introduces others which move the eye to one side as seen in Figure 10.8.

Visual strength

Each element in the elevation has its own visual strength or weight because of its size or colouring, resulting in some elements being more dominant than others. The roof and the wall are major elements, the windows less so. A balanced elevation will pay due regard to this fact by ensuring that the proportions of the elements are such that the composition is not unintentionally dominated by something of minor significance. For example, a very shallow pitched roof will appear mean when compared to a large area of wall. A window placed close to the end of the wall with little masonry cover appears weak, even to the point where the structural integrity of the building is in doubt. A bright, white rainwater downpipe and thick, white unplasticised polyvinyl chloride (UPVC) window frames will be dominant features in the wall so that they are seen much more positively in the elevation than anything else.

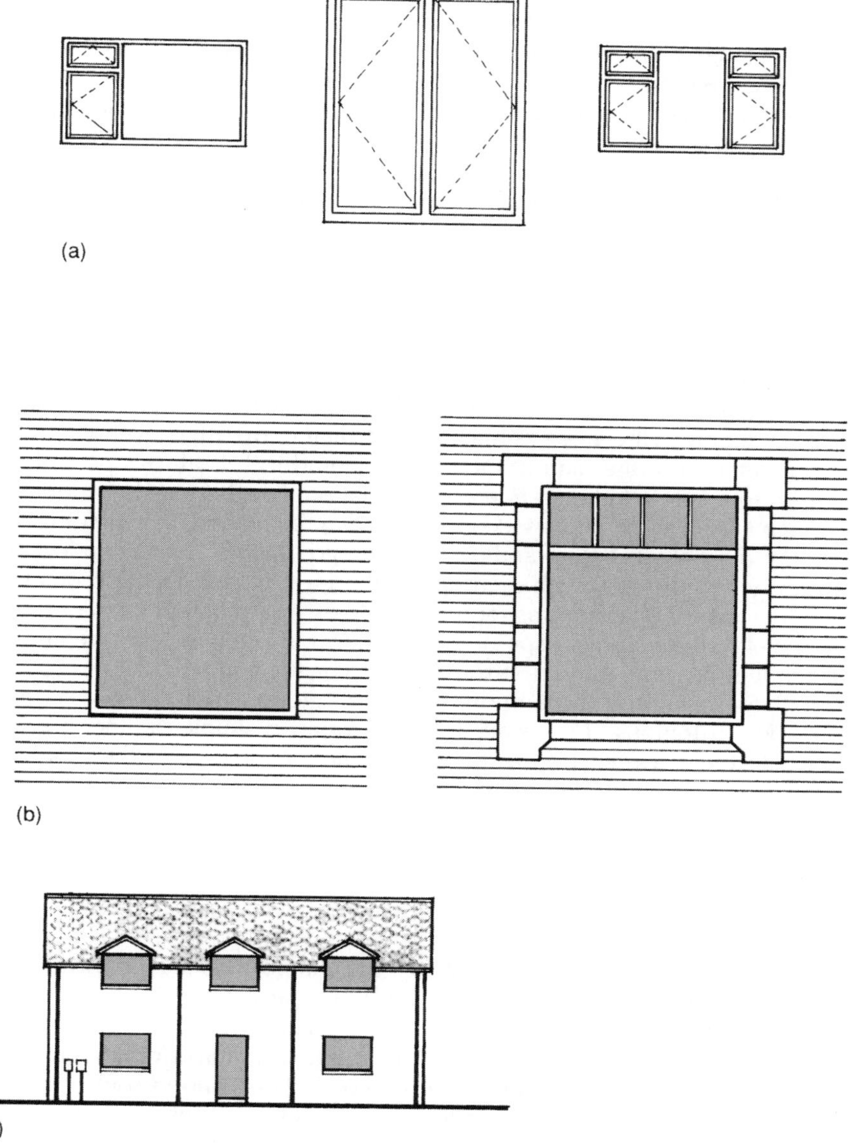

Figure 10.8 Adding detail to a window and its surround. (a) Arbitrary positioning of forms can destroy the visual balance. The introduction of detail in the example on the left upsets the visual balance of the composition. All the forms within the main forms can contribute to the visual balance, as shown on the right. (b) On the left is shown a plain simple window opening; on the right, detail has been added to the window framing and its surroundings. (c) A reasonably attractive balanced elevation in terms of wall, roof and openings. However, the introduction of essential rain-water downpipes between each dormer window, radically changes the actual appearance creating new unconsidered forms. Note also the electric and gas meter boxes which are ugly and unattractive and draw attention to themselves. Not what the designer had intended!

Visual force

As well as visual weight, some forms have a notional visual force. The square and the circle are truly static, but rectangles have a visual force which is their longest dimension, either horizontally or vertically. To remain balanced, or be in repose the sum of the visual forces of the elements should not over dominate the main visual force of the whole. For the simple wall used in this illustration, it can be seen that the size of the windows can be diminished or enlarged so that the wall is either a solid form with openings in it, or it is openings with a little masonry to hold them together. The proportion of solid to void significantly affects the way in which the wall is perceived as a form.

The design of the elevation may incorporate forms which can be deliberately manoeuvred to account for any or all of these issues. Elevations which are more complex require greater thought for the placing and dimensioning of the elements within them. The key factor to bear in mind is that the eye and the brain continuously seeks for comprehension and meaning. In many cases it is a desire to rest on simplicity through balance and harmony, to be able to receive the image and understand its logic. Images which are illogical or deny an expected balance or symmetry create doubts in the viewer's mind. In practical terms, the elevation of the wall in this example is generated from the plan of the spaces behind it. It is always possible to introduce subtle variations to create additional interest or variation, but they are best done by deliberate intent rather than merely appearing as a consequence of the planned layout. Once again, it cannot be stressed too highly that form and the function go together. If the windows on the gable end wall are to be symmetrical with the ridge of the roof, is their location practicable for the rooms which they serve? If the rooms require windows in a different position, what will they look like on the elevation? Comparison of the two positions requires consideration of both sets of demands at the same time if the best solution is to be created.

Consideration of form and proportion lead to determination of the *structure* of the composition. The illustrations of the wall show the forms in simple outline, but in reality they are materials with their own shapes, textures and colours adding a further quality of detail to their appearance as a result of their individual treatment in the manufacturing process and the way that they are put together in groups.

10.8 Detail

The form and proportion of all the elements within the composition may simply derive from basic functional and/or visual requirements, constructing and expressing each in strict accordance with its purpose. For example, a wall can be a simple rectangular panel of brickwork or rendered concrete blockwork painted in a single colour, and a window can at its simplest be just a rectangle, a hole in the wall filled with glass. There need be nothing else and in some circumstances this would be the most suitable design solution for these two elements of the building. Alternatively, the wall may be constructed in profiled steel panels or in random rubble stonework adding a richness of shapes, textures, and colours and the surround to the window opening highlighted with special, contrasting feature cill, head and jambs. The window itself can be subdivided into smaller units making its appearance more complex and interesting. The handles can be moulded into special shapes and even the screws fixing the handles to the frame can be considered for their visual appearance above and beyond their basic practical needs (see Figure 10.8).

Creating or adding detail to forms can be achieved in a number of ways depending on the constraints of manufacturing processes, the imagination of the designers or the costs and value of the results to the client. Some details may be disproportionately expensive and unjustifiable but others can be achieved by design at little or no extra cost. The options available include the selection of materials which already have their own inherent characteristics and qualities, the transformation of the shape of materials, components and assemblies or the articulation of forms in such a way that boundaries between or around them provide some means of contrast.

Materials

Every visible form in the building is made of some kind of material which in its natural state may be rough, smooth, shiny, dull, plain, coloured or patterned. Many materials currently in use are manufactured in relatively small units, so that as well as their individual unit qualities, assembled collectively they can make larger forms. Bricks, blocks, tiles and wood panels for instance can be assembled to create walls, roofs and floors. There are practical reasons for using small units in this way, but there are visual options and advantages too. A variation in the selection of bricks for example, can be used to divide a large wall into smaller panels, to delineate the changes between levels or to provide a datum linking together other common features such as window cills or heads. Bricks can be used to create patterning effects or to emphasise or soften the harsh line at the roof eaves and verge. A change in floor finish in a public building is a useful way of guiding people around the building, perhaps delineating public and private space.

Materials may be selected for the uniformity or consistency of their finish such as colour coated steel cladding or glazing. Every part of a wall constructed in these materials will look the same, although it is interesting that direct sunlight or shading can change the apparent colouring. It should be appreciated too that large areas of glazing may be quite dazzling on occasions, but they are sometimes used to advantage by reflecting their surroundings back to the viewer. This can create an interesting image, but may equally cause confusion and concern. Alternatively, the use of a brindle facing brick, or some types of stone such as marble for example, can introduce a mixture of textures and colours offering variety whilst remaining unified and harmonious. Beware though selecting too great a variety of materials and mixing them together, which can often be a mistake, destroying unity and harmony by being over fussy or too busy.

Transformation of shape

Transformation of shape can be considered both for the individual components and also their assembly into desired forms. Clay and concrete for bricks and tiles can be manufactured in special shapes, steel sheets can be pressed into various profiles including curves, glass can be moulded and shaped, and timber can be cut, planed and processed into many different forms. The unit material can be elaborated to match design intentions to make it simple or complicated depending on requirements. However, the placing of basic unit materials can also effect the detail quality of the form. For example, recessing or projecting courses of standard bricks changes the appearance of the wall, emphasised by the effects of shadowing. Using a different bond, mortar or pointing can radically transform the appearance of brickwork. A deep overhanging soffit and fascia at eaves level enhances the appearance of the roof and laying floor tiles in a herringbone pattern creates movement and interest as opposed to the static calm of a single coloured carpet.

There is a tendency to assume that all buildings are made up of straight lines because this is how most of them are. In fact curves can be created using standard materials with steel frames, masonry and glazing. Curves are visually attractive and stimulating, and can provide an excellent contrast to regular horizontal and vertical geometry. Curved walls, curved eaves and circular windows can be used to good effect to provide variety and additional vitality.

Contrast and enhancement

The choice of materials and the way that they are assembled influences the perception of the composition, particularly when it contains many forms. The composition is often improved if the individual forms have additional clarification or demarcation so that there is a better certainty as to what they actually are. In simple terms the most effective use of contrast and enhancement is at the edges between one form and another, acting as a visual stop so that the eye and the brain can better appreciate the point of change. Taking the building as a whole, the junction between the superstructure and the ground can be reinforced by a plinth or a band of darker or more textured material. This gives the building a visual anchor at its base, making it appear to belong in its position. The top of the roof seen against the background of the sky can be identified with a darker or specially shaped ridge tile, by a contrasting cladding flashing or by some other device such as a parapet wall with a coping reinforcing the transition point between building and sky. The roof colouring itself will act as a visual stop if it is darker than the superstructure of the walls below it.

The edges of the superstructure will benefit from clarification, particularly if seen against a confusing background. If a material is seen against a background which is more or less the same, it is difficult to distinguish the boundary line and therefore the forms become blurred. Stone or brick quoins are a traditional solution for masonry construction, but columns, cladding trimmings or changes in the colour of the materials would serve the same purpose. Openings in walls are nearly always improved by emphasising the immediate surround with a change of material or method of construction. Changing the basic horizontal of a standard lintel at the head of a window to a gentle arch softens the form of the opening and can make it more interesting and attractive. A recessed door opening with appropriate surrounding detail reinforces the visual importance of this feature in the elevation.

Elaboration and ornamentation

For many parts of the composition, both external and internal forms may be elaborated to a level beyond the demands of practicality and basic visual perception. The panelled door, the intricately moulded cornice, the elaborately scrolled banister and wrought iron balustrading become features which fit into the general scene, but which include details which are also enjoyable in their own right. There can be great pleasure in the way in which materials have been handled, the quality of craftsmanship and finish enhancing the potentially mundane feature. The extent to which any feature is elaborated perhaps comes down to 'taste' as mentioned at the beginning of this chapter. Taste and cost! The moulded plaster coving that covers the junction between a plastered wall and the ceiling is a detail well worth considering, whereas the external dummy timber window shutters at the side of all the windows could perhaps be said to be 'over elaboration' introducing additional unnecessary forms for the sake of it.

A mention must also be made in this section about the symbolic meaning and significance of decoration with respect to some building types, activities and to other cultures. The values associated with elaboration and ornamentation is clearly not universal and any design must respect the demands of its potential users. For example, the quality of detailing expected in the car-dealership showroom to attract and impress members of the public is likely to be of a far higher level than the paint spray workshop, which is for the staff concentrating on a single task.

The fourth important component of aesthetics is colour. This is a major topic in its own right which can only be briefly introduced in the space available in this book but an appreciation of the basics will perhaps help the interested reader to research the subject in greater detail elsewhere.

10.9 Colour

Colour only exists as a consequence of light; if there was no light, there would be no colour. Light is a form of electromagnetic energy which travels through space in waves. The unit of measurement of wavelength is the nanometre (nm) which is equal to 1 millionth of a metre. The electromagnetic spectrum is made up of waves of energy of varying length most of which are invisible to the human eye. Gamma rays, X-rays and ultraviolet rays at the short wavelength end and infrared, microwaves and radio waves at the long wavelength end of the spectrum can only be detected with equipment. Only the wavelengths between 380 nm and 750 nm can be seen and they are known collectively as the visible spectrum. The longest visible wavelengths are red and as they decrease in size become orange, yellow, green, blue, indigo and violet, the seven colours of the rainbow. When all the wavelengths are mixed together the result is white or colourless light.

In order to see colour, at least a little bit of light must be reflected off the surface of an object and be received by the eye. The light waves are received by the retina which has two kinds of cells that respond to them, rods around the sides which are sensitive to dim light and movement, and cones at the back which are sensitive to detail and colour. The information received by the retina is sent to the brain through the optic nerve for interpretation. The interpretation of colour depends on the amount and colour of the light absorbed or reflected by the surface of the object, so for example, when light strikes an object seen as yellow, it means that all the wavelengths in the light have been absorbed except for yellow which is reflected away.

Although each of the colours of the rainbow have their own wavelengths, some of them can be made by combining other different wavelengths together. Those which cannot be created in this way are known as primary colours which for light are *red, blue* and *green*. Mixing pairs of these colours together will make secondary colours, and subsequent combinations make all the other possible colours. Mixing colour as *light* will concern the design team in some circumstances such as buildings for the performing arts, special internal environments and the floodlighting of external facades, but is fundamental to the operation of film, television and computer monitors.

Of greater concern to the design of buildings is the mixing of colour as *pigment* for the selection of materials and colour schemes for the chosen finishes. Pigment is colouring matter used as paints or dyes and substances which give colour to materials, products and plants. When pigments are mixed together, the rules change because the primary colours are *red, yellow* and *blue*. Adding pairs together create the secondary colours of orange, purple and green, and subsequent combinations make tertiary colours and so on, offering the possibility of an immense range of alternatives. A mixture of all the coloured pigments will ultimately create black. It has been suggested that the human eye can detect at least seven million different colour variations, which to all intents and purposes means there are limitless options when it comes to selection. This becomes clear by looking through a window into the garden at the trees, shrubs, flowers and grass, which can all be described as 'green'. In fact, there are many different 'greens', some so dark that they are almost black, others quite light with yellows and white in them. The colours change depending on the brightness of reflected light, and how the light shines on the viewer, increasing and decreasing in intensity throughout the day.

In view of the range of options available, to simply refer to a colour as 'green' would be of little value, so before considering the use of colour in design, it is useful to understand how colours can be varied and how they can be described for communication purposes. There are three measurable ingredients in any particular colour.

(1) *Hue*
A hue is pure chromatic colour, containing no traces of any other colour, any black, white or grey. For example, red is a pure hue.
(2) *Chroma*
Chroma relates to the saturation or intensity of mixtures of pure hues. For example, yellow with 30% red will make a specific chromatic version of orange. Yellow with 70% red will make a different specific chromatic version of orange.
(3) *Value*
Value describes the effect of adding white, black or grey to any pure hue or chromatic variation. For example, blue with 50% white will be pale whereas blue with 50% black will be dark. Colours mixed with white are referred to as tints, mixed with black as shades and mixed with both, or grey as tones. The range of greys from white to black are referred to as neutrals.

Therefore, numerical reference to the percentages of ingredients in any particular mix will define its specific colour. There have been many attempts to classify colours for specification referencing in the construction industry, often reduced to a limited number of manageable standards. The full range of options is rarely necessary and a common understanding is often given based on British Standard reference numbers for a basic spread of choices. Although paint for example, can be obtained in many different colours, the choice of colour for a bathroom suite will generally be limited to the manufacturers standard range.

A more complex specification may be required for interior design work, for choosing decorative colours to match pre-coloured fixtures and fittings and for repair and maintenance work, when it may be important to try to match new colours with existing. Building materials are coloured in one of two ways, either as a result of an inherent natural finish belonging to the material, or by the application of an applied decorative finish. In both cases the colours must be selected to create the external appearance and the internal decor.

Colour as an element in design

Choosing and living with colours can be a very personal matter and enjoyment or dislike of colours is very much a matter of personal taste. Not withstanding the fact that some people are actually colour blind and unable to distinguish between certain colours, there is an emotional response to colours affecting the way that they are perceived. For example, red, orange and yellow are warm stimulating colours, whereas blue, green and violet are cool and relaxing. Colours have many associations which are commonly experienced and can create a mood or atmosphere in spaces which can affect people's health and well-being, sometimes at both ends of the emotional scale. In westernised cultures colours have the following associations.

- *Red* with health and well-being, with vitality, activity, the 'flame' of human spirit, yet also with anger, tension, warning and fire. A red light means 'stop'! Seeing red means being rather annoyed. 'In the pink' means feeling well; tickled pink means happy, looking through rose coloured spectacles means being unrealistic. Red letter day, red carpet, paint the town red are associated with celebrations; a red herring is a false trail.
- *Yellow* with the cheerfulness of the sun, a warm glow of contentment and yet also as a signal of danger and as a description of a coward. 'As good as gold' means being well-behaved.
- *Green* with the calmness of nature and growth, the soothing refuge of the theatre 'green room', and yet of also being unwell, inexperienced or green with envy. Green is used to discuss ecology and concern for the environment, energy conservation and sustainability. Yet it is also associated with money and a green light for 'go' and 'greener pastures' are always somewhere else.

- *Blue* with the calmness or tranquillity of the sea and the sky. Blue relaxes the central nervous system and is good for healing yet 'out of the blue' means unexpected and blue language is vulgar. A 'true blue' is a real friend, and yet blue also means sadness and longing; 'The Blues' is a form of emotional music associated with hard times.
- *White* with purity and cleanliness, yet also of being frightened, 'as white as a sheet' and the white flag of surrender. A white elephant is spectacularly unwanted and a whitewash is a total defeat.
- *Grey* with safety and security, but also non-involvement, concealment or doubt. Grey matter is brains and intellect, but ashen means illness or fear.
- *Black* with power and elegance, and yet also with death. Black suggests nothingness, extinction, the end. Blackmail, black list, black sheep, black market are all unfriendly or unpleasant terms.
- *Purple and violet* with mysticism, sensitive intimacy, charm, royalty and opulence. A purple patch means a successful period, purple haze is a state of confusion, purple prose is exaggeration.
- *Brown* with roots, hearth and family security.
- *Silver* with sleekness, elegance, modern, distinguished. Born with a silver spoon in one's mouth is a putdown.
- *Red, white and blue* suggests patriotism and national fervour.
- *Also,* how about marooned, bloody, crimson, bronzed, gingerly, honey, jaundiced, mustard, plum, rusty and sanguine which are all regularly used words based on colours.

The generally accepted theory for colour co-ordination comes from the work of AH Munsell, an American art teacher, who had the idea of linking the colours of the rainbow into a circle. The Munsell colour wheel has ten colours around its circumference, but further variations can be added in equal percentage steps as one colour shifts towards the next. The Dulux Colour Dimensions Circle includes 40 colours in the same overall format and is an excellent graphic representation of the association of colours. The typical colour wheel shown in Figure 10.9, has some interesting properties.

Analogous colours

Analogous colours, that is those which are next to one another anywhere around the wheel are unified by their common hue. They are harmonious when placed side by side and can safely be used together to create a restful atmosphere. The combinations of red + purple or red + orange, yellow + orange or yellow + green, blue + green or blue + purple all work together and can be extended further by adding variations on either side as for example, dark green + light green + yellow or violet + dark blue + light blue. In point of fact, the simple colour definitions given in this example are really too crude, as it is the subtle variation in chroma and value which work best in any combination. Many of the best unified colour schemes utilise a single hue, introducing change through variation in shades, tints and tones. For example, dark red, terra cotta, pink and cream could be the range of colours used for a room, all in harmony with one another.

Complementary colours

Complementary colours are the ones which are directly opposite one another anywhere on the wheel which have completely different hues. These colours placed together clash with one another, seeming apparently to vibrate, creating excitement and tension. Red and green, blue and orange, purple and yellow combinations compete against each other. This element of competition extends to cover the opposite third of the circle so that for example, yellow will clash with both red and blue. This situation is described by the term 'colour triad' and it

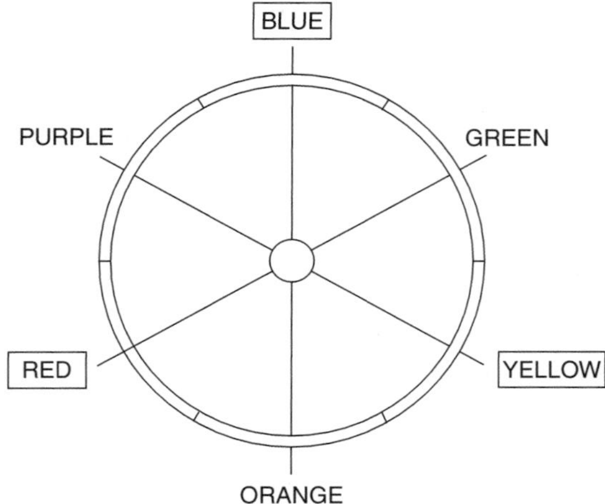

Figure 10.9 A colour wheel.

can create an image or an environment which is intense, hostile or simply uncomfortable. A complementary colour scheme can be softened by combining split complements together. This involves using a single hue with the two colours on either side of its complement. For example, yellow with a blue/purple and a red/purple. This is not such a confrontational mix as pure complements but retains elements of vitality and excitement.

Warm and cool colours

The colour wheel conveniently groups the warm and cool colours in opposing arcs facing one another; the greens, blues and purples on one side, which are cool and the yellows, oranges and reds on the other. Cool colours tend to move away from the viewer, making spaces appear to be larger than they actually are, whilst warm colours tend to do the reverse, moving towards the viewer, making spaces smaller and more 'cosy'. Rotating the selection around the circle can allow the colour scheme to have contrasts within it by for example selecting blues, greens and yellows, cool colours with contrasting warmth, or oranges, reds and violets, warm colours with contrasting coolness. Contrast can also be obtained by variations of background colour. A patch of colour will appear brighter in a dark background and dimmer in a light one, but a high level of contrast between colours can cause fatigue.

With respect to choosing colours for building materials, there are many criteria governing their selection, relating to cost, economy, availability and preference, always taking into account considerations for performance, strength, resistance to the effects of the weather, lifespan and ease of maintenance and replacement. No matter what technical reasons are used to support selection, their appearance is fundamental and the essential issue is whether or not they will be naturally finished or decorated. The implication is that a naturally finished material such as brick, stone, glass or timber should require little or no attention in the future with respect to its appearance. A naturally finished material will subtly change its appearance through weathering or through general usage but will essentially retain its integrity and its originally selected qualities.

An applied finish or decoration can be periodically refreshed or even altered completely at some time in the future. To a large extent the boundary between the two options in the UK occurs between external and internal conditions. The climate in the UK is such that every effort is now made to eliminate exterior decorative maintenance. The development of plastic coated steel sheet cladding and UPVC window and door frames, and fascias for example has lead to their widespread use on nearly every type of building constructed today. This is not the case in many other countries, where there is a greater sense of colour and flair in the external appearance of their buildings. Traditionally many buildings in this country were rendered and painted, but the onerous task of maintenance has largely made this type of finish redundant.

The possibilities for the use of colour are much greater with modern materials than was the case in the past, offering opportunities to create exciting and stimulating buildings, but equally imposing greater responsibility on those to whom it falls to make the choices.

10.10 Construction constraints

The elements of composition described in the previous sections *could* all be determined and mixed in the abstract as an artist might do in creating a free-form impressionistic picture, but it is rarely if ever possible to compose the appearance of a new building in a similar abstract way because the building is derived from functional requirements and the practicalities of satisfying them. For example, the floor area and height of the building are determined by the needs associated with the internal spaces, which will inevitably influence the location of structural elements such as columns and walls, and the position of door and window openings. The appearance of the building from any point outside or inside cannot be created without reference to the existence of these features, choosing to ignore them if their location happens to be inconvenient or does not quite suit the desired visual image.

The location of a window in an external wall, for example is determined by the need for natural light and views in particular location, and the practicalities of constructing it in that location, but its appearance is also part of the visual composition of the wall of which it is a part. Resolving the location of the window within the wall leads back to the dilemma of precedence or hierarchy of decision taking. Is the window a fixed element in the design so that other elements of construction must be fitted around it, or is there any flexibility to allow the window to be moved to a different location to fit in with other predetermined elements of construction? If the new building is to have a steel framed structure, what are the implications for an internal space whose only possible window location is at the point where there is a steel column in the external wall? Can the column be moved; can the space be moved; can the space function adequately without a window? There may be some tolerance but there is a risk that the design decisions which have lead to this situation were incorrect because the balance between form and function have not been fully accounted for at the analysis stage before proposing a solution.

This dilemma was discussed in Chapter 8 as a matter of design strategy. Is the frame to be visible or is it to be covered with some kind of cladding panels so that only the form of the cladding is visible? Should the framework within the roof be exposed or concealed? There are no absolute answers to these questions. They are posed to encourage reflection on possibilities, because although the structural elements impose constraints, they could be used as an advantage by making them visible adding a richness to the visual appearance.

With respect to the limitations of construction set against the possibilities of visual appearance, the two significant constraints are the nature of the design of the structural support system and the availability of commonly manufactured materials and components.

Structural support systems

The choice of structural support is largely determined by the size of the building in terms of both its horizontal and vertical dimensions. Continuously load bearing masonry in the form of brick and block walls is typical for small buildings like houses, where the height and span across spaces is relatively short, but larger buildings demanding greater spans are usually constructed with frames, either made from steel or reinforced concrete. A traditional load bearing masonry wall would normally be self finished in brick, block or stone, or sometimes rendered. The external appearance is the structure and the visual considerations are confined to selection of materials, bonding and pointing, together with functional matters such as expansion joints and daywork lifts. Unsightly daywork joints can often be avoided by having changes in the colour of the brickwork at appropriate levels or by introducing feature string courses to split the lifts into manageable proportions.

For the framed structure, the size of members, the columns and beams is determined by the work that they are having to do, the loading imposed on them. This does not prevent them from being larger than necessary for functional reasons, but will dictate minimum sizes. The spacings of columns is similarly determined which will have a profound affect on the appearance of the building both externally and internally. On one side of the wall, or the other there will be a bulge where the member projects beyond the face of the infill panels and is encased in concrete, brickwork or blockwork. The infill panels or cladding can be attached to the inside or outside face of the columns, or in between, so that either one side is smooth, or so that both sides have strong, regular vertical lines on them. The infill panels can be in masonry, subject to height limitations, or more commonly in steel cladding or glazing. In all cases there will be regular expansion and panel joints which are likely to be visible.

Other types of frame construction may be considered. Geodesic domes, space frames and high-tensile steel supported membranes may be practical solutions to the demands of covering large spaces, which each have their own characteristics with regard to appearance. Loads may be supported on domes, vaults and arches as well as by horizontal beams, affecting the appearance of spaces and openings. Cantilevers can be formed relatively easily in framed buildings.

Manufacturing limitations

Some materials and components are only available in predetermined sizes, either as a result of agreement between authorities or as the preference of the manufacturer. Bricks and blocks for example, are made in universal sizes, although variations and specials can be purpose made to order. Others can be made to size at little or no additional cost because they are manufactured to order anyway. For example, UPVC windows are not kept in stock, but are made to measure for each project. With regard to appearance, this is a significant advantage permitting the selection of components to suit the design rather than having to make do with what is available. Steel cladding and curtain wall glazing also offers this flexibility, although standard-sized units may be more economical because of the limitations of the manufacturers' machinery and processes. Large areas of cladding require covering trimmings at corners, eaves, ridges and junctions with openings.

Timber for components has dimensional limitations because of the nature of the material itself. Wood is a natural material with inherent instability, particularly when exposed to the effects of the weather. This dictates frame thicknesses and lengths, and the dimensions of boarding.

Plastics are also used commonly for rainwater goods. Surprisingly these apparently innocuous features are significant visual elements. Gutters, downpipes and soil stacks are very visible.

10.11 The effects of weathering

No matter how pristine the look of new materials on the outside of the building, exposure to the weather will, in time alter their appearance. Gradual changes in texture and colour will inevitably take place as a result of the actions of sunlight, warmth and cold, wind, rain, frost and general pollution in the air. The materials, colours and details of construction, carefully chosen to achieve specific visual effects may not stand up to initial expectations. Weathering can be beneficial, mellowing the surface finishes, adding to the qualities of the raw material. For example, the growth of lichens on concrete roof tiles can add considerably to the attractiveness of the roof, and the patina that develops on some sheet metals such as copper is often an improvement on the original appearance. This form of weathering softens the appearance, reducing harshness, brightness or uniformity of the basic material which can be rather overdominant, particularly in large areas like roofs.

Some other weathering effects are inevitable and must be accepted or tolerated, either left untouched or periodically maintained or cleaned. Most of the external materials will become discoloured through air-borne pollution, which will tend to progressively darken the surfaces. The client or the building occupiers would expect to clean areas of glazing on a fairly regular basis, but not so the remaining area of the external walls. Steel sheet panelling, brickwork and stone masonry will all accumulate grime, often in positions which remain damp for the longest period of time. A general darkening of surfaces would not be too obvious if it were over the whole area of the wall, but dark patches or streaks of staining can look very unsightly. The most common situations where this can occur are at ground level, where water splashes up against the building off surrounding hard surfaces, around and below window cills, where dirt is washed off glazing and below projections such as ledges, parapets and canopies, where rainwater on horizontal surfaces seeps down onto the vertical surface below.

The detail of the base of the building against the ground can be designed to minimise the visual effects of staining through the use of dark or brindle coloured materials, which are also sufficiently hard to prevent surfaces from remaining damp. Window cills and projections can be designed to include overhanging 'drips' so that water falls off rather than soaks into the wall beneath them, avoiding the concentration of grime and contamination that is washed off. The extent to which surfaces become stained will depend on exposure to wind, rain and sun. Some surfaces will be periodically washed clean by driving rain, others will not. The characteristic staining to one side only of columns or pilasters in an elevation demonstrates this effect, which in time will become a dominant visual feature, changing the appearance of the building. Providing it is practical to do so, this can be cleaned off to restore the original appearance.

Of greater concern to the design team and the client are weathering effects leading to transformation or breakdown of surfaces, or even in some cases loss of structural integrity. Discoloration is one thing, but serious damage can threaten to shorten the lifespan of the building. The most common failures are caused by the effects of water penetration and the cyclic effects of variation in external temperature. The solution to these problems rests with the selection of the materials and the detailing of the way that they are put together. It would not be practicable to outline all the potential problems and solutions in this book, but the general principles to consider include the following.

Materials and components

The selection of materials should be appropriate for the conditions in which they are to be placed, having regard to the degree of exposure to any of the elements during their anticipated lifespan. Manufacturers test data and recommendations should be followed carefully for matters such as absorption, tolerances, and potential expansion and contraction movements under different conditions. For example, some facing bricks will easily absorb rainwater, which in winter may freeze, expand and 'blow' off the surface of the brick. Water penetration into reinforced concrete may cause the steel reinforcement to oxidise and expand, blowing apart the construction. Any unprotected steel will rust, timbers may warp and twist, plastics become brittle. All these defects have significant visual consequences.

Assemblies

The elements of assembly which affect the appearance as a result of resisting weathering can include the pitch of roofs, the extent of projecting overhanging eaves, the size of window and door openings and the size of panels on framed structures. Construction should always be designed so that rainwater is shed away from the building rather than being left to soak in, form ponds or be uncontrollably discharged onto the ground. In the UK, rainwater disposal from roofs requires gutters and downpipes, which can be strong, often apparently random visual elements on any elevation. Consideration must also be given to the potential build up of snow and ice, which can slide off a pitched roof as it melts damaging anything below. The inclusion of snow guards to protect a conservatory for example, can be a powerful visual feature. Overhanging projections can offer shelter from driving rain, wind and solar gain, and can be developed or exaggerated for outside activities.

Materials which are likely to expand and contract such as sheet metal, plastics and flat roof membranes must be placed in such a way that anticipated movement can take place without causing buckling, tearing or creating gaps which water can pass through to the inside of the building. A framed building will have numerous details which allow for this movement, and lengths of brick, block and stone masonry must be subdivided to allow for thermal expansion if cracking is to be avoided. The choice of materials and construction details is intended to protect the qualities of the exterior surfaces and prevent deterioration of the structure and any of the internal surfaces and spaces.

10.12 The implications of wear and tear

The effects of wear and tear are the result of the way in which the building is used and the extent to which finishes or elements of construction become damaged over a given period of time. The pros and cons of selection of materials have been discussed elsewhere in the book, but as much as anything else they are a matter for the client to consider in relation to the anticipated lifespan and the costs of future maintenance and replacement. To this end, it should be clearly understood by all the members of the design team, and particularly the client, exactly what are the design criteria for acceptable wear and tear once the building is fully occupied. In purely visual terms, the issues concern the convenient, practical and economic cleaning of surfaces and the extent to which materials, components and assemblies are capable of resisting use without significant damage. To some extent, consideration of both is subject to the degree of wear and tear expected, the number of people using the building and the activities with which they are involved.

It may be thought that cleaning has little to do with design, that it is an inevitable, unavoidable consequence of the occupation of any building, which the occupiers undertake as they wish. In some circumstances this may be true, but consider the consequences of the design and specification of finishes and details for the reception space in the car-dealership. No buffer doormat to clean dirty feet at the entrance; a white carpet, which needs almost continuous cleaning; wallpaper which gets torn quickly; soft timber panelling which gets scratched easily; brass door furniture which needs continual polishing; high-level windows which cannot be easily reached for cleaning; out of reach light fittings full of dead flies; high-level ledges which accumulate dust. All are features which could have been designed differently to maintain their appearance with less effort.

With regard to deterioration, clearly design and specifications must provide circumstances which are suitable for the purpose for which they are required. The design team must assess the risks of likely damages against the capital costs and the costs of correcting failures. A regularly trodden on carpet in an entrance foyer will wear out more quickly than ceramic floor tiles. The appearance of the carpet will change much more quickly than the tiles, but it may be easier and more economical to replace the carpet from time to time than it would be to replace the tiles. The questions to consider are how easy will it be to replace, repair, maintain or redecorate surfaces, particularly if the design specification is for a *special finish* or materials which are finished in *special colours*? Will it be possible to obtain the same products or colours in the future? As a final thought in this section, if the development is undertaken in phases, will different suppliers and subcontractors be able to offer materials and finishes with the same appearance?

10.13 Project File content

The general aims at this stage are as follows:

- Completion of design proposals relating analysis to reality.
- Exploration of visual options.
- Assessment of the implications of horizontal layout against vertical appearance.
- Co-ordination of presentation material.

File material could include the following:
Illustration material

- Further presentation material based on the outline given in the previous Chapter's File content.
- Photographs showing views into and away from the site which may be useful to assess prominent vantage points.
- Photographs of similar building types for reference or inspiration.
- Sketch drawings of initial concepts with visual criticism, design concept drawings and finished presentation drawings.

Samples of materials and further trade literature

- Information about materials and forms of construction which must be incorporated into the design from a visual point of view.
- Samples of materials showing form, proportion and texture so that drawings are reasonably accurate.
- Colour swatches for standard materials and products showing the available range.
- Colour options for finishes which may offer greater variation such as paints and stains.

Colour schemes

- Details of proposals for external materials and interior decor, including fixtures and fittings, carpets and furniture, etc.

Cost assessments

- Checks that aesthetic considerations do not significantly alter the originally agreed target budgets.

10.14 Discussion points

(1) For many people, the 'kerb appeal' of new buildings is a swift judgement based on first appearances. What is the relationship between commercial buildings and their contents? How can designers create a favourable visual impression of a functional building such as a car-dealership? Is this an important aspect of promoting new car sales?
(2) How important are the UK regional variations in vernacular construction and use of materials? Should the standardisation associated with a company's style supersede local style? Should the Planning system be concerned with preservation of the status quo?
(3) The external appearance of some buildings is much more complex than others. What factors influence the degree of decoration of external facades in the UK? Why do building designers seek to change the appearance of buildings? Are designers able to claim visual expertise which overrides public opinion?
(4) Should building designers look backward or forward for inspiration for the appearance of their new buildings? What constitutes pastiche in the selection of materials; is there a good taste boundary? Should each generation of designers try to create innovative buildings belonging to their own time?

10.15 Further reading

Atherton J (2005) *Gestalt psychology*. www.learningandteaching.info/learning/gestallt.htm
Ching F (1996) *Architecture: Form, Space and Order*. 2nd Edn. Chichester: Wiley.
Cooper D (1992) *Drawing and Perceiving*. 2nd Edn. London: Chapman Hall.
Doyle ME (1999) *Colour Drawing: Design Drawing Skills and Techniques for Architects, Landscape Architects and Interior Designers*. Chichester: Wiley.
Hill R (1999) *Designs and their Consequences: Architecture and Aesthetics*. Singapore: Yale University Press.
Le Corbusier (1954) *Le Modulor*. London: Faber and Faber.
Le Corbusier (2000) *The Modulor: Modulor 2*. Birkhauser Verlag AG.
Pallasmaa J (2005) *Eyes of the Skin: Architecture and the Senses*. London: Wiley.
Parkin N (ed.) (2002) *The Seventy Architectural Wonders of Our World*. London: Thames and Hudson.
Patterson TL (1994) *Frank Lloyd Wright and the Meaning of Materials*. New York: Van Nostrand Reinhold.
Porter T, **Greenstreet** R and **Goodman** S (1980) *Manual of Graphic Techniques*. Volume 1. Oxford: Architectural Press.

Porter T and **Goodman** S (1982) *Manual of Graphic Techniques.* Volume 2. Oxford: Architectural Press.

Porter T and **Goodman** S (1983) *Manual of Graphic Techniques.* Volume 3. Oxford: Architectural Press.

Porter T and **Goodman** S (1985) *Manual of Graphic Techniques.* Volume 4. Oxford: Architectural Press.

Prizeman J (2003) *Houses of Britain: The Outside View.* London: Quiller.

Rattenbury K, **Bevan** R and **Long** K (2004) *Architects Today.* London: Laurence King.

Smithies KW (1981) *Principles of Design in Architecture.* London: Van Nostrand Reinhold.

Wakita OA and **Linde** RM (1999) *The Professional Practice of Architectural Detailing.* 3rd Edn. New York: Wiley.

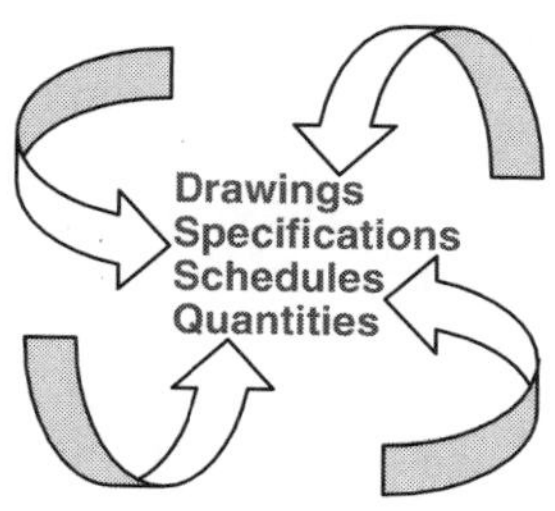

11

Construction information

11.1 Introduction

The earlier phases of the design development have been concerned with exploring ideas, leading towards a coherent presentation of an accepted possibility for the new building. The chapter on design planning shows the implications of critical decisions, taken by the development team which effectively move design work into the next stage. Although it is conceivable to step backwards, it is unlikely that design decisions will be radically altered once the team begins to prepare construction information. At this stage, the broad parameters associated with the size and shape of the building will be fixed and the design team will concentrate on identifying and defining all the elements of the new building so that they can be measured and costed in advance of construction commencing on site.

This chapter identifies the nature of the information needed and considers the production of working drawings, schedules, specifications and Bills of Quantity (BQ). The chapter concludes with a brief review of the use of supporting authority in the context of quality and cost control.

11.2 Making it work

So far, the outline of the new building has been described on typical sketches with limited detailed explanation. The plans, elevations and sections are essentially location drawings simply indicating the overall form and appearance of the building. It would not be practicable or cost efficient to fully describe, or even fully understand all the necessary construction details for elements which in any event, may still be subject to revision and which when fully considered will doubtless be refined. The extent to which the design team must develop and elaborate the details beyond this point depends on the terms and conditions relating to the procurement process that the client has adopted. Traditionally, the design team would begin to prepare working drawings, specifications and schedules so that the materials and labours needed could be quantified, with a view to inviting contractors to tender for the construction works. Alternative design and build or project management methods reduce the need for meticulous specification and description because the responsibility for some aspects of the work is taken directly by the contractor or the project manager. For example, the designer's drawings may indicate a brick wall of a particular length, height and thickness and include a note of which particular brick is proposed, but need not describe the requirements for good practice and workmanship in detail because this is an understood part of the contractor's service to the client, and any short comings are a matter for the contractor to deal with. It is of course important that the designer should be confident that the contractor is capable of offering work of a suitable quality before entering into such arrangements.

Changes can be expected and accommodated in the early sketches as ideas develop and information is obtained through research clarifying requirements. As decisions are taken and ideas are accepted and become well established, accommodating changes becomes increasingly more difficult as design work advances to meet programme commitments leading towards the time when fully costed proposals are ready for construction. The programme, described in Chapter 6 will usually include reference to critical points or 'design freezes' after which changes in design are unacceptable without having significant consequences on the costs and timing of the project.

The transition from the sketch level of detail to the preparation of construction information has two distinct implications.

(1) **The sketch ideas must be capable of being realised**

Loosely defined, indicative sketches should incorporate sufficient tolerance to accommodate the eventual demands of reality. For example, for the purposes of explanation to the client and the Planning Officer, a roller shutter door can be shown as a simple rectangle within the wall. To be meaningful, the simple rectangle must reflect that:

- its width and height are large enough for the purpose required,
- there is sufficient space within the structural elements of the wall to accommodate it,
- there are sufficient structural supports to hold it in place,
- it does not clash with other elements like steel framing or diagonal bracing,
- there is sufficient head room below roof eaves for the operating mechanism,
- there are no obstructions either inside or outside that would make the door difficult to operate.

If these issues are not considered in principle, then the sketched rectangle may be proved to be of little value.

(2) **Requirements must be defined to an appropriate level so that further design work and eventually construction can proceed**

Assuming that the general design issues are capable of being resolved, the specific definition of the element of construction must be clarified to sufficiently explain what is actually required under the circumstances. For example, the designer could simply specify 'roller shutter door' if the full specification is to be determined and agreed by others. At some stage a fully detailed specification of a particular door will be needed if the client is to obtain exactly what they want. Decisions must be taken by someone with regard to the roller shutter door's size, component parts, materials and colours in order to obtain costs for its supply, and also about how it is to be fixed and finished so as to determine the costs of its installation.

The implications of responsibility for detailed specifications will be discussed later but the designer's aim at this stage is to help to co-ordinate the detailing of the design framework drawings with the other members of the development team, integrating their requirements so that the ideas can be built.

Construction information is principally communicated in the following ways:

- Working drawings
- Specifications
- Schedules
- Bills of Quantity

11.3 Working drawings

The main factor distinguishing working drawings from any other type of drawing is the degree of accuracy indicated in this form of communication. Throughout the design development period, and sometimes during the construction process itself, sketches will be exchanged to illustrate ideas or intentions in principle, as a means of exploring and discussing options. Sketches can be rough, out of proportion and not drawn to scale, but they can still be suitable to communicate ideas well enough for the reader to understand. Once ideas are accepted though, sketches can be reconstructed accurately, in proportion and drawn to scale; representing reality. CAD drawings are automatically drawn to scale from the outset, using frameworks which can be enlarged or reduced to show the whole building at a small scale (say 1:200) or a detail of a window opening at a large scale (say 1:10).

In this way, the sketch ideas become working drawings which other members of the design team use as the basis for undertaking their own design work. They develop other drawings to illustrate their own contribution to the proposals. Details of construction are illustrated to gain formal approvals, to establish costs and to inform or advise contractors how to proceed on site, so that everyone can have confidence that all the elements required for the new building as described on the working drawings are appropriate, fit together correctly and that the building can be created as intended.

The working drawings cannot be regarded as finished articles at any one point in the process of developing construction information, because like the earlier design drawings themselves, changes will continue to be made and adjustments may still occur during the construction period, even up to the point of handover of the finished building to the client. The drawings are relevant to the stage of work currently in hand and it is common practice to keep sets of drawings on file as representing the development work at that stage. For example, the first issue of the designer's working drawings to the structural engineer may mark the point at which the engineer begins detailed design work for various elements of the structure, kept as a reference point for future work. Later on the package given to the quantity surveyor for measurement is a reference point for further changes as details and specifications are confirmed more accurately. The drawings issued for tendering, for commencement of construction on site and on completion showing 'as-built' records are additional points at which the working drawings will be updated to describe intensions for the new building at that time. The administration and management of working drawings is important as information is circulated to all members of the development team described previously in Chapter 3.

Working drawings are produced to illustrate information at three levels to explain:

- the location of spaces and major components;
- description of the separate individual components needed for manufacture or supply;
- details of the assembly of components for construction on site.

The distinction between each type of drawing was described in Chapter 3, together with the selection of appropriate scales, the importance of dimensioning and the meaning of lines. The content of working drawings is normally entirely 'technical', eliminating all the elements of 'artistic' interpretation. The drawings are produced manually with tee squares and set squares or electronically on computers. CAD is intrinsically accurate because it is necessary to insert mathematical definition in order to produce lines and shapes, but this does not mean that the drawings will automatically be co-ordinated, as the system still requires the designers or operators to check the figures that are inserted.

11.3.1 Working to a grid

To help to locate elements of construction or components on drawings, and in the process of setting out on site, it is useful to use a system of grid referencing, based either on arbitrarily fixed dimensions such as 10,000 or 5000 mm centres, or on the distance between repetitive elements like structural columns, which may of course be variable. For example, in a steel-framed building, the centre of each steel column along one side of the building might be referenced 1, 2, 3, etc., and A, B, C, etc. for columns at right angles, so that a feature within the building could be positioned by the reference 'B3' and an external door located as 3500 mm from grid 2, or at the early design stage, simply that it is located somewhere between grid 2 and grid 3. This simple referencing technique can be invaluable during the design process to ensure common understanding in discussion about a particular area or detail, especially in telephone conversations or in writing when the parties are not together.

The grid also creates a hierarchy of dimensions so that it is easy to separate the principle dimensions like the main structural elements from the minor elements fitted in between. This is particularly important when setting the building out on site, establishing the critical dimensions which can not be varied. For instance, depending on permissible tolerances, the pad foundations for the steel columns must be located accurately to within plus or minus a few millimetres, whereas positioning the external door may be a little more flexible, once the main structure is erected. The main setting out dimensions should be highlighted and clearly distinguished from others so that there is a minimal risk of confusion. If all the dimensions along an external wall were indicated consecutively in the same numbering style, there would be an increased possibility of a setting out error leading to a critical element being constructed in the wrong place.

11.3.2 Location drawings

Location drawings will be required to illustrate the site layout, all the floor plans of the new building, all the elevations and any number of 'typical' sections. Each drawing will have the purpose of communicating information, which for some of them can become very complicated. The site layout, for example can include hard surfaces, soft surfaces, drains, manholes, service ducts, parking space delineation and planting, as well as the building itself and the setting out dimensions needed to position them all. To illustrate everything on one drawing would be impracticable, and probably unreadable. Figure 11.1 shows typical working detail for part of the site. A single floor plan could show, walls, doors, foundations, drains, services, finishes and colour schemes, which again would be too much information for a single drawing. To this end, drawings are often split into information relating to a specific consultant or trade, prepared on copies of master drawings. In this way, drawings become 'layers' of information, selecting and describing one particular topic at a time, but great care is needed to co-ordinate the layers so that disparate elements do not clash. For example, the structural engineers' drawings for foundations, footings or ground beams must be compared to the drainage outlets and service intakes, which if

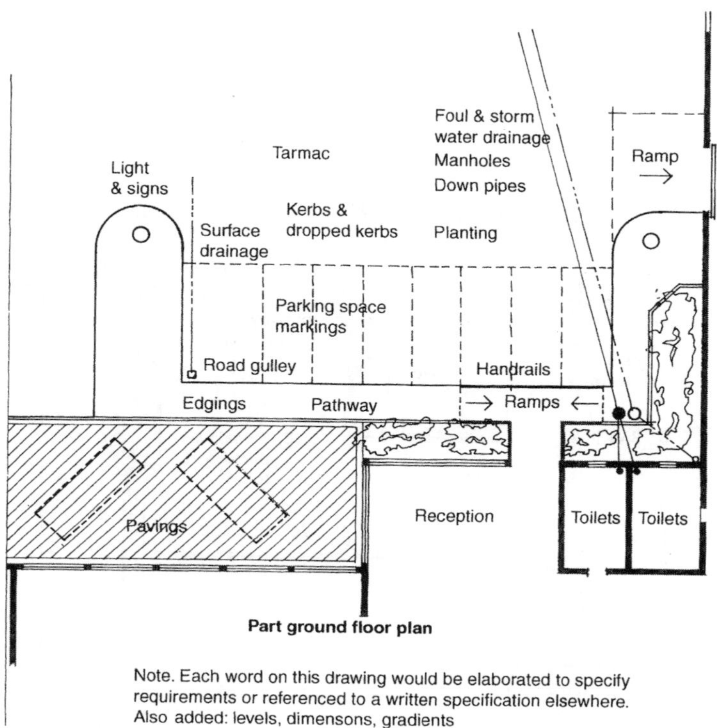

Figure 11.1 A site plan construction detail.

specified by someone else, may not be compatible. Although the sewer outlet may look convincing on the drainage plan, it will present considerable problems on site if it has been designed to run through the centre of a reinforced concrete pad foundation under a steel column. The location of a window on the designer's drawings and an internal vertical duct on the electrical drawings are not compatible. They must be positioned with respect to each other at the design stage. To discover the clash later on, particularly during construction is not an attractive prospect for the design team. A typical location floor plan is shown in Figure 11.2.

Layering of drawings also applies to buildings with multiple floor levels. The way that structural elements, drainage and service connections, staircases and lifts fit one above the another at different levels must be compared and matched on each plan. The lift engineer, for example will supply information relevant to every floor, including a ventilated housing at high level which may be above the roof, affecting the elevations as well as the floor plans. It is often difficult to comprehend layers mentally however good one's three-dimensional conceptual ability is. Comparing different drawings, styles and scales can easily lead to confusion. Once again, the grid referencing is an important link to check that dimensions at each floor level are consistent, and if drawings are prepared on film or tracing paper, they can be overlaid to check visually. Comparison of layers is an advantage with CAD drawings, but ultimately successful integration of all the building's elements is a matter of someone noticing or anticipating problems.

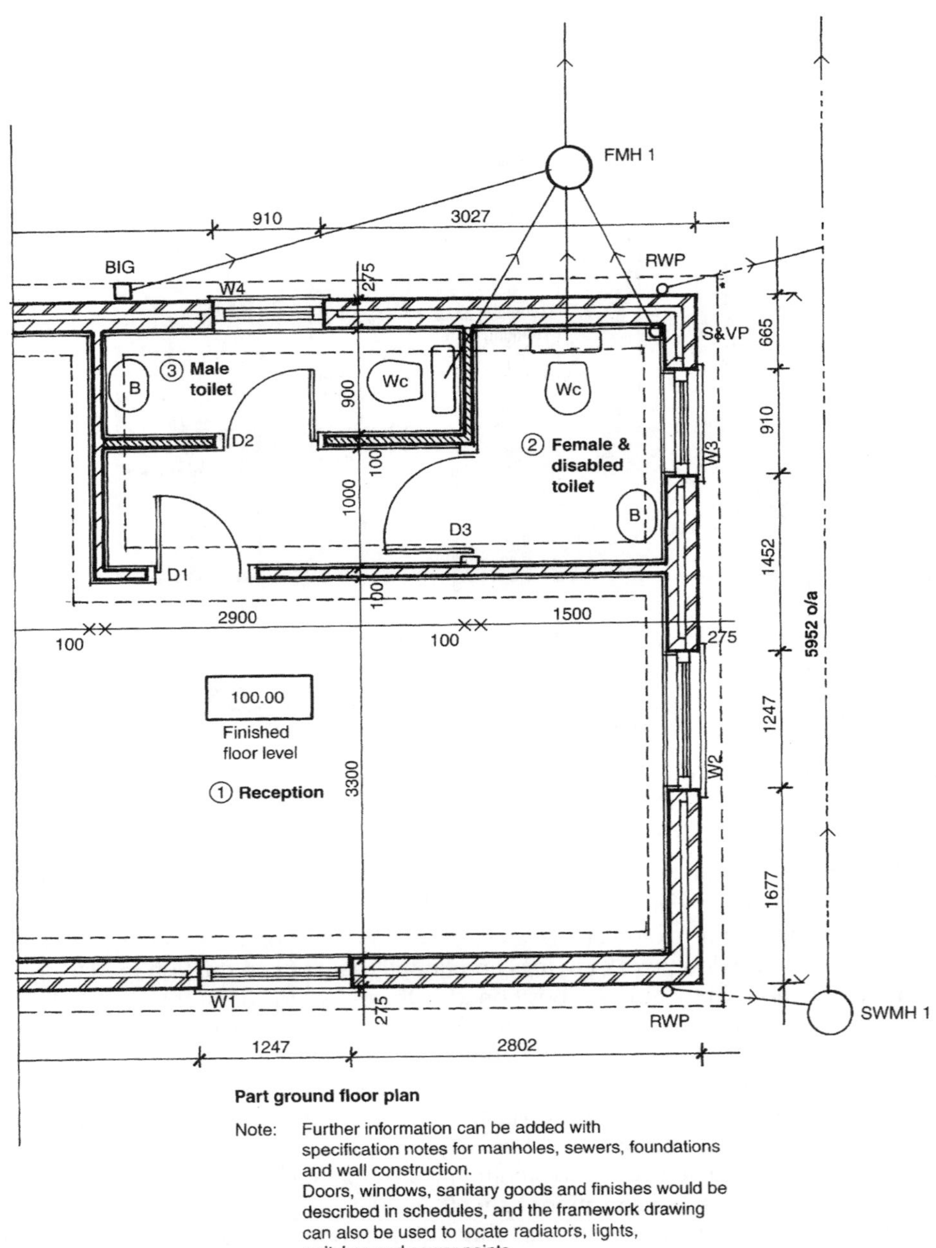

Figure 11.2 A floor plan construction detail.

The ease or difficulty of co-ordinating changes and revisions to drawings is a significant issue. If a location drawing includes too much detail, particularly for repetitive elements, the effort involved in preparing updates becomes substantial. If, for example, the encasement detail to every column on the location floor plan is shown in full, perhaps 20 or 30 times, not only is this pointless over-elaboration, but should there be a need to revise the encasement detail, then every single one must be changed. The use of CAD in handling repetitive elements eases this problem, but it can nevertheless take up time unnecessarily.

11.3.3 Component drawings

Component drawings are prepared to a much larger scale than location drawings to show the design requirements for a single element or a prefabricated set of elements. They would generally be used to show information about something which is non-standard or which must be made up before being delivered to site. Examples could include a staircase, an unusual moulding for a ceiling cornice or the requirements for a window, illustrated in Figure 11.3. Non-standard arrangements for a window or a staircase must show all the elements needed for their manufacture. It may be necessary to show plans, elevations and sections of each component to fully describe its content. It would be wasteful and difficult to give sufficient information on the location drawings for elements such as this, and would in any event be unnecessary, particularly if the component is repeated elsewhere in the building. If there were 20 'special' window configurations for example, it would be tedious and impossible to try to draw them all in sufficient detail for manufacture at a scale of 1:100 on the elevation drawings. A brief outline will suffice for location purposes, referenced to detail component drawings which can be prepared at a suitable scale of perhaps 1:50 or 1:20. This 'set' of drawings is then available for the manufacturer or supplier as a package, included with a specification and schedule.

As well as the designer's own drawings, the consultants, manufacturers or suppliers will also produce their own detailed drawings describing components. For example, the designer's rough sketch of a staircase with critical dimensions could be the basis for a joinery specialist to produce a full working drawing, which is then returned for comment and approval prior to manufacture. For some components, a specialist supplier or subcontractor will automatically prepare their own working drawings in any case. The supplier of steel sheet cladding for instance, will produce drawings showing all the panels, trims, flashings and linings at openings, fully referenced so that the correct items can be supplied and installed. These drawings will be based on the designer's plans and elevations, and brief specification of materials and colours. The designer retains responsibility for the overall concept, but the supplier should be able to assist with the preparation of the detailed information. The designer must still confirm that the supplier has interpreted requirements correctly and ensure that necessary amendments are incorporated before manufacture takes place. The suppliers working drawings become legally binding once approved, and liability for the consequences of subsequent discrepancies will be on the designer.

11.3.4 Assembly drawings

The assembly drawings explore and illustrate how the components are to be put together to create a satisfactory result. They are generally concerned with construction showing for example, how elements are to be securely fixed, how they can expand and contract, resist damp penetration, prevent heat loss, withstand frost damage or prevent fire spread. They can be used to appreciate the order of construction incorporating tolerances and safety issues. The example in Figure 11.4 shows how all the components surrounding the window are put together and fixed.

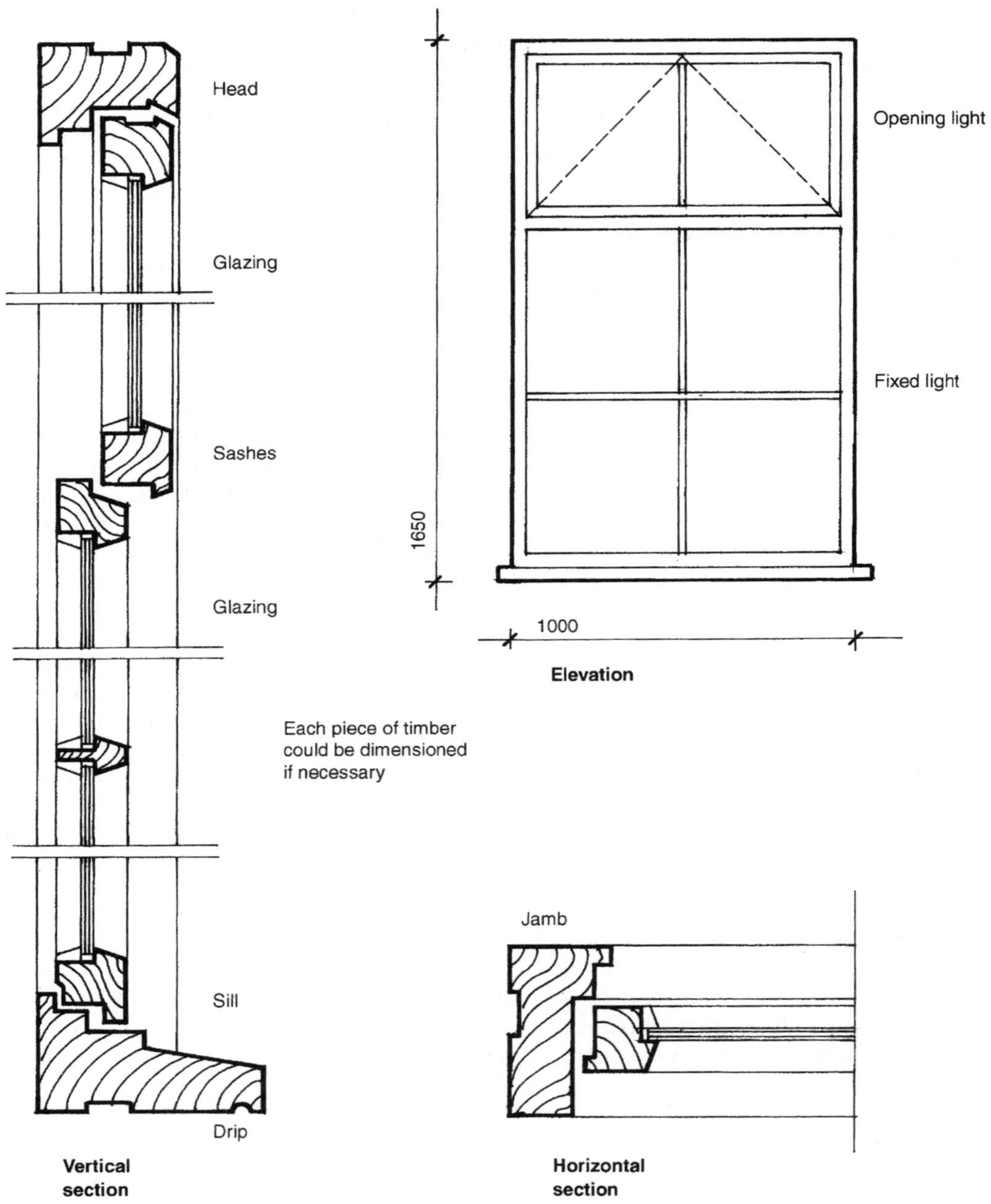

Figure 11.3 A component detail.

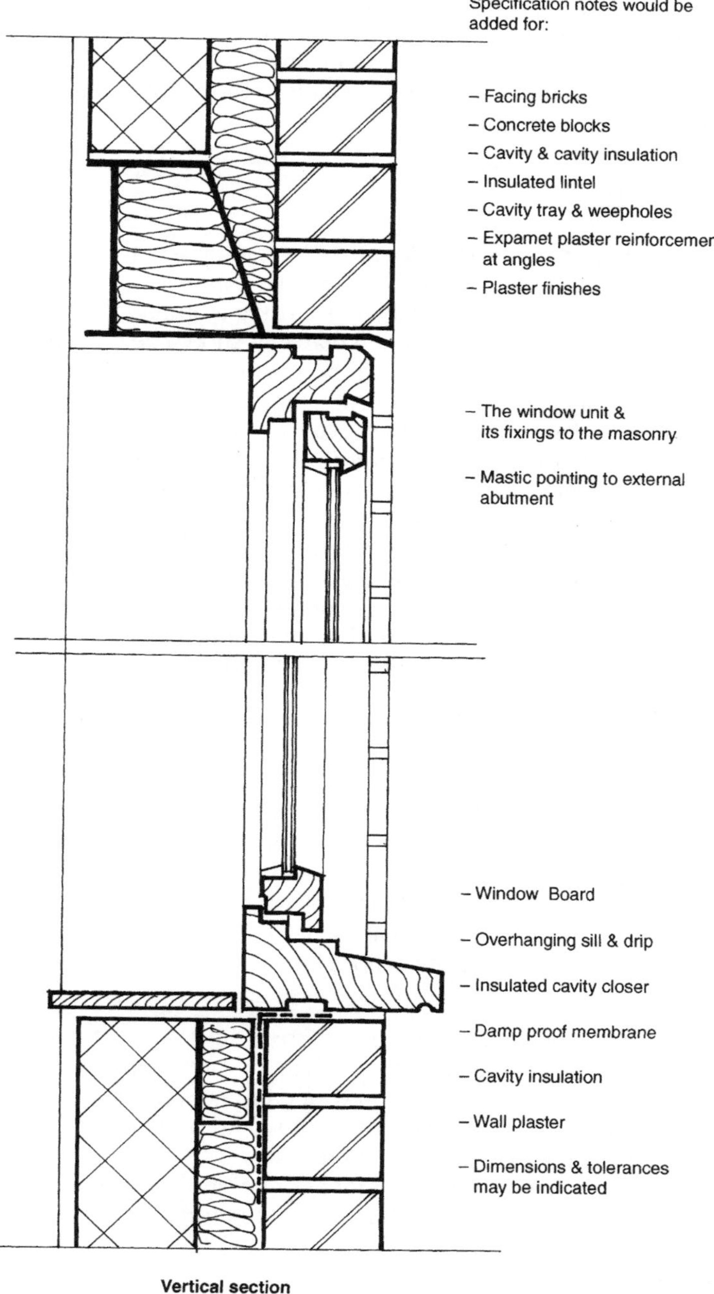

Figure 11.4 An assembly detail.

Notes can be added to drawings describing materials and workmanship, but they can rapidly become confusing. Methods of cross-referencing drawings to greater levels of detail elsewhere can be useful describing elements in a separate document. As a general rule, drawings should not become cluttered with too much written information. They should enable the reader to concentrate on location and visual definition.

11.4 Specifications

The purpose of working drawings is to identify and locate elements of the construction. It would be difficult, if not impossible to explain everything about the elements on the drawings themselves. The descriptive text could be confusing, and there is a risk that important information is obscured or unnoticed. In order to avoid these problems the drawings should be referenced to more detailed descriptions written separately in specifications. If necessary, details of the materials, labours and workmanship can then be fully described so that there is no doubt about what is required.

The extent of details included in the designer's specifications is very important and depends largely on the relationship between him or herself, the client and the builder. The traditional arrangement of precise designer specification, on behalf of the client is intended to tell the contractor exactly what to do. Different arrangements for procurement alter the designer's position so that the contractor may now tell the designer and client what they are going to do, or the client themselves will tell both designer and contractor exactly what they want them to do. Therefore, the degree to which the designer's documentation needs to be more or less specific is related to the understanding or expectation of all the parties as to what will actually be supplied and constructed. They all can take different views about what is an acceptable material or what constitutes acceptable construction.

Consider for example, the specification for bricks and brickwork. It is easy to presume that a designer understands how to select bricks and detail brickwork, that a manufacturer will know how to make bricks, that a builder or bricklayer will know how to lay bricks and anyway, everyone knows what brickwork looks like, don't they? In the event of the brickwork not being satisfactory, the client who is paying for the bricks is left in a vulnerable position if there is no formally agreed description of expected standards, which it can be shown have not been complied with. The specification for 'good' brickwork is no different in principle from that for a new car or television set, assuring the purchaser that the materials and workmanship are suitable for the purpose to which the article is to be put. The range or level of detail for specifications can include a simple statement as follows:

- *Brickwork*
  Whilst this is an example of a common item, there is little scope to achieve redress in the event of poor quality because of lack of any definition. Accusations of 'that is not what I expected or am prepared to accept' may fall on deaf ears, as even though it may be poor brickwork, it unarguably *is* brickwork. An improvement would be to state:
- *Facing brickwork*; Butterley Brick, dark brown brindle
  This is better, but does not specify bonding, mortar or pointing which affect visual appearance, or strength, density and cavity wall ties, which affect structural performance. The specification could be further elaborated as follows:
- *Facing brickwork; Butterley Brick, dark brown brindle, in stretcher bond with 1:1:6 coloured sand cement mortar with bucket handle joints*
  This is more precise but does not specify standards for the performance of materials or workmanship in putting them together. Permitted dimensional variations, daywork lifts, working in

cold weather and storing on site are all issues which can affect the quality of the finished work. The specification must define all these requirements in sufficient detail if the desired product is to be obtained. A more certain way to establish required quality would be to further add to the specification:

- Facing brickwork; Butterley Brick, dark brown brindle, in stretcher bond with 1:1:6 coloured sand cement mortar with bucket handle joints all in accordance with BS xxxx (British Standard for materials) and CP yyyy (Code of Practice for workmanship)

This represents a thorough description of requirements and providing that it is followed on site, guarantees as far as it is practicable that the completed work will be satisfactory. If the designer is in the position of specifying the brickwork then this level of description is essential. Even if the designer does not specify the brickwork, the client and the contractor would be well advised to take steps to ensure that they have a common understanding of brickwork if the finished product is to fulfil intentions without leading to costly disputes.

The position for the building designer is relatively straightforward. If required by the client to provide information, instruct others and approve the quality of their work, then specifications must be comprehensive. If on the other hand contractors, subcontractors and suppliers are offering materials or construction services directly to their client's, for which they are appropriately qualified, then it is they who take the responsibility for the correct supply of standard elements and methods of construction. Of course, the design and specification of something with which the contractor is likely to be unfamiliar such as a special effect, finish or detail must be very carefully described if the result is to be a success. In this case, all the information needed to achieve the end product must be made available. If things go wrong, it may be proved to be the designer's fault because of an inadequate specification, becoming liable either to the contractor for additional costs or to the client for a failure.

Typical specification clauses would be prepared for the following elements of work.

Demolitions, Alterations and Renovations; Groundwork; In situ Concrete and Large Precast Concrete; Masonry; Structural/Carcassing Metal/Timber; Cladding/Coverings; Waterproofing; Linings/Sheathing/Dry Partitioning; Windows/Doors/Stairs; Surface Finishes; Furniture/Equipment; Building Fabric Sundries; Paving, Planting, Fencing and Site Furniture; Disposal Systems; Piped Supply Systems; Mechanical Heating, Cooling and Refrigeration Systems; Ventilation and Air Conditioning Systems; Electrical Supply for Power and Lighting Systems; Communications/Security/Control Systems; Specialist Systems associated with fuels, etc.

Different forms of specification include the following:

- *Traditional specification*
  This document is prepared by the designer or quantity surveyor describing all the work involved in constructing the project. It would include preliminaries, preambles and definitions of requirements for materials and workmanship. It covers each activity or trade from excavation to decoration. It would not include measurements of quantities, which would be the contractor's responsibility for pricing and ordering materials.
- *Performance specification*
  Prepared by any of the consultants describing the way in which any part of the project will be expected to perform. It may apply to criteria for design, for example stating maximum and minimum temperatures required at different times of the day so that the M&E engineer or subcontractor can design an appropriate heating installation, or for supply, for example stating minimum floor loading requirements so that the flooring contractor supplies a system strong enough to meet it.

- *Standard pre-written specifications*
  Systems are available for the use of pre-written clauses such as the National Building Specification (NBS) covering every conceivable element in the building's construction. The clauses that should apply to the project can be selected and edited saving time in concocting new covering statements. It can be a useful system but still requires close attention to ensure that the correct information is made available.

Supporting authority
In the example given in the previous section, the specification for brickwork, or any other element in the new building, can be improved by reference to a recognised authority as a source of information about the element and the way that it can be used. The advantages can be summarised as follows:

- *Providing useful information at the design stage*
  There are many documents currently available for reference to help the designer to produce proposals and drawings which are realistic. Information about materials, components and recognized forms of assembly can be a sensible basis for a preliminary check that sketch ideas can be translated into a meaningful representation of the 'real thing'. There is also little point in struggling to redesign a 'standard' element which is already well documented.
- *Providing reassurance about the quality of specifications*
  Reference to accepted standards ensures that specifications are realistic, and offers the client assurance that by accepting the recommendations, the products and workmanship proposed for the new building should be of a recognised and understood quality.
- *Providing a source of quality assurance for elements of construction*
  The use of appropriate forms of supporting authority facilitates the communication of complex ideas in a relatively simple way so that everyone understands what is required and can better appreciate the demands and practicalities of construction, allowing for all the costs associated with the work.
- *Providing a reference for the quality of completed work*
  As materials are delivered and construction completed, the client or their nominated agent may choose to inspect and approve what is being offered, leading eventually to acceptance and hand over of the finished building. The basis of agreed standards of quality is likely to be much more certain if it is in accordance with a recognised authority, defining how the work should be undertaken and what the finished article should be. If for example, a CP for brickwork states how bricks should be stored on site, it provides the necessary support to require defects resulting from non-compliance to be remedied. If materials and construction are offered in accordance with an approved standard, then that is the best that can be achieved at the time.

The following sources are available:

- *British Standard Specifications (BS)*
  BSs gives details about the minimum criteria for materials and components. A BS relating to bricks, for example would give information about dimensional tolerances, permissible water absorption and anticipated strength, based on the results of exhaustive testing. A manufacturer is entitled to use a BS to describe their bricks on the basis that their products comply with the relevant standards. Some products carry a Kitemark as visible proof.
- *British Standard Codes of Practice (CP)*
  CPs refer to the appropriate qualities of workmanship needed to put the materials and components together. A CP for brickwork, for example will give information about building a

brick wall, including pointing and bonding, daywork lifts and protection from excessive water and frost.

- *Agrément Certificates*
  An Agrément Certificate can be used by a manufacturer or a supplier to verify a system of materials or components. For example, a flooring manufacture may offer a system of reinforced concrete beams, concrete blocks, polystyrene insulation and chipboard decking as a complete package. Although each material may be covered by a BS and a CP, if the performance of the unit is tested and approved as being acceptable, the award of an agrément certificate indicates the necessary quality assurance to specifiers and purchasers.
- *Building Research Establishment (BRE)*
  The BRE is an independent body which undertakes tests on many aspects of building construction. Their publications explain the results of the tests and can be used to support specifications.
- *Building regulations*
  The Building Regulations, described in Chapter 4 contain design criteria and details as examples of construction which is 'deemed to satisfy'. Compliance with the current regulations is of course essential, but it is also a useful source of reference for some design problems.
- *Government publications and legislation*
  A range of documents are available for design and construction requirements for different building types and users, including Local Authority Planning Design Guides, County Council specifications for highways and drainage, Department of the Environment Circulars, Home Office briefs for Magistrates Courts and Prisons and publications about design issues relating to children, old people and the disabled.
- *Manufacturer's current printed data*
  Most manufacturers and suppliers produce their own literature describing their products, the criteria for selection and use and recommendations for methods of installation, handling and storage. This information is essential as detailed working drawings are prepared and is an important means of obtaining quality assurance for finished work.
- *Trade publications*
  For some trades such as brickwork, lead, clay tiles and concrete pipework, manufacturers are represented by an independent organisation, which produces definitive information about the use of the product. This can be a valuable source of authority to ensure that specifications are consistent rather than subject to the individual interpretation of each manufacturer.
- *Technical press and reference books*
  Existing practice and current developments are well documented in books and journals. As with any of the reference sources described above, it is important to use information which is 'up-to-date' and currently applicable. Any material collected into a 'library' must be periodically examined and checked that the content is still relevant.
- *Forms of contract*
  Contracts between parties are a form of quality assurance defining the relationship between the parties and their expected behaviour and performance throughout the contract period. See Chapter 12 for details about contract documents.
- *Design guides*
  Section 2.8 introduced the idea of published design guidance. As well as general sources of information, local authorities sometimes produce specific guides about how buildings in their areas should be designed, or even for specific sites, setting out their expectations for the development.

- *Research papers*

  Higher education institutions are continuously publishing research information about the built environment. Specialist journals held in libraries throughout the world spread ideas which will influence design and construction in the future.

As a general point to conclude this section, if the designer assumes the responsibility of providing detailed specifications, it is sensible to be prepared to admit ignorance and seek specialist advice. There is no merit whatsoever in proposing or demanding action by others which cannot be achieved. It may also be to the building designer's advantage to ask specialists how something should be done. A bricklayer may be the best person to ask about bricklaying.

11.5 Schedules

Working drawings and specifications locate and define many of the elements of construction in principle, but whilst some of them are essentially repetitive they need additional definition because their specification varies in some respect. For example, although all the internal doors are indicated on the floor plans, the exact specification will vary from door to door with variations in size, materials and ironmongery, which must all be determined for each door so that they can be costed, ordered and installed. This information is assembled in schedules, collecting together all the elements associated with a single trade or supplier so that it is easy to see the extent of work. As well as confirming the exact specification of all the grouped elements, a schedule is the best way for communicating requirements for elements which are spread across a number of different drawings. For example, all the windows for the new building will be shown on the elevation drawings, which for a large building could be numerous; one A1 drawing may only show one window. By showing all the windows together, they can be illustrated on a single drawing, reducing the paperwork and avoiding the possibility of any of them being missed. Schedules can be used by the contractor for ordering the elements once construction has commenced, but should check their accuracy first.

Typical schedules would be produced to describe the following:

- *Steel reinforcement*
  Schedules are produced by consultants including bending schedules for steel reinforcement normally prepared by the structural engineer or by suppliers for steel cladding and trim variations. The designer will be required to approve the information contained in these schedules which are based on his or her detailed drawings.
- *Drainage, manholes, stop cocks, etc.*
  These schedules will include information about pipework sizes, depths and gradients, the size and depth of manholes and inspection chambers, and the location of control equipment.
- *Doors*
  A door schedule contains all the information needed for the supply of each door and could take the following format:

D1	Front entrance door to showroom as identified on ground floor plan
1800 mm × 2100 mm	Overall size of structural opening
Frame material	Aluminium

Leaf material	Aluminium
Colour	Blue frame, red leaf (accurate specified colour references would be given)
Glazing	6 mm toughened glass (and possibly details of beading)
Swing	Double swing (inwards and outwards) both leaves
Fire rating	Not applicable (NA) (in this case, but relevant for other doors in the building)
Vision panels	NA
Draught stripping	Specific details or 'as current Building Regulations'

- *Ironmongery*
  Ironmongery is normally the metal components associated with the operation of doors and windows, although in some cases other materials such as plastics may be included. The ironmongery schedule contains all the information about the fittings for each door and window as necessary including, for example, a list indicating requirements for hinges, locks, levers or handles, push plates, kicking plates, pull handles, automatic closers, letter plates, spy glasses, safety chains, bolts, indicator bolts (for toilets), etc.
- *Windows*
  The window schedule is similar to the door schedule identifying dimensions, fixed and opening lights and materials.
- *Lintels over openings*
  It is common practice to prepare a lintel schedule, specifying accurately the length, depth and materials for the lintel over each structural opening. Care should be taken not to confuse the structural opening dimension (say 1800 mm) with the lintel length (say 2100 mm) which allows for end bearings. This is a good example of a schedule which may not be suitable for ordering purposes, if the end bearing dimensions have not been added. It some cases, it may be appropriate to add the lintel specification to the door and window schedules rather than produce a separate document.
- *Sanitary goods*
  Sanitary goods in toilets, bathrooms and kitchens are collected together in a sanitary schedule including wc's, basins, sinks, baths, showers, bidets, urinals and disabled equipment. The schedule may include details of pipework and plumbing fittings.
- *Fixtures and fittings*
  A schedule may be needed to describe cupboards, benches, worktops and shelves which are to be manufactured off site and delivered complete, ready for installation.
- *Lighting fittings*
  A schedule of fittings may be prepared as indicative to assist the specialist consultants or subcontractors. It might include numbers and locations of items such as fluorescent tubes, spotlights, downlights, wall lights and emergency lights.
- *Electrical power points and switches*
  As with light fittings, an indicative schedule might locate socket outlets, fused spurs, cooker panels, single and two-way light switches, television and telephone points, etc.
- *Internal finishes, floors, walls and ceilings, ceramic tiles*
  Schedules will include materials, sizes, surface textures and natural self-coloured finishes.
- *Applied decorations and colour schemes*
  Schedules would normally be prepared for each space or each room where there are significant differences. Even if there are no differences, it is essential to explain to the decorator what is required in each space or room.

- *Signage and notices*
  In large buildings, each door may include room titles or numbers and there are a variety of other standard or special notices fixed to walls, including direction signs, fire notices and personnel identification.
- *Fire-fighting equipment*
  Schedules based on the advice and requirements of the fire officer will show the location and specification of alarm points, smoke detectors, fire blankets and extinguishers.
- *External surfaces and finishes*
  For complicated siteworks, schedules might be required to define areas of tarmac, concrete or concrete paving, block pavers and soil.
- *Planting*
  The details of the planting layout will appear on the siteworks drawing, but it may help to list the species type and number for each area.

Most schedules likely to be prepared by the designer are lists of materials and components. Occasionally, particularly for projects involving works to existing buildings, it can be useful to indicate detailed requirements in a *schedule of work*, defining the extent of work which cannot be numerically quantified or shown easily on drawings, such as replacement of damaged fittings or finishes.

11.6 Bills of Quantity

The information contained on drawings is often indicative or 'typical', supported by specifications for the materials and the way that they are to be used. Many of the drawings will be location drawings showing how components are placed together, illustrating each component with an example of construction at various strategic points around the building. For example, a single section through the external wall may sufficiently show all the relevant elements of construction, although it is probable that more than one section would be needed to describe all the conditions for all the walls. A limited number of typical drawings will communicate design intentions but they are not sufficient to determine quantities needed to establish costs, place orders and eventually undertake construction. The drawings will show a brick wall, and the specification will describe the brickwork, but the number of bricks and cavity wall ties, the volume of mortar and the quantity of insulation are not numerically defined on the drawings.

For small contracts, the cost of independently measuring all the components would be uneconomical and the task is normally left to the builder at both estimating and ordering stages. For larger projects, the client's chosen method of procurement will determine whether or not there is any advantage in measuring all the materials and labours involved in the new building in advance of construction. In many cases, contractors will not be willing to bear the cost of measuring the work themselves as they may not win the contract. Traditionally, the BQ are prepared by quantity surveyors as part of the process of obtaining competitive tenders from contractors. Normally, the measurements are contained within a single document, usually referred to as a BQ. With the other methods of procurement discussed previously, the BQ may not be required for purely competitive purposes although it may still be useful to the contractor for preparation of costings, and for both contractors, project managers and clients as a means to more precise cost control.

The BQ is prepared by the quantity surveyor based on the design drawings and specifications together with relevant information form all the consultants. The information is assembled in sections defining exact or provisional measurements. The mechanics of preparing the BQ should not be confused with the general cost advice given by the quantity surveyor throughout the design development period. In order to take off measurements the QS must have sufficient information, and it is a process which will not be repeated again and again to accommodate design changes. Prior to billing, the QS will be able to assist with decisions about sizes and volumes, construction systems and materials, finishes and decorations, but once complex elements of construction have been measured, the results will appear in the BQ. As the document is being prepared, all the members of the design team exchange information, explaining points or clarifying errors and omissions. For a large complex project, this is extremely helpful as individuals concentrating on their own work may easily miss or forget important items and elements of construction.

The advantages of the BQ can be summarised as follows:

- *Acting as a check list accounting for all the necessary materials and labours*
  The QS's methodical approach to compiling the bills minimises the risk of omissions.
- *Identification of expensive or uneconomic elements in advance of tendering which could be handled differently*
  As information is provided for billing there is an opportunity to consider the economy of specification
- *Assisting contractors to prepare realistic tender offers on a like for like basis with their competitors*
  Contractors may actually be unwilling to tender at all if they are put to the expense of preparing their own quantities knowing that they may not secure the contract.
- *Enabling evaluations of pre-contract and post-contract omissions and revisions on the basis of agreed rates*
  Should the client, designer or contractor wish to change any aspect of the works at any stage after the submission of tenders, the QS can value them in accordance with known costs.
- *Ensuring that stage payment evaluations are related to the value of work completed*
  When the contractor is entitled to payment, the evaluations of work carried out can be much more clearly identified and agreed.

The format of the BQ is generally as follows:

- *Preliminaries*
  Including a general introduction, definition of contract clauses and controls applicable to the project.
- *Preambles*
  Including a general specification of materials and workmanship for each trade. In this section the contractor is asked to identify percentage costs for profit and other on-costs such as plant and labour.
- *Bills*
  Including the measurement of materials and labours for each trade, by length, area, volume or number.

Typical bills would be prepared for the following:

Demolitions/Alterations/Renovations; Groundwork; In situ Concrete/Large Precast Concrete; Masonry; Structural/Carcassing Metal/Timber; Cladding/Coverings; Waterproofing; Linings/Sheathing/Dry Partitioning; Windows/Doors/Stairs; Surface Finishes; Furniture/Equipment; Building Fabric Sundries; Hard Surfaces/Paving; Planting/Landscaping; Fencing/Site Furniture; Disposal Systems; Piped Supply Systems; Mechanical Heating/Cooling/Refrigeration Systems; Ventilation/

Air Conditioning Systems; Electrical Supply/Power/Lighting Systems; Communications/Security/Control Systems.

- *Prime cost sums*
  For larger projects, it would be uneconomical to accurately cost some subcontract work such as the electrical installation, or the supply of special materials like the flashings to profiled steel cladding until the principles of construction have been established, either by competitive tender or by negotiation. The normal practice is to cover these elements of work with realistic estimates, which can be resolved later on. In the traditional procurement process, the use of prime cost sums also gives the design team the opportunity to directly nominate and control specialist subcontractors and suppliers rather than accept those offered by the contractor. The tendering contractor can include percentage additions to cover attendance and profit.
  The principle is similar with other forms of procurement allowing the client and the contractor or project manager to discuss, negotiate and agree costs based on estimates which can be confirmed later once the project is under way. Ideally, the prime cost sum included in the BQ, together with the contractor's percentage addition *should* be sufficient to cover the actual cost.
- *Provisional sums*
  Provisional sums are figures included for expenditure on elements of construction which can not be measured, or may be unknown until work commences, such as the cost of excavations for foundations or working in abnormally wet ground.
- *Contingency sum*
  A small figure may be included to cover unforeseen costs which can be deducted from the final account if not required. This can help the client to budget for unavoidable additional costs, and also gives the design team a little additional flexibility.
- *Summary*
  The summary is a statement of the costs for each section and a final total.

As an example, this is a selection showing a typical entry in the BQ, including the contractor's estimated evaluation.

Bill No. 3 Section D
Brickwork and Blockwork

01. Bring to site and remove from site all plant required for this section of the work.	Item	£1500.00	
02. Facing bricks as described pointed with neat flush joint as work proceeds. Half brick wall in skin of hollow wall.	586M²	£32.02/M²	£18 763.72
03, 04, 05, etc. Descriptions of other associated or incidental items	xxxM²	£yyy/M²	£zzz
To collection:			**£28 300.23**

Page 3/57

The addition of figures from all pages in Bill No. 3 are collected as a total cost for Brickwork and Blockwork which appears in the summary contributing to the grand total. Note that the figure of £1500 for plant (item 01) could appear in the preliminaries rather than the measured bill, and should be checked to avoid duplication. The exact cost for each element can now be seen, and forms the basis for decisions about any necessary revisions or changes. The mathematics can be checked, confirming that items have been added up correctly and that the total cost is correct.

11.7 Cost control

The significance of cost to the designer has been considered in terms of economy of design. The estimated cost of the scheme should broadly be in line with client's available budget, and wherever possible, proposals should avoid or eliminate features which are unnecessarily wasteful. At the construction information stage the design team are engaged in producing information to enable contractors to tender for the work, or are part of the contracting team organising themselves to start building. The framework of the building is set, and it may be too late to fundamentally change the design to achieve better economy, but the selection of materials, components, assemblies and finishes is still open to examination and determination, influenced by practicalities, appearance and cost.

In this respect, with regard to economy, decisions can be affected by the costs of supply and construction and the effects of 'cost in use'.

The costs of supply and construction

Requirements illustrated on drawings and defined in specifications are the starting point, but actually achieving or obtaining them may not be quite so simple. If the design of any element is 'non-standard' or 'special' it will almost always be more expensive to manufacture adding costs which may or may not be justifiable. Once work has started on site, it may be discovered that specified materials and components are no longer available, or that they cannot be delivered quickly enough. There is a risk of delay while alternatives are arranged adding extra costs for the contractor and client. A particular element of construction might be frustrated through lack of suitable labour or equipment or through particularly inclement weather conditions. These issues must all be considered as specifications are determined, attempting to predict possible difficulties which may increase costs.

The effects of 'cost in use' for the foreseeable future

The design and specification of some elements in the new building can be governed by consideration of costs over a period of time of use. The period could be the predicted lifetime of the building, 30, 60 or 100 years, or whatever figure is agreed as appropriate. The period could be quite short, 2, 5 or 10 years, a more easily understood target based on present experience. It is becoming increasingly more difficult to predict changes in standards and technological advance which could revolutionise building design. The notion that a building should last for 100 years, even 30 years may soon no longer be as important as it once was. Whichever figure is selected, the initial capital cost can be related to the potential running and maintenance costs to assess the possible financial benefits of more or less investment at the beginning.

For example, consider the design and finish of an internal wall. It could be an expensive, high quality, pre-finished facing brick or stone construction which over a period of say 30 years requires no maintenance or decoration. Alternatively, it could be a very much cheaper concrete block and plaster construction which will require maintenance and redecoration every 3–5 years. The capital and maintenance costs can be compared over the period to see which represents the better value for money. The costs of borrowing must also be included in the equation, adding interest for higher expenditure at the start or saving interest on lower expenditure, but spending money periodically in the future. The assumption in this example of course, is that the wall is to be maintained in its original location and condition. For many commercial buildings, the occupants may wish periodically to reorganise the spaces within the building, or change the appearance of the wall. In such a case, the expenditure on a higher quality of construction and finish may be a disadvantage and represent a waste of money. It may be much more economical to specify a cheaper system which can be easily modified to suit future demands. It could equally be argued that in the light of the need for flexibility, it could be appropriate to use a demountable system for the wall, which could be more expensive than facing brick or stone, but which can be taken apart and reused in the future. It is only by considering the costs over a projected period that a decision can be taken about which one represents the best 'value for money'.

Other elements will have a more rapid pay back time. For example, an expensive but more efficient heating system can pay for itself within a very short time by reducing energy consumption. The choice of cheaper equipment may not only represent higher running costs, but may lead to more regular maintenance and discovery that it wears out sooner, requiring earlier replacement. The introduction of an automatic door to the workshop, which is an expensive piece of equipment may not only lead to a warmer, happier workforce, but save so much money by eliminating wasted heat that its cost is recovered in savings within 2 or 3 years.

As well as the running and general maintenance costs typical to all buildings, the client and future occupants may be faced with additional costs because of the difficulty of gaining access to carry out maintenance because of the way that elements have been positioned. For example, high-level windows are much more difficult to clean or replace than those which can be reached easily. Changing a light bulb in a fitting on the ceiling at the centre of a large open space could require the installation of scaffolding, and replacing a broken part in a heater located in a tight corner with insufficient working space may take much longer than anticipated. Consideration should be given to the costs of subsequent dismantling and demolition as required by the Construction Design and Management (CDM) legislation described in Chapter 4. The focus of the legislation is the protection of future occupants and workers from risks to their health and safety, and the cost implications can be significant. For example, the selection of hazardous materials such as asbestos, frequently used in the past is a major problem to any building owner who now finds that the material is in their building and must be replaced.

11.8 Project File content

The general aims at this stage are as follows:

- Translation of sketch ideas into viable forms of construction.
- Selection of appropriate and economical forms of construction and materials.
- Preparation of essential detailed information needed for costing, measurement and construction.
- To control the development of detailed design work in accordance with agreed initial budgets and cost planning.
- To maintain communication between all the members of the design team co-ordinating detailed design development work.
- Incorporation of all mandatory statutory requirements.
- Work towards maintaining the design development programme.

File material could include the following:

Notes, letters and minutes

- Recording exchanges of information between the members of the design team, including sketches of construction ideas and preliminary framework drawings.
- Details of discussions with statutory authorities reflecting requirements or changes to the design as detailed work proceeds.

Checklists

- Question and answer lists exchanged between members of the design team to confirm revised details and specifications.
- Details of client approval to detailed information incorporated into final design work.
- Details of costs of proposed design details and materials, including evaluation of relative economy and implications of changes to design as work proceeds.

Working drawings

- Examples of location, component and assembly plans, elevations and sections showing exactly how elements of construction are to be handled.
- Site layouts for surfaces, drainage, services, external lighting, signage, landscaping and planting, parking arrangements, walls, fences and boundary treatments.
- Floor plans showing location grid and space referencing, floor levels, dimensions, foundations, forms of construction, fixtures and fittings, sanitary goods, fire safety measures, stairs and lifts, services arrangements for heating, power, lighting and communications, internal drainage and plumbing, etc.
- Elevations showing lengths and heights, openings, gutters and downpipes, pipework, extract grilles, overflow pipes, materials such as brick, block, cladding and glazing.
- Sections showing internal dimensions, methods of construction and details of abutments and overlaps.
- Examples of superseded drawings retained for historical reference only.

Specifications

- Examples of comprehensive description of materials, products and assemblies including sizes, qualities, colours and workmanship.
- Reference sources for quality assurance, BSs, CPs, etc.
- Examples of colour schemes and specialist layouts such as ceiling tiles, floor tiles or wall finishes.

Schedules

- Examples of schedules for repetitive elements such as doors, windows, ironmongery and manholes.

Details of drawings issues

- Lists of principle drawings needed for tendering, details of drawings and documents supplied to nominated subcontractors and suppliers for estimates to be included as PC or provisional sums.

- A register of all drawings prepared by the building designer and other members of the design team including, structural information, power, lighting, heating and ventilation and communications.
- An example of a drawings issue slip sent to members of the design team to confirm transfer of information.

Measurement

- Details of information supplied to the QS for inclusion in the BQs including full specifications, Prime Cost Sums, provisional sums and contingencies.
- Query sheets from the QS requesting clarification of points of information.
- A preliminary BQ for approval prior to printing.
- A BQ ready to be sent out to tender.

CDM issues

- Risk assessments for elements of the design.
- Risk assessments for materials and construction processes.

11.9 Discussion points

(1) What are the design variables for a simple product such as a concrete roof tile? Are there any variables which the building designer need not specify accurately? What are the implications for changing variables as a means towards innovation?

(2) What is the useful life of a building and how could it be reasonably extended? Why do some buildings survive for longer than others? How do the current Building Regulations attempt to prevent future building defects?

(3) To what extent are detailed product specifications governed by statutory legislation? How have statutory controls influenced product design throughout the twentieth century? Do statutory controls inhibit innovation?

(4) What is holistic design? Why is 'life cycle' costing becoming increasingly important in the design of new buildings? How can building designers best keep up-to-date with such rapid developments taking place in technology?

11.10 Further reading

British Board of Agrément. www.bbacerts.co.uk
British Standards and British Standard Codes of Practice. www.bsi-global.com
BRE: Building Research Establishment. www.bre.co.uk
Ching F and **Adams** C (eds) (2001) *Building Construction Illustrated.* 3rd Edn. Chichester: Wiley.
Chudley R and **Greeno** R (1999) *Construction Technology.* 3rd Edn. Harlow: Longman.
Cox P (1994) *Writing Specifications for Construction.* London: McGraw-Hill.
DTI construction statistics. www.dti.gov.uk
DTI Constructing Excellence. www.constructingexcellence.org.uk

Emmitt S and **Yeomans** DT (2001) *Specifying Buildings: A Design Management Perspective.* Oxford: Butterworth-Heinemann.
Grundy JT (1991) *Construction Technology*. 3 Volumes. London: Edward Arnold.
Gorse C and **Emmitt** S (2005) *Barry's Introduction to Construction of Buildings.* Oxford: Blackwell.
Lyons A (1997) *Materials for Architects and Builders.* London: Edward Arnold.
NBS National Building Specification. RIBA Enterprises. www.thenbs.com
Norton B and **McElligot** W (1995) *Value Management in Construction: A Practical Guide.* Basingstoke: Macmillan Press.
Rosen H and **Heineman** T (1999) *Construction Specifications Writing: Principles and Procedures.* 4th Edn. Chichester: Wiley.
Seeley IH (1995) *Building Technology*. 5th Edn. Basingstoke: Macmillan.
Specification 3 Volumes (1999) EMAP Architecture.
Styles K (1995) *Working Drawings Handbook.* 3rd Edn. Oxford: Heinemann-Butterworth.
Trade Association Publications *Bricks, Concrete, Lead,* etc.
Trade Literature *Windows, Roof Tiles, Drainage, Ironmongery,* etc.
Willis CJ and **Willis** AJ (1997) *Specification Writing for Architects and Surveyors,* 11th Edn. Oxford: Blackwell Science.
Yeomans DT (1997) *Construction Since 1900: Materials.* London: BT Batsford.

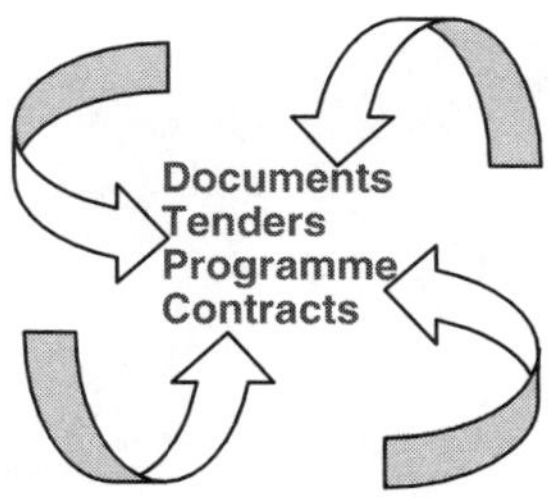

12

Pre-contract administration

12.1 Introduction

The title of this chapter refers to the organisation and arrangements needed before work begins on construction of the new building on site. The approach described here is understood as the 'traditional' method of procurement, where the consultants secure the services of builders on behalf of the client. There are other forms of procurement, other relationships and mechanisms, but the traditional relationships enable useful discussion about the professional responsibilities, forms of contract and tendering arrangements, where consideration of tender figures or evaluation of tenderers' offers, or simply an appreciation of value will be needed.

This chapter will also look at construction planning and the construction period programme, pre-contract issues such as health and safety planning and insurances, and general site establishment getting ready to start building.

12.2 The traditional route

The previous chapter described the transition from sketches and ideas into working drawings and specifications. It is a progressive change, as shown in Figure 12.1, adjusting and refining the framework by adding more and more details, so that the proposals for the new building become realistic and comprehensive. The original plans, elevations and sections are the basis for the work of all the members of the design team, who create their own new drawings for their own purposes, or who sometimes provide information that can be added to the originals. The quantity surveyor will offer advice on the cost implications of design decisions and prepare the Bills of Quantity (BQ), summarising the materials and labours needed to construct the building. The relevant statutory authorities will have been involved with the elements of the design for

Figure 12.1 Ideas to reality.

which they are responsible and will either have given their approval to the proposals, or will be in the process of doing so.

Setting aside the qualities of the design, and the apparent comprehensiveness of all the descriptive information assembled so far, the pre-contract stage is perhaps the most nerve wracking one for the designer involved in the traditional procurement method because realising the construction of the new building is now dependent on finding suitable contractors, subcontractors and suppliers who are willing and able to build it within the target budget costs expected by the client.

No amount of effort by the members of the design team can guarantee that a contractor's assessment of the costs of all the work will match the budget. It is usual to invite a number of different contractors to submit competitive quotations or tenders for the work in the anticipation that at least one return will be close to, or less than the target figure. Sometimes this does not happen, and the design team may need to consider changes to the design or possible reductions in the quality of some of the specifications to make savings, if any building work is to proceed at all.

This is one of the problems currently causing concern in the construction industry leading to the increased use of other methods of procurement. The problems facing the construction industry worldwide are universal: the differences can be explained by reference to differing approaches to project organisation. However, appointment of a contractor by tender or negotiation is only the first hurdle. As construction proceeds it will become clear as to whether or not the design team's documentation is sufficient to obtain the quality of the work desired. It is not uncommon to find construction contracts running into difficulty, as for a variety of reasons additional work is required because some elements were overlooked, inadequately specified, or unforeseeable. Clients and/or designers may change their minds about the extent or quality of works. Additional costs or extras, can amount to a considerable sum over and above the original contract figure. For most clients, this is a problem which they would obviously prefer to avoid.

As the building designer may play little or no part at all in negotiations between contractor or project manager and client with respect to the costs of construction, this chapter focuses on the traditional route to procurement, arranging for the appointment of a builder or general contractor, leading to an exchange of contracts with the client. Although for many larger projects this is becoming a minority approach, it is still useful to consider the traditional role in order to be able to fully appreciate the differences. The designer may be involved in choosing which contractors should be invited to tender for the work, and may then help to assess their response and advise the client on who should be employed. The aim is to see that a competent builder is employed at a reasonable cost so that construction work, once started will proceed smoothly and amicably to a satisfactory conclusion.

12.3 Professional responsibilities

Reference was made in Chapter 2 to the building designer's professional responsibilities occupying the traditional role as the client's agent. In this situation, acting as 'director' or 'controller', the designer must also act as an arbitrator endeavouring to ensure that all parties are treated fairly, and that they comply with established conditions and agreements, particularly once legally binding contract arrangements have been accepted and exchanged. For projects where the designer is not required to fill this traditional role, the responsibilities for administering contracts are not the same; other members of the design or construction teams take the leading role.

With any procurement method, the building designer must supply the necessary information to the design and construction teams at the right times, but it may be the contractor or the project manager who is responsible for liaison with the client to confirm decisions, arrange for payment for completed work and to monitor the quality and progress of construction in accordance with the agreed programme. Negotiations between the client and their project manager about financial arrangements and management systems may essentially be private matters to which the designer is not required to contribute. If this is the case, then the designer will not need to obtain competitive tenders from general contractors, but could be called on to help with the selection and appointment of subcontractors or suppliers. The designer may also be part of a design and build team, attempting to secure work in competition, so an understanding of the broad principles of tendering is important.

Preparing and submitting tenders involves considerable time and expense, which for all but the successful tenderer cannot be recovered. If tenders are to be truly competitive, those involved must feel that they are competing with their rivals on an equal footing, and that the results will be evaluated and compared on a like-for-like basis. The person inviting the tenders should check and verify that selected tenderers are:

- Interested in offering a competitive tender within the time available.
- Able to meet the defined quality standards.

- Capable of undertaking the work within an agreed period of time.
- Prepared to enter into a contract should their submitted tender be acceptable.

Other considerations include the following:

- The selection of tenderers should be related to the value and complexity of the work required. Competition between a small company and a large company who's experience, management structure and overheads are not the same may lead to a significant variation in offers which can be difficult to evaluate.
- Tenderers should be aware that their offer carries no legal weight until contracts are exchanged and that submission of the lowest tender does not guarantee appointment but that in principle, submission of the most attractive tender is likely to result in the award of the contract.
- Tenders should be based on the design team's drawings and written specifications, referred to as the *contract documents*. Additions or revisions to the contract documents may occur during the tender period, which can be accommodated before submission of tenders, providing that all tenderers are advised in writing that their tenders should include the modifications.
- Tenderers should be asked to notify the design team of any obvious shortcomings in the contract documents, if they are aware that some important element has not been included. Discovery of additional works and additional costs once construction work has commenced can be acutely embarrassing and disastrous to the client's cost planning arrangements. Sometimes a tenderer may add a rider to their offer, clarifying their interpretation of some element or elements.

After due consideration, one of the tender offers may be accepted, leading to a formal exchange of contracts and appointment to undertake the work.

12.4 Forms of contract

A contract is said to exist between someone who offers payment to someone else in return for goods or services. The construction of the new building is based on a contract between the client and the contractor, formalising their relationship by exchanging signed copies of the contract documents. The contract is intended to ensure as far as is practicable that both parties treat each other in a fair and rational manner. As with nearly all legal documents, however, words are subject to interpretation, and they may have to be examined very carefully in order to determine their exact meaning in the event of disputes.

The Joint Contracts Tribunal (JCT), established in 1931 produces a variety of standard forms of building contracts, guidance notes and other documentation for particular types of development. Their general purpose is to define what is, or is not to happen, and to act as a reference in the event of disputes, to preclude the expense of independent arbitration. Typical documents include the following.

Main contracts

- Standard Form of Building Contract; Standard Form of Prime Cost Contract
- Major Project Form
- Standard Form of Building Contract with Contractor's Design
- Standard Form of Management Contract
- Intermediate Form of Building Contract
- Agreement for Minor Building Works

Subcontracts

- Nominated Subcontract, Domestic Subcontract
- Prime Cost Nominated Subcontract
- Subcontract for Major Project Form
- Standard Form of Sub-subcontract

Agreements and warranties

- Client/Construction Manager Agreement
- Main Contractor Collateral Warranty for a Funder
- Nominated Subcontractor Collateral Warranty
- Fluctuations Supplement for use with JCT 98

The interested reader can research detailed forms of contract to suit their interests, but the typical, broad content is as follows:

- *Definition of development personnel, their obligations and powers*
  Including names and contact arrangements for all the members of the design team with brief descriptions of their involvement throughout the contract period. It is important to clarify their responsibilities and the authority which they have to instruct the contractor or approve work as it takes place and when it is offered as completed. For example, if the designer is to be responsible for approving the quality of constructed brickwork, the contractor must know that the instructions of others, including the client carry no weight, particularly if additional costs are likely to be incurred. Details of interested external authorities should also be included in this section so that everyone is made aware of their responsibilities throughout the contract period.
- *Description of contract documentation, on which the tender is to be based, and which will govern subsequent actions throughout the construction period*
  Lists of drawings, bills and specifications which apply to the project.
- *Standards for materials, goods and workmanship to be priced and used by the contractor*
  References to printed sources of quality assurance with which the contractor is expected to comply.
- *The requirements for site supervision and minimising risks of injury*
  Descriptions of the contractor's obligations to maintain staffing levels to see that site operatives and subcontractors perform as required, and comply with all statutory health and safety duties. It is essential that the contractor is represented on site by someone who is available to meet the client and the design team, and who can respond to instructions.
- *Procedures for dealing with variations to the tender documents and the expenditure of provisional sums; the implications of value-added tax (VAT), and any other statutory costs and taxes*
  The management procedures for instructing or approving variations and accounting for all associated costs.
- *Possession, programme, practical completion and defects liability*
  The time framework from start ('possession' is the moment the contractor assumes responsibility for the site) to finish ('completion' is the moment of formal handover of the finished building to the client) which the contractor is deemed to accept when contracts are exchanged. This is particularly important to the client with regard to occupation of the new building, but also to the design team who must maintain the flow of essential information to avoid delaying the contractor, and to the contractor who must plan resources to meet critical target dates. Defects is a term used to describe avoidable failures after completion and

occupation. For example, a warped door leaf or an ineffective boiler which may result from poor materials or workmanship. The terms of the contract ensure that the contractor corrects defects for a limited period of time beyond completion.

- *Employment controls, fair wages, assignment, certificates, payment*
  These sections oblige the contractor to implement nationally agreed standards for the employment of site personnel and subcontract labour. Assignment refers to the wholesale transfer of the works or parts to other independent contractors. The contract may require the issue of formal certificates confirming instructions and valuing completed work. This is an area discussed in an earlier chapter that is related to formal qualification to practice. For example, the RIBA and CIAT have different certificates for situations which may arise during the construction period. They will not be specifically identified in this book.
- *Fluctuations*
  Fluctuations are defined in the contract because some variation in the cost of materials or labours can occur during the construction period, which are beyond the control of the contractor. Inevitable price rises to cover inflation or nationally agreed wage rates, for example, may be permitted as described.
- *Nominated subcontractors and suppliers*
  The conditions associated with the design team's choice of subcontractors and suppliers.
- *Insurances*
  Confirmation of the extent of insurances to be carried by the contractor to cover damages and injury as a result of undertaking the works. The relationship with the client's own insurance cover is also described.
- *Damages for non-completion, extension of time*
  This section covers the penalty costs which can be imposed on the contractor for failing to complete the works within the agreed period of time. Some situations, such as undertaking additional work as unforeseen variations, or because of poor documentation, or at the request of the client will take longer to complete than predicted, and the contractor may be granted an extension to the contract period at no penalty.
- *Loss and expense*
  Additional works described above will involve the contractor in additional personal expense, such as keeping staff, hired site accommodation or machinery in place for longer. In this case the contractor may be recompensed for the necessary additional costs.
- *Determination of the contract*
  There are circumstances which may lead to either the client or the contractor withdrawing from the contract. For example, failure to comply with contract conditions, unreasonable demands or non-payment are reasons for the contractor to stop working, and poor workmanship, failure to meet the programme or bankruptcy could result in the client seeking to employ someone else to complete the works.
- *Outbreak of hostilities and war damage: antiquities*
  General provision for unforeseen events beyond the control of any party to the contract describing how responsibility and additional costs should be handled.

Each form of contract will include or exclude clauses as appropriate to the development, and elaborate the implications in different ways depending on the value of the contract or the way in which it is to be carried out or executed. The purpose is to attempt to clarify issues so that all parties understand their responsibilities and so that disputes can be resolved amicably. This is not always possible and in the event of deadlock, the contract allows for some independent source to be appointed to adjudicate or arbitrate. Their conclusions are normally binding on the parties, but arbitration should be regarded as a matter of last resort.

12.5 Tendering arrangements

In the traditional method of procurement, the design and specification of the new building is based on an assessment of likely costs, but until a contractor or subcontractor makes a firm offer to build, the figures are simply theoretical estimates. Construction, like the sale of most goods is a matter of supply and demand, and prevailing market forces will usually determine the real costs of the work. One way of testing the market is to invite independent quotations or tenders from interested companies. This has traditionally been seen as the way to obtain value for money, but it is not necessarily the case. The element of competition in an active market may mean that submitted tender figures are deliberately reduced in order to secure employment, in the knowledge that additional costs will be recovered once work has commenced. Alternatively, tender figures may be deliberately inflated by heavily committed companies, not particularly interested in securing further contracts, but happy to do so at a premium rate. The reasons for possible additional costs concerning the building designer mainly arise through errors or shortcomings in the design team's specifications and contract documents. Consequently, although a tender figure for example, may appear to be 10 per cent lower than the budget estimates, by the time the work is completed the final account may have risen up to the budget limit, or even beyond. Some additional costs such as extra deep foundations in subsequently discovered poor ground, may be unavoidable and legitimate. Others like an improved quality of light fittings in the showroom, may well be contentious if the original specification is shown to be inadequate. The client may expect more from the contractor and the contractor expect more from subcontractors than had been allowed for in the tenders.

This problem, together with uncertainty about construction costs up the point of receipt of tenders, is leading both clients and contractors to the conclusion that other methods of establishing costs may be preferable. Negotiation based on actual costs plus agreed profit margins and overheads, for example, can prove to offer better value for money in the long run. However, should the client choose to test the current market by inviting tenders, the design team will assemble and issue relevant drawings and documents so that tender returns can be compared on a 'like-for-like' basis. Variation in tender returns will inevitably arise because each tenderer will value risk elements and profit in a different way. Some of course, may also have greater purchasing power enabling them to offer more competitive rates for the supply of materials and labour.

There are a number of different ways that tendering can be managed, as described below.

Open invitation

This system offers the opportunity to any interested party to submit a tender for the works. It can take the form of an advertisement in the local press or in national trade journals, inviting a response from suitably qualified companies. It is not a particularly satisfactory proposition for the design team because supplying the necessary information to a potentially high number of applicants would involve considerable time and effort, and may yield responses from organisations about whom little is known. It is difficult with this system to be certain about the quality of work that is being offered, particularly if very low tenders are received. It is not a good system for the tenderers either who have no assurance about the equality of their competitors.

Selective list

It is better to limit the number of possible tenderers to a reasonably sized list of companies who's work is familiar to the design team. The normal practice for projects of the size of the car-dealership would be to select up to six general contractors from the list and up to three subcontractors or suppliers for individual elements of work. Examples of completed work are reasonable indicators of the general ability of the tenderers, but further research may be needed to confirm that their inclusion on the tender list is appropriate at this time. Many more tenderers could be asked to submit offers for the work, but there would be little to gain by widening the list as competitiveness is unlikely to be improved, and the tenderers will be less interested in making offers as there would be less chance that they will secure the work.

Sometimes, the selected tenderers' quotations may all exceed the target budget costs, which is an unsatisfactory outcome for everyone. The possibility of obtaining other tenders offers no guarantee of success. The design team may have to reconsider the proposals and change elements in the design or the client may be left with no choice but to find additional finance. The probability is that the whole development will be delayed and costs will escalate, which is exactly the situation which other methods of procurement are attempting to avoid.

Negotiation

For many larger developments, clients are choosing to procure their buildings by negotiation. By this method they can invite a contractor or subcontractor to prepare a budget quotation for the work based on limited information. They can select their preferred contractor and develop the scheme jointly, considering design and construction methods together so that difficulties or undue expense can be avoided before work starts on site. This can be particularly advantageous for commercial developments that must be started and finished quickly, minimising the delays associated with the competitive tendering process. Negotiation requires both client and contractor to be honest with one another about true costs and profit margins.

Traditional tendering arrangements for the general building works are normally handled by the designer or the quantity surveyor. Tendering for specialist subcontract works or the supply of materials may be dealt with by the appropriate consultant, such as heating and electrics or mechanical hoists. Costs for these items are covered in the BQ as prime cost or provisional sums as described in the previous chapter.

12.6 Tendering procedure

The general procedure is as follows:

- The tender documents are issued to the tenderers who should be asked to confirm their receipt. The drawings issued for tendering will be a selection of the most important as it is not usual to issue all the detail drawings at tender stage. The documents issued to the tenderers include the main location drawings, indicative component and assembly drawings, specifications, BQ, the form of contract, a letter of invitation, a tender form and a tender return envelope.
- The return date is given, usually a minimum period of 4 working weeks from the date of issue.
- The contractors will visit the site to check any specific constraints.
- They will split up the drawings, specifications and bills as necessary to obtain prices from their domestic or preferred subcontractors and suppliers.

- If the construction period has not been defined already, the contractor will indicate starting availability and how long they anticipate the work will take. This can be used as an element for assessing tenders, discussed in the next section.
- The contractor may ask for clarification of matters of uncertainty and additional information made available for inspection if requested.
- Queries from individual contractors can be useful to highlight errors and omissions in the contract documents. The resolution of any points must be communicated to all tenderers to maintain fairness.
- Tenders are returned in identical sealed envelopes on the date agreed. The envelopes are issued with the tender documents and should be numbered so any missing at the time of return can be identified. The contractor should understand that the inclusion of conditions with the tender offer may lead to its disqualification.
- The tenders are opened and the return figures are assessed.

12.7 Evaluating tenders

When the tenders are received, they are normally simple gross figures. Assuming that one of the offers is attractive, the company will be advised of the position and supplied with an additional copy of the specification or BQ into which they are required to insert all their detailed prices and rates. There are two reasons for doing this. Firstly, so that the mathematical addition of all the priced elements can be checked to see that the gross tender figure is correct. If errors are discovered, the tenderer has the opportunity to confirm or withdraw the offer, but either way it is unfair and unwise to proceed without advising the tenderer, who may have omitted a significant sum which will cause financial difficulties later on. Secondly, once construction has started it is essential to know how much has been allowed for elements which may be deleted, added or altered so that accurate cost savings or extras can be determined. The second most attractive tenderer may also be notified to stand by in case problems are found with the lowest one.

It is normal to include a statement in the contract documents to the effect that there is no obligation to accept the lowest, or indeed any of the submitted tenders. It may also be that the tenderer's anticipated construction period is an element of the tender itself. This allows scope to consider the viability of tenders, which must be carefully assessed and compared before the client decides to enter into formal contract arrangements.

Evaluation may not be all that straightforward. Consider the following notional returns:

- **A** £950 000 in 16 weeks
- **B** £1 038 000 in 12 weeks
- **C** £966 000 in 14 weeks
- **D** £835 000 in 21 weeks
- **E** £1 111 000 in 25 weeks
- **F** £975 000 in 18 weeks

They can be evaluated as follows:

- **E**

  This is the most expensive offer, requires the longest construction period and can be discounted as uncompetitive.

- **B**
  This is the second most expensive offer, but in the shortest period of time. Compared with the next shortest period, this offers the client the possibility of 3 weeks additional commercial operation. However, this can be discounted if the additional capital cost is greater than the benefit of earlier occupation.
- **D**
  This is by far lowest offer, but over a relatively long period. Although this could be realistic, it may be that the tender figure has omissions or that it is deliberately low in order to secure the contract, which may mean difficulties later on if and when money runs out. This tender is worth consideration, but may be unrealistically low.
- **A, C & F**
  These three figures are relatively close together for gross capital cost and duration of construction, possibly indicating the true value (closest to commercial reality) of the project. F is the least attractive because it is the most expensive of the three and requires the longest construction period. C is more expensive than A, but offers 2 weeks additional occupancy time, which may be of greater value to the client than the additional capital cost. A is the lowest cost of the three, but involves the loss of 2 weeks occupancy time.

Having weighed up the pros and cons of the various offers and determined which one represents the best value, the client can decide how to proceed. The successful tenderer will be notified, contracts prepared, signed and exchanged. The unsuccessful tenderers should be notified and given a list of their competitors and a list of the tender figures received, separated to maintain confidentiality.

12.8 The construction programme

The design development programme, described in Chapter 6, shows how the time required to reach the various stages throughout the design process is organised, leading to commencement of construction on site. The time needed to complete construction to the point where the building is handed over to the client and all the work is completed may be shown as a single block at the end of the design period programme, setting the target date for occupation by the new buildings users. This period of time must itself be subdivided and structured to organise the succession of activities that the contractor is involved with as construction proceeds. There is another critical path showing the route from start to finish, fitting in the organisation of all the materials and trades showing what must happen if the work is to be completed on time.

The detailed content of the construction programme is normally the responsibility of the appointed contractor who will assess the implications for themselves at tender stage. A typical construction programme is illustrated in Figure 12.2. It is useful to discuss the probable overall length of the construction period with at least one of the contractors in advance of tendering so that demands are realistic, particularly if there are difficult site conditions or constraints associated with the construction works which should be taken into account. There is little value for the client asking for the works to be completed within a period of say 9 months if this is impracticable and in reality, 12 or 15 months will be required.

The way that the contractor constructs the programme will depend on their own company structure and the way that they view the tasks and the risks associated with completing them.

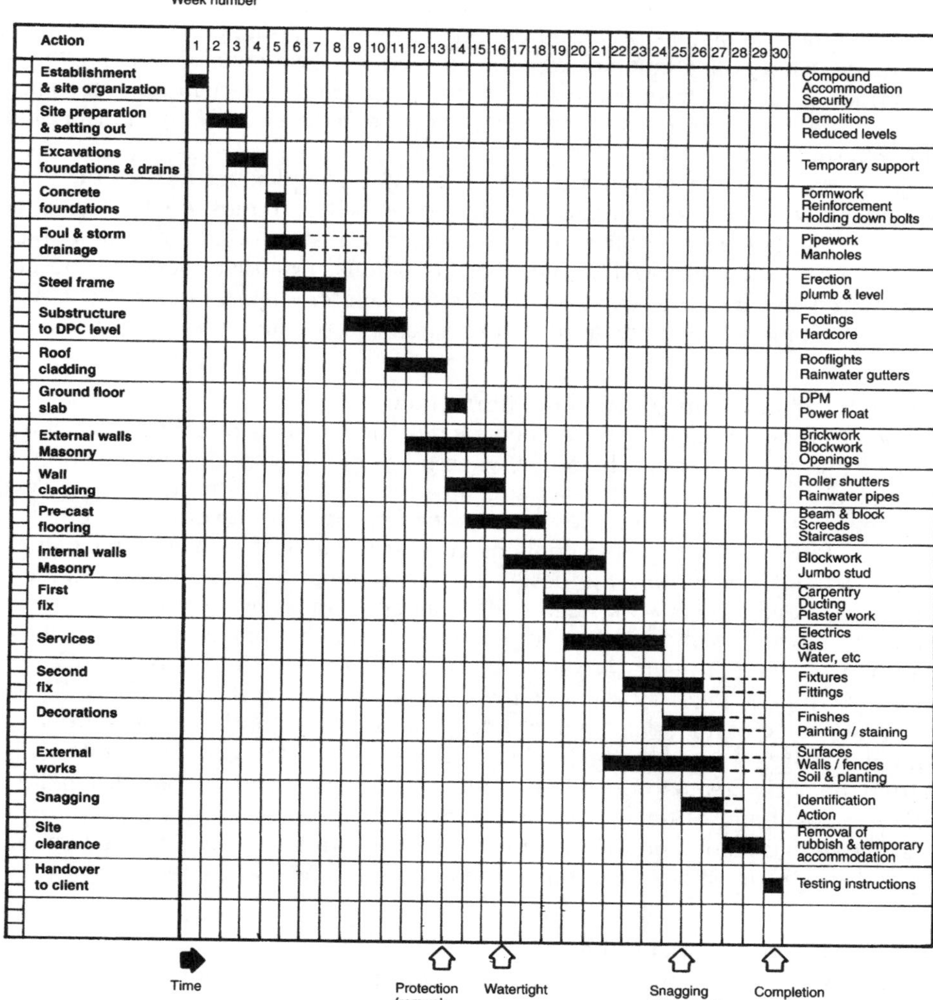

Figure 12.2 A typical construction programme.

The issues concerning the contractor's management structure, employment arrangements, supervision and ordering mechanisms, and risk strategies are beyond the scope of this book and are matters which can be referred to elsewhere. The contractor's purpose is to plan the work so that they can reach completion without penalty and operate in such a manner as to obtain the profitability from the contract that was anticipated at the tender stage.

With respect to the design team's involvement, the construction programme includes the following considerations.

Establishing a realistic construction period

The design team must be confident that the demands being placed on the contractor are reasonable or that the contractor's proposed period of time is realistic, particularly if it is a factor in accepting the tender offer. In either case, the consequences are likely to be that the standard of work is poor because it has been rushed or the completion date *will be* exceeded and the work finished later than agreed at the outset. It is far better to finish early on a longer programme than late on a short one.

The projected construction period must also reflect the complexities of construction and the time needed to obtain special materials and components. Regard must be given to factors such as the ease or difficulty of access to the site, the space available for administration and storage of materials and the likely prevailing weather conditions. Extensive works in the ground in January may well take longer to complete than they would in August. Bricklaying in a month of heavy rainfall will inevitably be delayed if the desired end product is to be achieved.

Ensuring the 'flow' of communication of information

The contractor's programme will show when each of the elements of construction are planned to take place and can be used to highlight when relevant information must be confirmed and issued. Final decisions about the exact specification of materials, construction details and schedules may not have been completed at the tender stage, or indeed by the time that work begins on site. Providing that the BQ contains the correct sums of money for the elements, this presents no difficulty, permitting the design team additional time to resolve requirements in due course. However, once work begins the contractor must be in possession of sufficient information to proceed without delay. For example, the facing bricks needed for the external walls can be described at the tender stage in principle, covered in the tender offer as a 'cost per thousand'. The contractor's programme will indicate at what stage this must be translated into an actual choice of the brick, so that orders can be placed and the bricks arrive on site at the moment that are required by the bricklayers. To this end, the contractor will request information from the design team through Information Required Schedules, formalising the position regarding the possibility of delay if decisions remain outstanding. They can be regarded as helpful to the design team, and the client because they are then fully informed about exactly when final decisions must be made if the programme is to be maintained.

Assisting supervision of work on site

At site meetings, one of the principle issues for discussion and monitoring is the progress of the works. The contractor's programme can be used to check that intentions are being fulfilled or to show how additional labour might be needed to recover lost time. It will help the design team to understand when important inspections should take place so that required quality standards can be established for materials or workmanship and when approvals are expected for completed work. The programme may also be a useful guide with regard to any arbitration needed in the event of disputes, particularly over claims for extensions of time resulting from works being undertaken out of sequence by different subcontractors, or delays caused by inclement weather.

The programme will also show when inspections are required at various stages by the representatives of the statutory bodies such as the building control officer, health and safety inspectors, and the fire officer. Although liaison with the building control officer on site, for example, may be the responsibility of the contractor, the design team must be in a position to know that the necessary approvals have been given.

Projecting cost planning

The programme will show when major elements of construction are likely to be completed or when expensive components will be delivered which will help the client to appreciate when significant payments will be due. Payments to the contractor are usually based on the value of work completed and of materials on site, assessed on a monthly basis throughout the construction period. The value of payments will vary substantially from month to month affecting the client's cash flow arrangements.

12.9 Pre-contract arrangements

There comes a point at this stage in the creation of a new building when a contract is formally exchanged between the client and the contractor, committing both parties to its contents. It is a document signed by both parties confirming the contractor's offer to build and the client's agreement to pay for the work. The term 'contract' is also often used to describe the period between starting work on site and finishing the completed building. It is generally understood to cover the contractor's involvement from 'day 1' once work has actually commenced. Hence, pre- and post-contract can refer to operations on site as well as the moment of exchange of agreements.

Before the formal contract document is exchanged a variety of issues must be checked, as follows:

- *The contract figure is correct and agreed by both parties*
  The agreed contract figure must be accepted by both parties, ensuring that all associated costs have been accounted for and are mathematically correct. The figure must include the sums for prime cost and provisional works, and all relevant contingencies, leaving no room for doubt as to what the contract figure represents. The client in particular must understand what is included in the contract figure and what is not, including professional fees, statutory authority fees and charges, and any costs associated with their own operations, such as transferring existing machinery, equipment and furnishings, or financing new ones which may have not been measured. The costs of their own staff time in meetings, inspections and commissioning the building on completion can be considerable.
- *The clauses in the contract have been completed and are acceptable to both parties*
  The form of contract used will contain a number of clauses into which relevant figures or descriptions are inserted for the particular project, including significant dates, criteria for assessing additional costs and penalties for failure to comply with the contract. One of the principle issues to agree before exchanging contracts is the date for possession and the date for completion. It can be useful to have a formal meeting to ensure that all parties understand their rights and responsibilities as defined in the contract.
- *The requirements of Construction Design and Management Regulations (CDM) legislation have been met*
  See Chapter 4 for more details of the responsibilities of all parties at this stage.
- *All necessary insurances and bonds are in place*
  It is essential to check that the client and the contractor understand the extent to which they are responsible for insurances for damages to property and injury to site personnel and visitors. The points of transition of responsibility as the contractor 'takes over the site' at the outset, and when it reverts to the client at 'handover' are critical with regard to exposure to risk.

- *Essential formal approvals are in place*
  The design team must be certain that the fundamental approvals have been formally obtained before any contracts are signed. For example, Planning Permission must have been granted before the parties are legally committed to the project. For other approvals which may take place once work has commenced, the design team should be in the position of being confident that they can and will be obtained.

On the assumption that the formal agreements are in order, a pre-contract meeting will be arranged with the client, the consultants and the contractor to discuss the project and resolve all relevant issues before work commences. Although some of the issues referred to may have been fully documented at the tender stage, it cannot always be assumed that they have been understood, appreciated or accounted for. It is better at this stage to be sure that construction work will start, and continue smoothly. The issues to consider include the following:

- *Introduction of management personnel*
  The design and construction teams will introduce the personnel who will manage and supervise the contract, and confirm their powers for matters including procedures for supervision, requirements for approving sample materials and construction, making and receiving instructions, confirming variations, and preparing and agreeing valuations.
- *Ensuring that the contractor has sufficient information to start work*
  The contractor should be issued with a full package of drawings, schedules and specifications covering all the necessary information for construction. This is a formal set of documents defining and describing the works at this moment in time. It represents the point at which subsequent revisions, omissions and additions can be judged and measured. Additional information will be issued as it is needed but the design team must now be fully aware that variations to the issued contract documents will represent costs, either as savings or as extras.
- *Examination of the contractor's proposed construction programme*
  The contractor will outline their strategy for undertaking the work and this meeting is an opportunity for all parties to begin to better understand their part in the construction process.
- *Confirmation of nominated subcontractors and suppliers*
  The design team may be in a position to advise the contractor of their selected nominations or agree when this must be done to meet the programme.
- *Approve the contractor's 'domestic' subcontractors and suppliers*
  It is a usual practice to require the contractor to indicate who they will be employing to carry out certain aspects of the works. It may be that the experience of any member of the design team could lead to objections about specific individuals or companies who they would prefer not to be involved with the project.
- *Discussion about the contractor's proposals for site planning*
  Agreement should be reached about the location of temporary access roads, secure compounds, site accommodation for offices, stores, canteens and toilets, and storage arrangements for materials and waste. If not already illustrated, the contractor should be advised about where signboards and advertising hoardings may be sited. This may also be a subject of interest to the Planning Authority.
- *Dilapidations survey agreement*
  The condition of existing features should be clearly recorded, preferably on a series of photographs of boundary walls and fences, trees and any existing structures, materials or property, even though it may be adequately described in the contract documents. This should include records of the condition of other buildings adjacent to the site boundary where

potential damages could occur. The value of the dilapidations record is explained in Section 7.8. The agreement should also include descriptions of how important features will be protected throughout the construction period. It is quite common for designer and contractor to make their own records to ensure full coverage.

- *Agreement of a date for the first site meeting*
  It is usual to agree when the first site meeting is to be held after work has commenced as soon as suitable accommodation on site is available.

12.10 Project File content

The general aims at this stage are as follows:

- Assemble packages of information which are precise, accurate, clear, easy to read and understand.
- Provide sufficient information to establish realistic costs of construction.
- Arrange tendering procedures which are fair to competing contractors and which will result in the client being able to rely on the quality of work being offered.
- Ensure that all necessary statutory approvals are in place prior to commencement of construction on site.
- Co-ordinate all pre-contract matters between all parties prior to commencement of construction on site.

File material could include the following:

Notes, letters and minutes

- Further exchange of information between the members of the design team.
- An agenda and issued minutes relating to the pre-contract meeting.

Query lists

- Clarifying points of information with the client for inclusion in the contract documents.

Form of contract

- Information about the client's preferred form of contract, including copies of the appropriate documents.
- An example of a contract document.

Tender documents

- Drawings and associate information to be issued to tendering contractors, including forms of tender, letter of invitation and return envelopes.
- Details of revisions during the tender period, circulated to all contractors.
- Details of queries from contractors pointing out errors in the tender documents or asking for further clarification of points of information.
- Examples of tender returns from a number of contractors.

- A priced BQ or specification.
- Examples of a contractor's analysis of the project, including considerations of risks and programming.

Evaluations

- Details of enquires about possible contractors and interviews to confirm their suitability.
- Details about elements of risk included in the Tender Health and Safety Plan.
- Consideration and analysis of tender returns.
- Advice forwarded to the client about contractor selection.

Letter of appointment

- Details about the client–contractor agreements before signing contracts.
- A letter of authority from the client confirming forthcoming exchange of contracts so that the successful contractor can begin to make plans for construction work.

Contract documents

- Details of the drawings, specifications and schedules which form the basis of the contract between the client and the contractor.
- Nomination documents for subcontractors and suppliers if not to be selected by the contractor.

Pre-commencement meeting

- Including details about the proposed construction programme, the contractor's lead in and set up timings, proposed start date, extent and location of the contractor's site signage.
- Details of the drawings and other information to be issued at this stage with an indication of the extent of further information which will be produced in due course.
- Contractor's information required schedules issued to the design team requiring the formal issue of information and instructions in accordance with the construction programme.
- Details of nominated subcontractors and suppliers.
- Details about the contractor's domestic subcontractors and suppliers.
- Discussions about special aspects of the works to clarify requirements, or highlight issues which the contractor may not have fully appreciated.
- Dilapidations records and details of proposed protections for vulnerable features.

Site Health and Safety Plan

- Compiled by the Principal Contractor showing how risks highlighted in the Tender Health and Safety Plan will be addressed.
- Any relevant permits or certification which must be publicly displayed including employment, health and safety, insurances, etc.

Completed formal documents and approvals

- Including Planning Approval, etc., which are critical prior to commencement, and advisory notification such as Building Control which must be deposited prior to commencement of construction on site.

Completed business and financial arrangements

- Including tax certificates, VAT, cash flows, bonds and insurance agreements, etc.

12.11 Discussion points

(1) How do different forms of contract cater for special conditions? Do contracts encourage confrontation or help to create better working relationships? Can design quality be guaranteed by contractual obligation?
(2) How can designers be certain that tender offers fully reflect their requirements? What can specifiers do to minimise the possibility of misunderstandings? How can a client be made aware of the distinction in levels of quality and performance at the tendering stage?
(3) How are the theoretical discussions about Health and Safety reflected in the reality of actions taken on site? How can building designers work towards improving the poor health and safety record of the UK construction industry? Could all accidents be avoided by design?
(4) Which formal authority approvals are required before commencing work on site? What are the risks of starting work on site before detailed design work has been completed? What arrangements need to be in place to transfer responsibility for the site to the general contractor?

12.12 Further reading

Aqua Group (1999) *Tenders and Contracts for Building*. 3rd Edn. Oxford: Blackwell Scientific.
Ashworth A (2001) *Contractual Procedures in the Construction Industry*. 4th Edn. Harlow: Longman.
Chappell D (2000) *Understanding JCT Standard Building Contracts*. 6th Edn. London: Spon Press.
Chappell D and **Powell Smith** V (1999) *The JCT Design and Build Contract* 2nd Edn. Oxford: Blackwell Scientific.
Connaughton JN and **Green** SD (1996) *Value Management in Construction: A Client's Guide*. London: CIRIA.
Godfrey PS (1996) *Control of Risk: A Guide to the Systematic Management of Risk from Construction*. London: CIRIA.
Hackett M and **Robinson** I (2003) *Pre-contract Practice and Contract Administration for the Building Team*. Oxford: Blackwell Scientific.
JCT: Joint Contracts Tribunal. www.jctltd.co.uk
Kwakye AA (1997) *Construction Project Administration in Practice*. Harlow: Longman.
Lavender S (2000) *Management for the Construction Industry*. Harlow: Pearson.
Murdoch J and **Hughes** W (2000) *Construction Contracts, Law and Management*. 2nd Edn. London: Spon.
Ndekugri I and **Rycroft** M (2000) *The JCT 98 Building Contract: Law and Administration*. London: Butterworth-Heinemann.
Seeley IH (1997) *Quantity Surveying Practice*. 2nd Edn. London: Macmillan.

SECTION 3 Construction Period

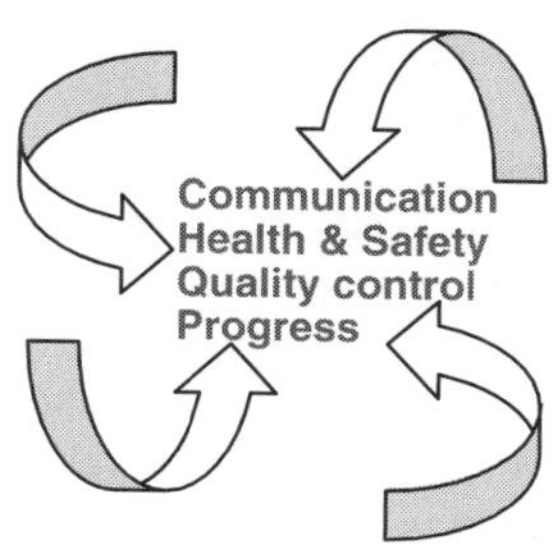

13

Construction supervision

13.1 Introduction

The third part of this book is concerned with the undertaking of construction, completing the new building and handing it over to the client for use by its occupants. The process of construction will take a considerable length of time, but the building designers' direct day-to-day involvement on site may be quite limited. Regular meetings will be held to monitor progress and ensure the flow of essential information. The consultants and other specialists will continue to produce and elaborate detail to inform and guide the contractor, subcontractors and suppliers. The formal mechanics of contract supervision and certification will not be discussed at any length in this book. The legalities of practice can be studied in existing reference sources elsewhere. The intention in the following chapters is to briefly outline the designer's relationships with the design and construction teams as work proceeds showing how supervision affects the quality of the finished building. The focus will primarily be on the purpose of the powers of supervision which are available and the process which can be adopted to create good working relationships throughout the construction period.

This chapter explains how and why information is exchanged, updated, revised and corrected as construction takes place. It reviews the significance of supervision and inspection of work on site, together with the part played by the building designer in maintaining the target programme, the budget, a safe working environment and friendly relations with other members of the design and construction teams.

13.2 The implications of reality

Throughout the design development process, during the period described as 'pre-contract', the new building has been created 'on paper', incorporating and illustrating the demands of the client, the expertise of the consultants and the constraints of the authorities. The costs associated with this exercise are relatively small in comparison with the eventual total cost of construction and commissioning, and at any point the process could change direction, be delayed or even be halted. Such an eventuality becomes increasingly more difficult the further the process has developed, but the client's commitment, limited to lost fees and any compensation which may be applicable could be said to be manageable, and in any event the process could be put 'on hold' for a limited period and started off again if and when conditions were more favourable.

However, once work commences on site (see Figure 13.1) the costs of withdrawing become considerable. The formal exchange of contracts imposes obligations on the parties, and both client and contractor begin to be committed to the regular expenditure of large sums of money. Although the project could still be abandoned, the creation of elements of construction becomes a reality which cannot be easily changed or reversed, and it would be very costly to reinstate the site to its original condition. The financial commitment drives progress, and the actions of everyone must be clearly focused on the end target, defined and limited by the information contained in the contract documents. The client has budgeted for the building described in the contract documents, and expects to take possession of a finished product within the period illustrated in the programme. The contractor has also budgeted in his tender to provide the building within the time allowed.

To the design team, the fundamental fact is that intentions, whether well documented or not, must now become realities. The contractor, and all the people involved with them on site, must understand every element of the building. If work is to proceed smoothly to the agreed

Figure 13.1 The excitement of starting to transform a run-down, derelict site into a useful developement. Peel Street redevelopement for Nottingham Trent University.

completion and handover dates, the contractor must be in possession of sufficient, accurate information so that materials can be ordered and delivered, skilled labour can be employed and all necessary equipment brought to site at the right time. Notwithstanding the quantity and quality of the specifications and defined standards in the contract documents, it is essential to establish the standards of workmanship expected and the design team must form working relationships with the contracting team based on understanding and trust, which can be easily destroyed if misinterpreted or neglected requirements are not discovered until the work has been completed. Sample materials or examples of acceptable workmanship should be discussed and agreed as the standard required for the project and once approved, left in a prominent position for future reference.

At the commencement of construction, the site temporarily becomes the property of the contractor, who is fully responsible for activity within the boundaries. This lasts for the duration of the contract up to the date of handover of the new building to the client. The contractor must arrange for the necessary insurances to protect property and personnel on site and implement effective security measures to protect against unauthorised access. The designer, the consultants and the client become 'visitors' who, like anyone else involved in the construction works, must submit to and comply with the contractor's own management procedures with regard to health and safety issues, and security measures.

The design team's interest during the construction process is broadly as follows:

- That the building is completed and handed over to the satisfaction of the client or its potential users.
- Construction is completed as designed, as anticipated, *'fit for the purpose'*, on time and to the agreed cost, in accordance with the defined standards, samples and testing.
- Sufficient detail is issued to meet the requirements of the programme and any necessary additional information is supplied promptly when requested.
- Regular meetings are held to monitor progress.
- Form and finishes are as described in the contract documents.
- Changes and alterations are approved and instructions confirmed.
- Contract clauses are adhered to, and not varied without formal approval.
- Situations of confrontation are avoided wherever possible and disputes resolved fairly without favouritism.
- Health and safety issues are dealt with in a responsible manner as described in Chapter 4. The implications of the Construction Design and Management Regulations (CDM) are described in Sections 4.6 and 4.7.
- Construction proceeds in accordance with the requirements of interested statutory authorities.
- Payment is made to the contractor for works completed.
- Accounts are settled by the client promptly as recommended.
- Final accounts are agreed and settled amicably.

The powers that the design team rely on to undertake supervision are defined in the form of contract. As discussed in the Introduction of this book, supervision or management of a contract carries legal implications which cannot be undertaken lightly as the consequences of mistakes can be substantial. The position of contract supervisor demands a high level of skill, knowledge and experience, and for a project the size of a new car-dealership requires the services of someone with suitable qualifications to act in this capacity. It is probable, although not a certainty that one of the professionally qualified members of the design team will be the contract supervisor. It is they who will take the formal responsibility of issuing instructions, approving work and sanctioning payments, but their associates, perhaps as yet without formal qualifications can still play an important part as construction proceeds.

13.3 Communication

During the design development period, information was regularly transferred between the client and the other members of the design team through an exchange of drawings and various written documents. Although the design period programme fixed target dates, there was some flexibility to research and refine the ideas before they became commitments to future design work. A delay caused by reconsideration of some significant factor may actually have lead to the design of a better building, presenting advantages which the client could understand and accept. The cost of delay could arguably be offset by the gains made by a more economical design. However, once construction starts on site, this flexibility disappears because the construction programme is much more specific in its defined target dates and carries much more significant implications for costs. Changes in completed construction cannot be reversed and materials once positioned cannot easily be reused.

In an ideal situation, the contractor would be in possession of all the necessary information for construction at the beginning so that building work proceeds smoothly. In this scenario the contractor would know exactly what to do and the new building would appear without any further effort from the design team. In reality, this is rarely the case as even the best set of drawings, specifications and Bills of Quantity (BQ) will leave some gaps for interpretation or clarification, and there are always some construction details which have not been sufficiently considered or illustrated. Even if they are, the contractor and subcontractors may not fully understand them. Conditions on site are not always as anticipated and unforeseen findings may require a change in the design of some elements. Decisions about the final selection of materials and components are often delayed until the point when they must be confirmed to meet the construction programme, highlighted in the Information Required Schedules mentioned in the previous chapter. Sometimes agreed elements can be changed at the request of the client, the consultants or the contractor, and Prime Cost and provisional sums must be resolved and nominations confirmed. The possibilities for construction to vary from the defined contract documents are considerable, representing a significant difficulty in successful project supervision.

One of the issue is communication with regard to effectiveness and authority. The contractor must know exactly which materials to order, where they are to be placed and how they are to be constructed. None of these factors can be easily altered or 'improvised' once work has begun. If the end product is to be satisfactory, it can only be on the basis that a single clear instruction is given and received. Arriving at the instruction is a matter of the co-ordination of the design team and disseminating the instruction is a matter for the contractor's management. In this respect, the point of contact must be limited to its simplest level and the usual practice is for the supervising designer or the project manager, acting as the client's agent to issue all instructions, and the contractor's agent, either the contract manager or the site foreman to receive all instructions. This ensures that both parties are represented by someone who can take an overview of the matter which may have wider implications.

For example, consider the specification of a facing brick for an external wall. The exact specification may already be given in the contract documents, in which case the contractor is certain about what is required. Alternatively, the specification may be based on a provisional cost in the BQ, following discussion with the Planning Officer who has approved the selection.

During a visit to the site, the client sees the brick and tells the subcontract bricklayer that it is not acceptable, demands that the work is stopped and suggests an alternative brick. This situation is untenable because the client's own choice may be technically unsuitable and considerably more expensive, and the subcontract bricklayer is responsible to the contractor for the costs of any deviation from the agreed specification. The client and the subcontractor are not in a position to fully appreciate the consequences of their actions in relation to everything else. Their spontaneous response can be very disruptive. The implications may include the following:

- *A change to design intentions*
  Ideas about the colour, texture and style of the originally specified facing brick may be closely related to the choice of other materials specified for the new building or related to existing materials on other buildings around the site. The visual appearance of the whole building could be radically transformed for instance, by the use of a predominately dark coloured brick instead of a lighter one.
- *The need for Local Authority Planning Approval*
  The original brick may have been specified at the Planning stage, or agreed subsequently with the Planning Officer as a condition of approval. Even if the alternative brick is acceptable to the Planning Officer, there will be a delay whilst the matter is considered and any work undertaken in the mean time is at risk if approval is not forthcoming.
- *The technical performance of the new brick may not be suitable for the location, which could lead to construction failures, poor weathering and loss of manufacturer's guarantees*
  The specification of facing bricks is variable depending amongst other things on the loading requirements and the degree of exposure to the elements. An incorrect specification may lead to a rapid decline in the brick's performance with no possibility of seeking redress against the manu-facturer or supplier. For example, an absorbent facing brick placed in a position where it will be regularly saturated such as a window cill or a parapet wall will inevitably suffer frost damage, which after a relatively short time will lead to the face of the brick being 'blown' and the surface will disintegrate. Other bricks will stain more easily or support undesirable algae growth.
- *The alternative brick may be more expensive than the originally specified brick which has capital cost implications on the agreed tender figure*
  Not only might the brick be more expensive to purchase, but it may also be more expensive to handle and lay in the construction of the wall. The dimensional tolerances and colour variations for example, may lead to more waste or the need to take greater care when mixing batches before laying. These costs may be difficult to estimate in advance and only become evident once the work has been completed.
- *The contractor may be delayed whilst new orders are placed, possibly leading to a claim for an extension of time to complete the building*
  The delay may also represent a loss to the client as a result of later occupation of the building, affecting business operations.
- *Loss of labour*
  Depending on the method of employment, the subcontract bricklayers may decide to work elsewhere while the new brick is being considered. There is a possible further delay if the contractor is unable to obtain the services of new bricklayers to match the revised programme.
- *Possible affects on sequence of works*
  The delay associated with resolving the new specification could lead to some elements being incorrectly constructed out of turn which may become damaged by exposure to the elements.

Construction of later elements may be rushed and poorly constructed in an attempt to recover the lost time.

A similar example may arise if the structural engineer instructs the site foreman to reposition a steel column. This decision may be for entirely legitimate technical reasons, but the repositioned column may interfere with the positioning of the heating engineer's ductwork and the client's hoist, or change the appearance of an attractive open plan entrance foyer which had been designed as an important feature. As far as the structural engineer is concerned, the steel frame may be more economical, showing cost savings when compared to the contract documents, but the other consultants may face difficulty redesigning the installation of their equipment resulting in considerable potential additional costs as compared to the contract documents. The effect on the open space foyer may be irredeemable, remaining as an irritation to the future occupants for many years to come.

This shows the significance of pre-planning, the importance of sticking to it and how easy it can be to make mistakes by focusing only on a limited number of elements. Simplified, direct communication between the design and contracting teams is essential to protect and reinforce the information already agreed and to control work on site so that the desired result is achieved at the agreed cost. At least there is then an opportunity to assess the implications of an instruction and resolve the matter amongst the design team before issuing it, and for the contractor to consider how the instruction affects the management of the building work on site. The design team and the contractor are concerned with achieving their own ends, but they are all interlinked and at this stage in the creation of the new building must all be very much aware of the possibility of escalating costs. Communication is a two-way process and although the design team will attempt to control the flow of information at the right times, the best contractors will assist by asking for information at the right times.

13.4 Supervision and inspection

The word 'supervision' is sometimes applied to the administration of a construction contract in the sense that the 'supervisor' has an interest in seeing that the terms of the contract are complied with by the parties who are involved. This is not necessarily a continuous activity, but rather a matter of periodic checking that stages are reached as intended and that a range of conditions have been met. Supervision can also be taken to be the direct overseeing of the activity of others, directing, inspecting and approving their work on a regular or even continuous basis. In adopting the title of 'supervisor', or agreeing to undertake 'supervision', the member of the design team must distinguish between these two positions and clearly establish their meaning with regard to the extent of their involvement on the site.

The 'day-to-day' supervision of the contractor's site personnel is a matter for the contractor's own management relating to their own methodology and practices for undertaking and completing tasks. The contract documents should require the contractor to maintain suitably qualified staff on site at all times to control their own workforce and any independent subcontractors and to respond to directions from the design team and the statutory authorities. Supervision of the entire project is the responsibility of the contractor's site foreman, but other personnel are also likely to be present, including trade foremen for concrete, bricklaying or plumbing and even gang foremen for individual groups of workers who may be operating on different parts of the site. The contractor must establish the chain of command to ensure that there is consistency in performance between these groups.

The designer's, or other members of the design team's supervision is different. Assuming that the designer is not based on the site, and that he or she is involved with other projects elsewhere, contact with operations on site will be limited to occasional inspections and formal site meetings. The significance of this situation is that the designer is not in a position to see the works on a day-to-day basis, and will only be aware of progress at various stages. For example, at one visit the foundation trenches may have been excavated ready for concrete to be delivered. At the next meeting the foundations may have been placed, footings constructed and the reinforced ground floor slab completed. Although everything may appear satisfactory, the quality of compacted hardcore, the positioning of reinforcement steelwork and the adequacy of the damp proof membrane are matters which cannot be seen and the designer cannot confirm that they have been constructed correctly. The designer can either take a risk and assume that the work has been done properly or make arrangements to be represented on site on a more regular basis. Figure 13.2 shows typical construction work taking place.

For a project of the size and complexity of the car-dealership, such inspection may be required every day as work proceeds, requiring ideally the presence of an experienced **clerk of works**, who's duty is to see that the building is constructed in accordance with the contract documents, and who is available to witness every action by the contractor and subcontractors as they carry out their work. The clerk of works will also be able to help the contractor to understand the complexities of construction, to resolve any difficulties as they arise and to ensure that formal inspections by external authorities such as the Building Control Officer take place as required. The clerk of works is an important link between the design team and the contractor with respect to making sure that sufficient information is issued and maintained on site and that design requirements are practical. A particular detail on the drawings may be inappropriate or could be improved in the light of the clerk of work's experience, which is a further check that proposals are realistic.

Even if the project has the benefit of a resident clerk of works, the contract should include a clause to the effect that supervision cannot be taken as absolute in the event of subsequently discovered failures. It is still the contractor's responsibility to comply with the requirements of the contract documents and even the most diligent inspectors cannot be expected to see everything. For example, the designer may approve a sample panel of brickwork by specific inspection, defining the desired effects. It is straightforward to compare the appearance and workmanship of the construction of the new walls with the sample panel, but the inclusion of all necessary cavity wall ties in accordance with documentation cannot be verified without seeing every tie installed. This would be an impossible situation for any form of supervision. Even the site foreman must trust that the individual bricklayers install the cavity ties correctly. However, if the wall collapses because of insufficient ties the contractor's responsibility must not be diminished because of the acceptance in any form by any members of the design team that the work was completed correctly. Approvals must be restricted to elements which can be seen and which are known to have been installed correctly.

The designer must decide how often to visit the site and when inspections are necessary. It is common practice to hold formal meetings on site at least once a month. This is a reasonable length of time to judge progress and ensure that all relevant matters are dealt with as work proceeds. The organisation and recording of meetings was outlined in Chapter 3. The findings or conclusions reached following any meetings on site must be communicated and confirmed as quickly as possible if the programme is to be maintained.

It is important to be aware of the conditions on site and try to appreciate the difficulty involved in achieving the end product. Although the contractor, the subcontractors, the bricklayers and the labourers are there to 'earn a living', they will have as much pride in carrying out their work as

Figure 13.2 The new building begins to come out of the ground. Sites in Notttingham.

the design team have in creating the design. They may have a better understanding of the possibilities of creating a good end product than the designers, and should always be made to feel involved and encouraged to offer advice and suggestions which may be invaluable. Working relationship with all the site personnel should be positive and encouraging. It is far better to discuss and resolve potential problems in advance than to turn up 'out of the blue' and condemn something which has been worked on patiently and conscientiously for 2 days in the pouring rain!

13.5 Quality control

Reference to aspects of quality control or quality assurance have been made in earlier chapters in this book in relation to the design team's performance and to the use of recognised standards for design work and specifications. Sources of quality assurance play an important part in supporting the design team when supplying information and making demands on the contractor at the start of the construction process. They define materials, components and assemblies either in principle or in absolute terms so that the contractor can understand what is required. They can be regarded positively as forms of instruction, used to tell the contractor what to do and also negatively as a form of defence, as protection in the event of a construction failure. Of course, it could be said that defence is positive, particularly when offering professional advice or opinions, but it is an 'after the event' form of defence which is not a guaranteed form of protection. A legal resolution to a construction failure is not certain to go in the designer's favour and may in any event result in a substantial loss of money and prestige. No one involved in creating the new building would welcome a construction failure or a disappointing result, particularly the client.

With this in mind, everyone must attempt to make sure that the materials and components supplied and the assemblies constructed do not lead to failure or disappointment. This may seem to be an unduly obvious statement, but it highlights the need to act conscientiously in advance to see that anticipation is matched by reality. Ideally, samples of all the materials to be used should be made available for inspection and approval before ordering, particularly those which will remain visible on completion like facing bricks, roof tiles, fair-faced concrete blocks and ceramic wall tiles. Materials which will be hidden like sand, cement and hardcore should be checked that they comply with the contract specification. Setting aside the possibility that materials have been supplied fraudulently, there may have been an error in ordering or at the time of despatch. Pre-mixed or 'ready-mixed' concrete for example, is always tested on delivery and after a period of time to confirm that the supplied mixes were correct.

Components can be checked by visiting the manufacturer and examining processes and quality control methods to see that the finished products are satisfactory. If the new building is to have pre-fabricated aluminium window frames for example, they can be seen being made in the factory to confirm that the appropriate standards can be achieved and maintained.

Samples of assemblies must also be considered to check workmanship. A sample panel of brickwork for example sets a standard for bricklaying; the thickness and straightness of horizontal joints, the verticality of perpends, the colour and style of the mortar and pointing. Further discussion will be needed about colour mixing of delivered batches of bricks to avoid unsightly lines between batches, or dominant patches of darker or lighter bricks which have not been mixed carefully. Sometimes the bricklayer, or the tiler for roofing is so close to the work that such changes are not noticed until the work is complete.

Perhaps the most difficult area of quality control is the achievement of satisfactory finishes, particularly applied decorations. Painting and staining timber or masonry can leave much to be desired if standards are not agreed on a small scale before large areas are completed. Similarly, the quality of pre-finished products such as kitchen units, sanitary goods and floor coverings can deteriorate rapidly by the time the building is ready for handover if they are not carefully protected.

13.6 Variations

Variations are changes to the design as illustrated on the contract drawings, or defined in the contract specifications which may occur at any time during the construction period. There are a variety of reasons why changes do, or should occur but perhaps the most significant issue associated with them is whether or not they have an impact on the cost as agreed in the contract documents. They may represent a cost saving, but in many cases lead to additional costs which must be accounted for. Determining the responsibility for any additional costs depends on the reason for the variation. The common alternatives include the following:

- *Changes made at the client's request*
  Providing that the request is practicable, changes can be accommodated and the costs either added to or subtracted from the contract figure. The client must be made aware of any associated costs such as causing a delay to the contractor or suppliers as a result of making a change, but the increased costs in this case are attributable to the client.
- *Changes to the design because of error of specification*
  If the contractor undertakes work in accordance with the design team's drawings and specifications then any changes or corrections to work already constructed will be at no cost to the contract. That is to say that the client is not expected to finance the error which may well be attributable to the designers. Changes may also be required to account for errors, omissions or simple variations to definitions in the BQ. Changes on a 'like for like' principle are not uncommon based on the rates included in the bills.
- *Changes as a result of incorrect construction or materials supply*
  Errors in construction or supply are the responsibility of the contractor and the relevant suppliers who would be expected to correct their mistake at no additional cost to the contract. In some circumstances such errors could be accepted by the client because they are equivalent in quality or may be accepted on the basis of an agreed discount.
- *Changes following unforeseen findings on site*
  Changes in design and specification may be forced by the discovery of some factor on the site which had not been foreseen or anticipated. The site analysis described in Chapter 7 should reveal most problems, but this is not always the case. It is a common practice to include contingency sums in the BQ to cover such eventualities. Expenditure is ideally confined to the sum included, but if exceeded would normally be the responsibility of the client. If expenditure is not required, the contingency will not be needed and there will be an overall saving in the final account.
- *Changes enforced by statutory authorities*
  The statutory requirements for the new building may be revised during the construction period or the inspector for the authority decides to interpret some issue in a different way. This cannot always be predicted, and will invariably be unavoidable. Associated costs will be the responsibility of the client, providing the designers and the contractor can show that they have not been negligent in their actions to date. In most cases, changes in legislation cannot be enforced retrospectively once formal approval has been given. For example, a change in the Building Regulations to require an increase in the thickness of insulation in roof spaces will not affect building work if the authority has already approved the construction proposals.

13.7 Valuations

For a large project such as the car-dealership, the contract figure will be a considerable sum of money, running to some hundreds of thousands of pounds. It would be unreasonable to expect the client to make all the money available at the beginning or for the contractor, subcontractors and suppliers to finance the cost of the development until completion. To assist the cash flow of both parties, periodic valuations are prepared as a method of arranging stage payments from the client to the contractor at regular intervals based on works completed and materials on site at certain specified, regular times. This is usually a lump sum paid to the contractor who distributes payments to the subcontractors and suppliers. The details of this arrangement are not described in this book as specialist matters about finance can be studied elsewhere.

Normally, the quantity surveryor (QS) will prepare the formal paperwork to be presented to the client and will check the value of work completed on site together with the value of any materials or components delivered and awaiting installation. The objective of this arrangement is to fairly recompense the contractor for legitimate expenditure but specifically to prevent too much money being claimed in the early stages of construction which may lead to problems later on.

The main interest to the design team in this process is the handling of variations or 'extras'. Some changes as described in the previous section are easy to value because of the descriptions in the BQ. A change in specification for the facing bricks is straightforward based on the figure or the rate in the BQ. It is either more or less and can be simply calculated. However, if the change demands some work which is not fully described in the BQ, like for example removing some previously unknown asbestos insulation from an existing building or demolishing a newly constructed partition to make space for a larger than envisaged car ramp, then valuing the work is much more difficult. The contractor may not be able to give an accurate estimate of the total cost of the work until it is completed. The normal practice in this situation is to record the activity on a daily basis, referred to as 'daywork' including the involvement of labour, the extent of materials used, plant and transport needed and legitimate overheads. Added to this is an agreed percentage for the contractor's profit, established in the contract documents. This method allows the supervising designer, or the clerk of works to check the extra work that is being carried out and to verify the legitimacy of additional payment within a reasonable limit. In this way, the contractor is unable to claim excessively high costs for a simple piece of work, but equally cannot be denied recompense for legitimate outlay.

The valuation payments to the contractor would normally be made at least monthly for a project like the car-dealership, and include an element of retention, held back until the work is fully completed. This acts as an incentive to the contractor to ensure that the work is completed satisfactorily at which point part of the retention is released, the remainder being released after correction of any defects identified at the end of the maintenance period described in the next chapter.

13.8 Progress

The contract period is defined in the contract documents, and the contractor's construction programme illustrates how this target date will be achieved. At the regular site meetings and by occasional inspections it will become apparent as to whether or not progress is 'on programme'. The costs associated with a delay can be considerable for all parties to the contract and although the design team cannot assist the contractor with the physical work on site, discussions about progress and determining responsibility are an important part of supervision. As with apportioning responsibility for the costs of variations, the reasons for the delay of the contract must be clearly established. The form of contract should include provisions for circumstances that can lead to claims for extensions of time.

The reasons that the contract may be delayed include the following:

- *The client*
  The client's requests for changes or variations may result in extra work which cannot be absorbed within the contract period. Sometimes the client may employ their own labour for elements of the work which can interfere with the contractor's programme causing a delay.
- *The designer or consultants*
  Design and specification variations may add work, prolonging the contractor's involvement and all members of the design team must be aware of the implications of failure to issue sufficient information at the right time or confirm instructions quickly, which will prevent the contractor from proceeding as planned.
- *The contractor*
  Some factors will cause the contractor problems which will affect the progress of the works. Unavailability of labour or shortage of materials are factors which are beyond the contractor's control, but in other cases such as lack of diligence and organisation, or undertaking remedial works because of poor quality materials or workmanship are issues to do with the contractor's own management of the works.
- *Unforeseen circumstances*
  In some cases, delay results from the actions of other people or events which are outside the contractor's control, such as a strike, a fire or war. The most common problem for many contracts is exceptionally inclement weather, such as prolonged frost or very heavy rain lasting for a long period. Depending on the stage of construction, this can cause a serious delay over which the contractor has little control. If the persistent heavy rainfall takes place just after the site has been excavated and foundation trenches dug, the site may become almost inaccessible for a time and further work inevitably delayed.

The consequences of a delay can incur penalties to either client, contractor or both. A legitimate extension on the grounds of inclement weather would enable the contractor to avoid penalties and possibly be recompensed for the additional costs of maintaining staff and machinery for a longer period. However, this is effectively a penalty against the client, who must delay occupation of the completed building. A claim for additional time from the contractor because of poor programming or the need to correct poor workmanship would be rejected and result in the contractor facing a claim for liquidated damages as defined in the contract documents as well as the costs of continuing on the site until the work is completed.

13.9 Disputes

It is not coincidental that this section is reserved to the end of this chapter. The focus of everything that has gone before has been on the establishment of formal and informal relationships based on a careful definition of requirements so that the new building is created to the satisfaction of all the parties involved. Notwithstanding the care taken by the client, the designer and the consultants in the preparation of their information and fixing of the contractual controls, it is still possible for things to go wrong and positions to be reached which can be described as disagreements. As already mentioned at the beginning of the book, the process of construction of a new building can be quite confrontational, and it is not uncommon to find the client looking to reduce the final account or the contractor seeking ways of increasing it. The client may insist on discounts for some perceived omission or inadequacy and the contractor demand payment for extras which were apparently never included in the original tender offer.

The supervising designer can find him or herself at the centre of the argument, being required to defend the client's interests on the one hand, but to protect the contractor from unfair criticism on the other. For example, the client may express dissatisfaction with the quality of workmanship of a brick wall, or deny that an instruction for a more expensive floor finish was approved. The contractor may claim that additional costs were incurred through a lack of information from a particular consultant at the right time causing a delay, or that more materials were required than were included in the BQ. On most occasions the form of contract used for the development should include clauses which allow responsibilities to be determined, providing a solution to the difficulty which should be acceptable following reasonable consideration. This is easy to say but not always so easy to implement and it is not uncommon for loopholes to be exploited by either side.

The options for dealing with disputes which cannot be resolved amicably are essentially as follows:

- *Litigation*
  Litigation is a process of bringing a dispute to court by suing for damages. It is an expensive business which may cost more to undertake than is recovered in the event of success.
- *Arbitration*
  Arbitration is a cheaper alternative whereby the parties present their case to an independent inspector or arbitrator and agree in advance that they will accept the decision whichever way it falls. It would normally be used in cases which are complex and where significant sums of money are at stake.
- *Adjudication*
  Adjudication is similar to arbitration but applied to simpler disputes which can be resolved quickly. It has been developed from the Latham Report produced in 1994 to be included in the form of contract as an attempt to reduce confrontation in the construction process.

In all three cases, the decision reached will be based on the facts of the case and the applicable law, viewed impartially and must be accepted as binding. In all three cases, there will undoubtedly be costs involved in the process and the parties must appreciate the consequences of their actions. Everyone will be best served by a genuine attempt at resolving difficulties without recourse to external powers.

13.10 Project File content

The general aims at this stage are as follows:

- Maintain the flow of information from the design team to the contractor in order to maintain the construction programme.
- Focus all parties on the need to comply with accepted contract documents.
- Establish standards for workmanship and finishes.
- Record progress as an aid to resolving disputes following any delays.
- Account for changes to the contract documents.
- Value work completed and arrange for payments to the contractor, subcontractors and suppliers.

File material could include the following:

Notes, letters and minutes

- Notes about issues discussed informally on site.
- Letters confirming points of information between members of the design team.
- Copies of minutes recording discussion and outcome at site meetings with the design and construction teams.
- Formal documents displayed by the contractor regarding Health and Safety, site management and insurance arrangements.
- Examples of the contractor's records for ordering, receiving deliveries and checking the quality of materials and components.

Information required schedules

- Typical lists of information needed by the contracting team if progress is to be achieved and maintained as per the agreed programme.
- Details of subcontract information requiring confirmation, such as profiled steel cladding or windows, often in the form of manufacturing drawings which must be checked and approved prior to manufacture.
- Details of information issued during the construction period confirming outstanding matters such as colour schemes and planting specifications.
- Details of information to be resolved by the client such as signage, fixtures, fittings, equipment and furnishings.

Revisions

- Examples of revised drawings, specifications and schedules, updated as work proceeds.
- Variation instructions describing how variations are to be incorporated and accounted for as construction proceeds.
- Revisions to the construction period programme showing how changes will be accommodated to maintain the completion target date.
- Any information which should be added to the Health and Safety File as required by the CDM legislation.

Daywork sheets

- Examples of the clerk of works' records of additional activity including materials, machinery and labour required on site.

Valuations

- Examples of the method for valuing completed work and advising the client that payment should be made.

Samples and test results

- Details of approved samples with record photographs.
- Details of typical concrete cube test results.

- Information about the performance of materials including manufacturers' recommendations.
- Test results for the performance of equipment such as heating and ventilation systems.
- Details of observations made by any authority's inspectors.
- Clerk of work's observations about any aspect of his or her inspections.

Progress photographs

- Record photographs as work proceeds for interest and for use in displays on completion.
- Record photographs of elements of construction to illustrate conditions on site at the time which may help to resolve disputes later in the event of claims for an extension to the contract period, or that work has been carried out incorrectly.

13.11 Discussion points

(1) What is the relationship between building designers and site operatives? How can building designers best use the knowledge and skills of those undertaking construction work? Do designers sometimes ask for the impossible?

(2) How can a designer check that specifications have been correctly carried out on site? What precautions are required for approving work which is unseen or covered up before inspection? What are the implications if specified materials are not available at the time that they are required on site?

(3) How can building designers ensure that construction information is kept up to date as construction proceeds? Do CAD drawings and specifications encourage unnecessary design changes? Does electronic communication help with site management?

(4) What can designers do if their client is unhappy with something on site? Does the client as a 'customer' have the right to expect exactly what he or she wants? How can the design team ensure that their client understands what they are going to get?

13.12 Further reading

Bielby SC (1992) *Site Safety Handbook.* London: CIRIA.
Clarke RH (1984) *Site Supervision.* London: Telford.
Davies VJ and **Tomasin** K (1996) *Construction Safety Handbook.* 2nd Edn. London: Telford.
Kelly J, **Morledge** R and **Wilkinson** S (eds) (2002) *Best Value in Construction.* Oxford: Blackwell.
McCabe S (2001) *Benchmarking in Construction.* Oxford: Blackwell.
Schaffer T (1983) *'Site architect's guide', Architect's Journal,* 20th April, 27th April and 4th May.
Thorpe B (2004) *Quality Management in Construction.* Aldershot: Gower.
Ward HNE (1986) *Building Technicians and Clerks of Works' Handbook.* Braunton: Merlin.
Watts JW (1983) *The Supervision of Construction: A Guide to Site Inspection.* London: Batsford.

Motability

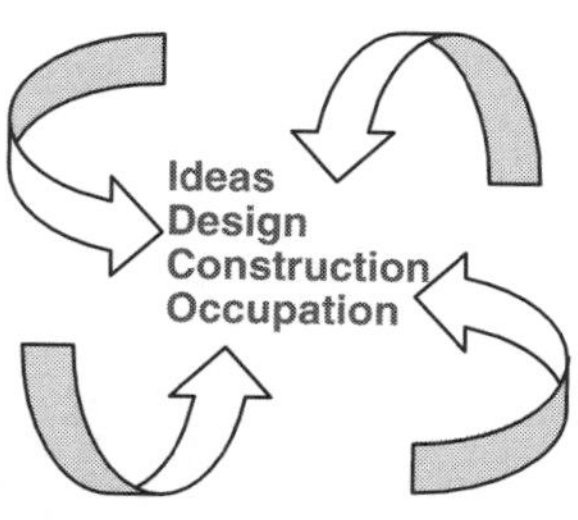

14

Completion

14.1 Introduction

Finishing work on the new building is identified on the construction period programme as a target date for handing over the completed building to the client and users. Completion is really a process in itself, usually spread over a period of time as everything finally comes together. It is reassuring, and often quite surprising to see the results when scaffolding is taken down, mess cleared away, decorators move out and protecting covers are peeled off. The building emerges from its cocoon of chaos ready for imminent occupation. For a large and complicated building like the car-dealership, there will be much to check and test before the end is actually reached. Inside and outside spaces, surfaces, fixtures, fittings and equipment will be carefully inspected, identifying any remedial work needed to leave them in excellent working order.

Once agreement is reached that everything is as it should be, or was expected to be in the contract documents, the building is handed over to the client, who begins to furnish and equip the premises ready to commence business activity. The finished building enters a transition period where maintenance may be needed and where the building occupants have time and opportunity to discover if there are any problems requiring attention before the contract is completed. Identified latent defects will receive attention later on before agreeing the final account and a post-occupancy review will collate feedback on client and customer satisfaction. There will be lessons to be learned for all members of the design and construction teams before starting with the next project.

14.2 The end is in sight

Creating new buildings can be time consuming and occasionally frustrating but when it is finished there is the relief that the bulk of the hard work is over and the satisfaction that it has resulted in a tangible end product. The design team's intentions for the structure, finishes and all the workings of the new building are at last realised as illustrated in Figure 14.1. The quality of the design, its

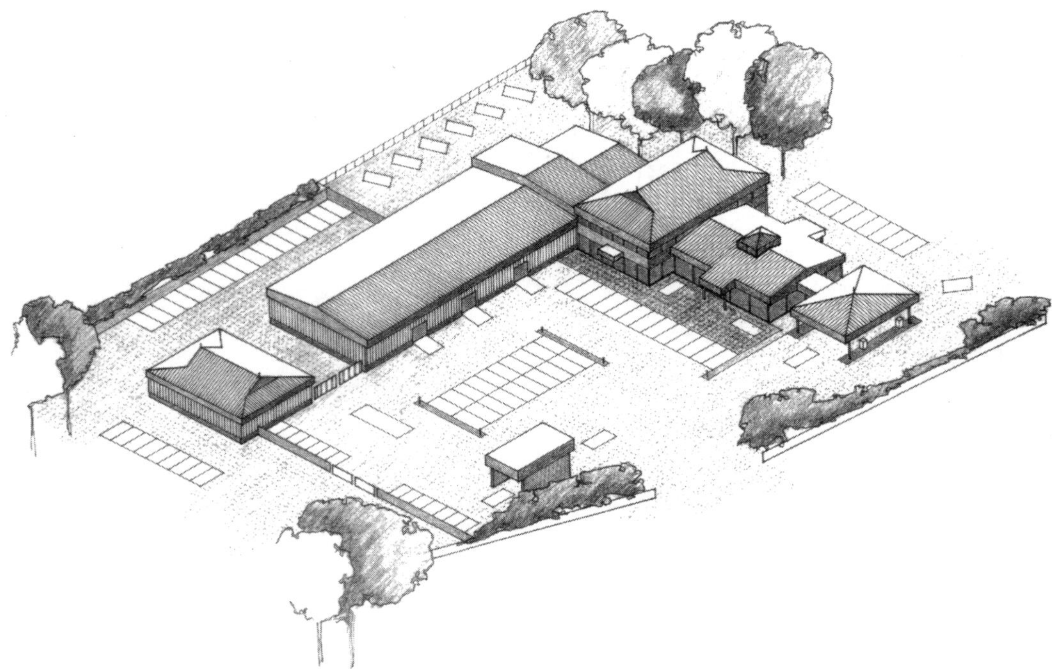

Figure 14.1 The finished building.

layout, appearance and resolution of all the details is visible and the building (good, bad or indifferent) is available to be praised or damned by the client, the building's users and any other critics.

For many of those involved in design and construction, completion on site can be the most satisfying stage in the whole process, but it is susceptible to a fundamental difficulty. It is always easier to start a task than it is to finish it off and whilst the broad sweep of work at the beginning, motivated by enthusiasm drives progress, attention to the minutiae of detail at the end can be diminished by weariness, and previous good work may be spoiled by a lack of care at the end. This can be true of the design process itself, resulting in poor detailing through lack of full attention. With respect to the construction process, finishing off is an element of work that can be very hard to control. Contracting is essentially a messy business, often undertaken in difficult conditions by staff and operatives who suffer the ravages of cold and wet conditions with remarkably good grace, but the skills needed to construct the envelope of the building in the face of severe external environmental conditions are not always compatible with those needed to complete fine detailing and perfect finishes.

The design team's role in this respect is to see that the standards defined and agreed in the contract documents are achieved on site so that the completed building is handed over to the client in the best possible condition. It is, however, at this point where difficulties may be experienced and great care needed to remain unbiased in not demanding the impossible or overlooking the obvious. There can be a great temptation for the contractor to rush the finishing stage of the works as cost limits are reached. The fixed costs associated with preliminaries such as staffing, temporary buildings and insurances may be consuming anticipated profits, and there is a risk that work will not be satisfactorily completed as a result. There is often considerable pressure to relax standards in sympathy with the contractor's situation resulting in compromises which may not meet with the client's approval. It should also be said that there may be considerable pressure from the client to demand an unreasonable level of perfection over and above the accepted normal standards as defined in the contract documents. It is very important that the client should fully appreciate that sufficient time must be included in the programme to attend to snagging so that practical completion and handover can be achieved amicably.

At this final stage, the target is the formal acceptance of the building as being complete and ready for occupation by the client as intended. Practical completion and handover mean that the new building becomes the responsibility of the client who takes over the necessary security and insurances. The completion certificate releases due payments less a small percentage held back as a retention to cover the costs of any subsequently necessary remedial work to correct defects. All the support documentation and instructions, and records are presented to the client for future reference.

14.3 Snagging

The first stage before accepting the new building is to agree with the contractor when it will be ready, either as already understood from the construction programme or otherwise as progress has actually been achieved. The building may be completed earlier or later than indicated on the contract programme, and the client will need to know about any revisions in order to make arrangements for occupation. The term 'snagging' is used to describe the identification of minor defects in construction and finishes which should be attended to or corrected prior to presenting the building to the client as being completed. The process normally takes place once the finished building has been offered as complete by the contractor, who will invite inspection of the works to see that everything is satisfactory. In essence, the inspection is to check that the construction complies with the contract documents and any other formal instructions regarding variations which have arisen during the contract period. At this point, there should not be any major deficiencies as they should have been identified, considered and dealt with already. The issues involved should be of a relatively minor nature such as missing screws from ironmongery, localised poor paint finishes or components which have not been thoroughly cleaned. Incomplete or unsatisfactory work is defined on a *snagging list* issued to the contractor for immediate attention.

Snagging undertaken by any of the members of the design team should not take the place of, or replicate the work of the contractor's finishing foreman or any of the other trade foremen. Obvious discrepancies should be dealt with by the contractor prior to offering the works as completed. This is the contractor's responsibility, and part of the contracted tender offer to hand over the building in a perfect condition. It is always sensible to discuss expectations with the contractor in advance of snagging so that issues are clearly understood. This is done throughout the job as standards for materials and workmanship are agreed, but as finishing approaches it is always useful to emphasise once more the exact qualities that will be required for the building to be acceptable. Care should also be taken that any formal snagging list is comprehensive before it is issued, so that once completed, there are no further outstanding matters or the need for further improvements before handover. The contractor must be in a position to know what the target is and when it has been achieved. It is not reasonable to prolong completion unduly in the search for ever-increasing standards.

The snagging inspection will also include checking the proper operation of moveable parts like doors and windows, together with the testing and commissioning of all services equipment and systems such as power, heating and ventilation, hot and cold water systems, and foul and surface water drainage. Once all the items included on the snagging list have been dealt with, a further, final inspection will be carried out to agree with the contractor that everything is satisfactory and ready for handover.

14.4 Handover

In most cases the finished building will be handed over to the client in one go when everything has been completed. Sometimes, particularly if spaces are clearly distinguishable it may be practicable to operate a partial handover, offering the client the use of sections of the building development in succession. For example, if the car-dealership was located in several separate buildings on the site, it could be that a single part could be made available for use while the remainder is being completed. Handover can also be conditional, based on a formal agreement that certain outstanding works will be completed as soon as possible after handover. Such items could include completion of planting and incidental external works, replacement of damaged materials or components and erection of signs and notices. In this case, the situation regarding completion, insurances and payments must be clarified carefully. There is undoubtedly a risk that the contractor may remove personnel and equipment from the site when they are actually still needed in order to finish the works.

The formal handover is an acknowledgement by the client and the design team that the building is completed to their satisfaction. Outstanding defects may still be apparent and will be formally noted. The works now enter the maintenance period, which is a specified time limit for normal occupation, which allows the client to discover signs of latent defects. For most elements in the building this period is 6 months, but for mechanical equipment and systems it is usually extended to 12 months to ensure that it has been used through a full climatic cycle. Subject to the client's agreement, formally noted defects can sometimes be left until the end of the maintenance period when latent defects are dealt with, permitting the client to occupy the building and commence trading as soon as possible. For example, if incorrect push plates had been fitted to door leaves, or the wrong plastic covers placed over fluorescent lighting tubes, replacement may be more convenient later on when other matters receive attention.

At handover, the contractor is expected to have removed all temporary accommodation, equipment, materials and waste. All services systems will be commissioned and in full working order and the contractor's costs for fuels and water finalised by taking meter readings. The transfer of insurance responsibilities should be checked and confirmed, and the client presented with all essential keys.

Most importantly for the design and construction teams is the issue to the client of all the 'as-built' records relating to their own input. Drawings and specifications in particular must be revised to take account of any changes that have occurred in construction so that the client has an accurate record of materials, components and assemblies as they actually now exist. The package should also include the operating instructions for equipment and systems, all necessary instruction about maintenance and cleaning to materials, components and finishes, and a complete, updated Health and Safety File as required under the current Construction Design and Management Regulations (CDM) Legislation.

14.5 Latent defects and final account

The term 'latent defects' is used to describe problems which only appear after a period of occupation and use. They would not have been apparent at the time of handover, or were not discovered at that time. Depending on the exact form of contract used to cover the construction works, provision is normally made for latent defects to be corrected by the contractor at no additional cost to the client, providing that the defect is consequential to the contractor's work, and not caused by others or by normal wear and tear. Whilst the terms of the contract will specify the applicable time limits the position is nevertheless a matter of the supply of goods which should be fit for their purpose, and there may be other statutory rights applicable in this respect offering the client the prospect of longer-term remedies.

The usual procedure is that an inspection is carried out with the client to identify any works requiring further attention. This is done in advance of the time limit so that a new list of defects, including any previously noted at handover is issued to the contractor for remedial attention on or shortly before the expiry of the maintenance period. Typical issues likely to arise would include warped door leaves, cracked plaster and damaged finishes resulting from poor materials or workmanship. The contractor's remedial action can include replacement, repair or redecoration. For services equipment and systems the client may require immediate assistance with defects or faults, and the contractor would be expected to provide a degree of 'after sales service' to ensure for example, the effective, smooth running of the heating system or attending to leaks in pipework and valves. Such problems would not be left for 12 months before receiving attention. The client may choose to employ the contractor to undertake a facilities management service, looking after the building for a period to make sure that everything is functioning correctly.

Ideally, remedial work is arranged by agreement with the contractor, based on the original contract documents, but in some cases disputes may arise, which can only be resolved by reference to the form of contract, or through the intervention of an arbitrator. Should the contractor fail to attend to latent defects as required, the contract should allow the client to employ another contractor to undertake the work, offsetting the costs against the original final account by deducting retention monies or outstanding payments. Depending on the costs involved, there could even be a situation where additional compensation can be obtained from the contractor.

The works included in the defects list should be attended to within an agreed period of time, but at the convenience of the client and the building users. Once completed, a final inspection will take place to confirm that everything has been attended to satisfactorily. Completion of this element of the works will lead to release of all retention monies, and settlement of the final account. The final account is the last payment and marks the end of the contract between the builder and the client. The final account includes all the costs related to the contract including any work which has been re-measured, and the value of any additions and savings. For the car-dealership, the quantity surveyor will normally administer this stage of the works determining the final payments to be made by the client, ideally with the agreement of the contractor. All the members of the design team will eagerly anticipate final payment of their outstanding fees.

14.6 Customer satisfaction

At the beginning of this book, the creation of a new building was described in terms of it being a product, made for the benefit of its users. A number of judges were referred to including the client, the building's occupants and users, passers-by, the authorities, the critics and of course, the building designer, the other members of the design team and the contractor, and anyone else involved with the project. Each have their own criteria for regarding the development as a success or a failure. For those who have been employed success may be pride in their achievement, recognition and awards, or simply making a profit on their efforts. Critics may describe the building in glowing terms and authorities be delighted that the development is a safe, attractive addition to the built environment in their area. Passers-by may enjoy the occasional glimpse of the building and the goings-on around it as seen in Figure 14.2, but ultimately customer satisfaction is about the client and the building's occupants and users who have paid for the development and must 'live' with it for many years to come. They will either benefit from all the careful thought that resulted in 'their' building or soon begin to discover the lack of it.

Customer satisfaction can perhaps be measured in two ways. It can be seen as fulfilling a contract to the 'letter of the law' between parties in strict compliance with its terms. In absolute business terms, this is a crucial point ensuring that the parties give and receive an understood level of quality. But equally importantly it is about providing a service which genuinely satisfies the customer's needs, which is sometimes beyond the strict definition of contract clauses. Quality control of design is not a simple concept and personal behaviour within any relationship is difficult to guarantee. The fact that a building's user is stimulated by their new environment or regards occupation as a pleasure is not a statement which could be easily assured in advance.

Throughout this book the designer's relationship with other members of the design and construction teams has been explained with a view to achieving the best possible results for the

Figure 14.2 Finished buildings in use as seen by passers-by. Various car-dealerships in Nottingham.

client and the building's users. The ability of the designer to consider and anticipate outcomes is at the heart of customer satisfaction. Other issues can be more accurately measured such as efficiency of occupation with respect to running costs, long-term maintenance and replacement costs, and how well the building stands up to human use and the effects of the weather. For the car-dealership, the success of the client's business operations may be a judgement on the quality of the new building, which has been made better, more efficient and profitable as a result of the design team's efforts. The client's staff may be happier, more contented and enthusiastic because working conditions are so much better than they had previously experienced.

Customer satisfaction is crucial for the designers because their relationship with the new building does not cease once the final account has been settled. Liability for failure continues beyond this point and can yet come back to haunt even those who would claim to be the most professional and conscientious.

14.7 Project File content

The general aims at this stage are as follows:

- Complete construction to the standard as described in the contract documents.
- Complete construction to the satisfaction of the client.
- Resolve defects prior to handover of the new building to the client.
- Reach agreement about settlement of final accounts.
- Compile all as-built records for handover to the client.

File material could include the following:

Snagging lists

- Pre-snagging details setting standards and identifying key issues.
- Prepared snagging lists for the contractors attention.
- Defects maintenance list of items requiring attention before settlement of the final account.

Practical completion

- Documentation confirming that works are complete and settlement of the bulk of outstanding accounts.

As-built records

- Information handed to the client on completion including as-built drawings, corrected specifications and schedules, details of finishes, instructions for operating equipment including maintenance schedules.

Health and Safety File

- The completed Health and Safety File containing all the information described in Chapter 4.

Photographs

- Photographs of the completed building for interest and record purposes.

Opening ceremony arrangements

- Consideration of proposals for a formal opening ceremony including publicity.

Final account

- Completion of latent defects works.
- Agreement between all parties that the building is finished.
- Agreement of final valuation of works undertaken.

14.8 Discussion points

(1) What are the remedies for client disappointment about the quality of finishes inside a new building? What should the building designer's attitude be to poor workmanship on site? Who should be responsible for agreeing the snagging of unfinished or defective works?

(2) How do forms of contract deal with extensions to the contract period? Is it worth considering an extension of time if there is likely to be an improvement in quality? Should a contractor be penalised for being late even if the quality of work is excellent?

(3) Is it realistic to have partial handover of new commercial business? What are the business risks of unfinished accommodation? How are insurance liabilities transferred back to the client?

(4) How should the new building be opened? What kind of celebrations could be planned to mark the finish of the works on site? What constitutes customer satisfaction with respect to the finished product?

14.9 Further reading

Holroyd TM (2003) *Buildability: Successful Construction from Concept to Completion*. London: Telford.

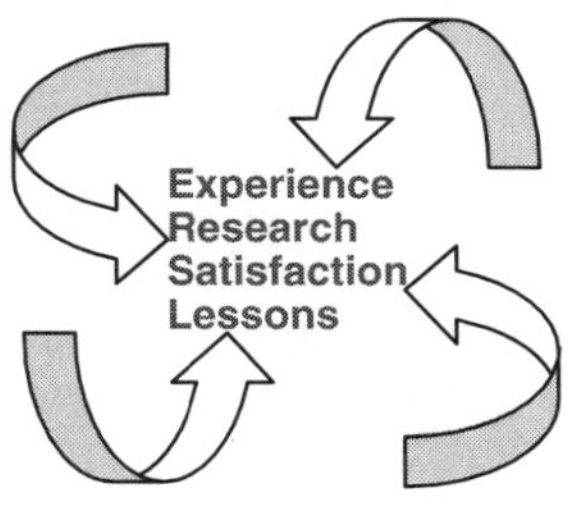

15

Feedback

15.1 Introduction

One of the professional responsibilities of the designer and all the other members of the design and construction teams described in Chapter 2, is to keep up to date with current practice, developments and legislation by at the very least maintaining and wherever possible, improving personal knowledge and experience. The draughting and communication tools of the designer's trade are changing, and will doubtless continue to do so in response to available technology and the demands of the other people involved in creating new buildings. The electronic environment is one which new designers will take for granted enabling them to produce ever more complex and swift answers to design problems. Further discussion about the mechanics of electronic communication is almost meaningless because the pace of change is so rapid that a new and better system will be available before the last one has been mastered. The traditional relationship between designer, client and contractor described in this book will continue for some time for many construction projects, but the commissioning of large and expensive buildings will not always follow the same route. Political initiatives are continually being considered with respect to ways in which the process can be made more economical and less confrontational.

The way that new buildings can be developed with regard to their function, construction and appearance is unlikely to ever be fixed or frozen in time as it once was. The invention of new materials and structural systems, and changes in lifestyles and business practices will continue to revise expectations about what the nature of the built environment should be. Today's society is governed by change in almost every area of life, none more so than in the expansion, replacement and refurbishment of the building stock. Whilst the design team must attempt to create the new building by examination of 'true' purposes and resist the influences of previous and possibly poor experience, they must also remain fully aware of current trends and movements, of commercial and political forces and of the desires and aspirations of all those involved.

Building controls and associated legislation do not remain static either. Changes incorporating additions and refinements can be expected periodically, reflecting concerns about many issues including health and safety, accessibility, working practices and the effects on the community at large. Where formal approvals are required, it is relatively easy for the design team to keep up to date, because there is opportunity to discuss requirements and correct proposals in order to comply with demands. In other areas, which may presently be advisory only or matters of 'good practice', there is considerable onus on members of the design team to collect current literature or attend seminars and presentations to ensure that advice is available about standards which might apply to the design and construction of the specific project.

On the assumption that the designer will be involved with future projects, perhaps the most important way of increasing knowledge and experience is to examine the project just finished to see if lessons can be learned about successes or failures which could be repeated or avoided when the process is repeated for the next project. It can be argued that the best way to learn is through experience, by an awareness of the relationship of theory to practice.

15.2 Lessons for the development team

Feedback is useful when it gives an overview of the project, reflecting on the success of the venture in the widest possible terms. For some designers, professional critical acclaim is a good indicator of success, but only in terms of the standards applied by the judges. It does not always follow that the client and the building's users see the finished building in the same light, and the consultants, authorities and the contractor may all have different views about the finished product. Their experience of the process may be of real value to the way in which subsequent projects are approached. A post-occupancy review conducted by someone uninvolved with the project can be an excellent way of understanding how the design and construction process worked, and if the result represents value for money. Such a report might include a brief history of the project, comment on the briefing process, an assessment of the accommodation provided, a review of the tender and completion costs, and a summary of the views of anyone involved with the project, notably the building users, through the use of 'secret' questionnaires, inviting their honest views about how the new building performs.

For the building designer, feedback also comes through informal visits and casual discussion. Glaring design and construction failures will soon become apparent, perhaps demanding further attention by the designers and contractors, but minor difficulties or inconveniences are often overcome or accepted without further comment. Elements of good or successful design may not be mentioned at all, apparent only from the way in which spaces and materials perform, or by the ease of everyday use by the building's occupants. Information about the good points may only be noticed by paying visits to the new building to observe the way in which it is being used.

As well as considering if all the working relationships between the designer and the other members of the design and construction teams were satisfactory during the development period, the designer may benefit from their experience by asking questions about their performance, either addressed directly to the individuals concerned or by reflecting on their involvement.

The client

The client has been, and will be, involved with the new building for the longest period of time from first thoughts through many years of subsequent occupation. In this respect, immediate feedback may be useful, but a better insight may be offered once the building has been used for some time when qualities or deficiencies are fully understood:

- Did the client perceive the working relationships between client/designer, client/consultants and client/contractor as satisfactory?
- Was their contribution acceptable and appreciated?
- Did the briefing arrangements result in the building as anticipated?
- Were the client's real needs understood by the design team?
- Was the contract completed on time within agreed costs?
- Would any elements of planning or specification be changed if the process was repeated?
- Are there any aspects of the new building that they did not appreciate until occupation?
- Is the standard of workmanship in line with expectation?
- Is the function and appearance of the site works, the building and all its elements satisfactory?

The consultants
The consultants supplied specialist information at various stages throughout the design and construction periods:

- Were their working relationships satisfactory with others involved with the project?
- Was their contribution essential?
- Was their initial advice given at the right time with respect to developing detail proposals later?
- Was their performance satisfactory and on programme?
- Were communication systems adequate?
- Did they co-operate in accommodating design changes?
- Were their proposals in line with budget planning?

The authorities
The authorities approve design and construction work, and offer advice and recommendations about good practice:

- Were informal discussions easy to organise and was information received quickly enough?
- Were applications processed smoothly and rapidly?
- Were inspections and approvals forthcoming as programmed?
- Did their involvement result in additional post-contract costs?
- Did they adhere to the 'letter of the law' or was there any flexibility in their approach?

The contractor
The contractor and subcontractors constructed the building:

- Were their working relationships satisfactory with the client, consultants and authorities?
- Was sufficient information made available at the right times?
- Were drawings and specifications accurate and correct?
- Were any construction details incorrect or inappropriate?
- Could any details have been different making the work easier, safer or more cost efficient?
- Was work conducted diligently in accordance with agreed programmes?
- Is the quality of construction appropriate as required?
- Was the contractor's management structure well organised?
- Were health and safety requirements properly implemented and adhered to?
- Was the project profitable?

- How did the contractor respond to instructions and attend to defects?
- Did the contractor feel that they had been treated fairly with respect to supervision and the application of contract conditions?

The designer

The designer helped everyone else to create the new building:

- Were the design and construction processes managed efficiently?
- Were the consultants and the contractor well chosen and co-operative?
- Could improvements be made in management and communication to increase profitability?
- Could elements of layout planning have been considered in a better way?
- Does the building work as intended?
- Were the specifications for materials and components correct and appropriate?
- Were any construction details difficult to achieve or unrealistic?
- Are there elements of risk which could lead to subsequent failures?
- Have the elements of the new building stood up to use as anticipated?
- Is there merit in submitting the scheme for awards?
- Can the scheme as a whole, or details within it be used for publicity?
- Are there any lessons from the design and construction of this type of building which may be relevant to any other?

Consideration of the answers to all these questions with analysis of the implications will add to the understanding of issues related to design and construction in general terms, and the best way that the individual designer should approach each project in the future. The designer must have the ability to produce proposals which are appropriate and practicable, and which comply with statutory requirements. Of equal importance is the designer's ability to be conscientious and professional so that proposals are realistic and that relationships are established quickly which lead to maximum co-operation throughout the design and construction periods. Perhaps it is the feedback about the quality of relationships which will be the most significant factor in achieving future success.

15.3 Discussion points

(1) How can useful feedback be obtained about the performance of a new building? What criteria should be used for judging the success of building design? Whose judgement is most important with respect to 'lessons' for the future?

(2) Should building designers listen to their critics? Is subjective criticism always likely to mean widespread dissatisfaction with current professional activity? Should designers confine themselves to satisfying public opinion?

(3) Should building designers stick with specifying what they know and understand? How important is innovation in design and construction? Is innovation legitimate for its own sake regardless of necessity?

(4) Are the differences between the opinions of design professional, users and the general public about the qualities of finished buildings understandable? Can the appearance of a new building be judged without reference to its surroundings? What responsibility to future users do building designers have for their completed buildings?

15.4 Further reading

Preiser WFE and **Vischer** JC (2005) *Assessing Building Performance.* Oxford: Butterworth-Heinemann.

van der Voordt DJM and **van Wegen** HBR (2005) *Architecture in Use: An Introduction to the Programming, Design and Evaluation of Buildings.* London: Architectural Press.

16

Conclusion

The design and construction processes that I have described in this book have been related to the creation of new buildings based on my own experience of the construction industry during the past 30 years. I have also tried to reflect my experience in higher education, and the current changes in technology and ideology with respect to development of the built environment today. The selected case study, the car-dealership, may of course be a building type which any designer only encounters once. Some of the analysis and discussion have been specifically focused on this particular type of building, but in general terms the attitude of the building designer should be broadly the same for any new building. Creating a house, a shop or a church involves the same degree of research and analysis to understand the elements of the building before synthesising and proposing a solution, even though the client, the users, the location and the moment in time are different and one designer's ability and experience will not be the same as another's. The designer's responsibilities will vary with the method of procurement adopted by the client, but I believe that the analogies used in the context of the car-dealership can be related to other building types as well.

Although each project is different, the design and construction processes are essentially repetitive. Each project can be viewed as an opportunity to be tackled in eager anticipation of a new challenge, or to be 'processed' by 'going through the motions'. Any designer can get used to doing things in their own way, forgetting that every project needs careful, individual attention if the best possible results are to be achieved, and it is all too easy to make assumptions which do not reflect the true needs of clients and users, or take into account the changes taking place in business and management procedures. This can tend to lead to complacency and reliance on 'authority', in the sense that the 'professional' must know best, and therefore should not be challenged. As discussed in the earlier chapter about design, there is strong evidence to suggest that if designers are told, or expected to work in a particular way, they probably will do so. They may become unwilling, or unable to challenge the *status quo*.

As a simple analogy, when I was first invited to teach at Nottingham Trent University, I put a great deal of effort into the preparation of my initial lectures and seminars. Addressing each topic for the first time was exciting and slightly dangerous, being unsure how it would be received by the students. The material was new to me, and I was anxious that the content of my presentations should be as comprehensive, and as interesting as I could possibly make them. In the second year, faced with repeating a lecture, I knew what I was going to say, but I found it more difficult to generate that spark which comes from something fresh, and on a number of occasions neglected to explain some things as well as I might have done, because I took them for granted. By the time it came round to the fourth year, I realised that I could not just regurgitate the same material again and again, but must try to rethink my purpose and intentions, changing things for the benefit of my students, and myself. The essential content remained but the presentations became more focused towards what I began to appreciate the students needed to know, rather than what I had thought that they ought to know. I began to better understand the importance of their role in the process, to help ideas and discussions to emerge around basic themes, and was often helped by the students themselves, who sometimes asked me questions for which I was unable to give an immediate answer.

I attended a seminar of senior academics and practising designers at which we discussed among other things, the nature of the content of courses in building and design. We debated at length the issue of problem solving on the basis that students should be able to give answers, and we carefully examined teaching rationale and the anticipated learning outcomes, well documented by both academic and professional institutions. I suggested that an important one was missing, namely that students should be able to recognise when they do not know 'the answer' and cannot do what they are being asked, but are able to go and find out.

Of even greater importance, they should be able to analyse problems and propose sensible, rational solutions which they can justify. The building designer will find him- or herself in this position on many occasions as new buildings are created. The increasing interest, concern and direct involvement of clients, other professionals, project managers and contractors in the procurement process are leading to a continual re-examination of procedures and responsibilities. Consequently, the process must be a team effort, sharing experience and knowledge in an attempt to achieve the best possible results. In this respect, the part played by any one person depends on their understanding of the process with which they are involved and how they may be able to contribute to it.

Professional organisations associated with the built environment

ABE: **Association of Building Engineers** (www.abe.org.uk)

Promotes the advancement of knowledge in the technology of building, building engineering and the built environment, and aims to encourage and facilitate co-operation between the construction professions.

ACE: **Association of Consulting Engineers** (www.acenet.co.uk)

Represents the business interests of the consultancy and engineering industry in the UK, lobbying government, major clients, the media and other key stakeholders to promote the contribution engineers and consultants make to the nation's developing infrastructure.

ACPO: **Association of Chief Police Officers** (www.acpo.police.uk)

The national policy-making organisation for policing, but notable for their 'Secured by Design' initiative to encourage the construction industry to incorporate anti-crime measures into development of all kinds.

APM: **Association for Project Management** (www.apm.org.uk)

Aims to develop and promote project management across all sectors of industry and beyond.

ARB: **Architects Registration Board** (www.arb.org.uk)

Protects the consumer and safeguards the reputation of architects by being concerned with establishing and maintaining standards of education and professional practice.

BDA: **Brick Development Association** (www.brick.org.uk)

Promotes the use of clay bricks and pavers in all forms of construction.

Beddington Zero Energy Development (www.bedzed.org.uk)

An energy efficiency model for a living/working development in Beddington, London.

BGS: **British Geological Survey** (www.bgs.ac.uk)

The world's longest established national geological survey and the UK's premier centre for earth science information and expertise, provides expert services and impartial advice in all areas of geoscience.

BIFM: British Institute of Facilities Management (www.bifm.org.uk)

Facilities management is the integration of multi-disciplinary activities within the built environment and the management of their impact on people and the workplace. It is concerned with both strategic and operational management of resources, and activities to create safe and efficient working environments.

BRE: The Building Research Establishment (www.bre.co.uk)

BRE provides a consultancy, testing and commissioned research service for most aspects of construction and the built environment, and associated industries. It can provide advice and certification for materials and systems to UK, European and international standards, as well as consulting engineers (CE) marking and product approval.

BSRIA: Building Services Research and Information Association (www.bsria.co.uk)

Provides independent and authoritative research, product testing, consultancy, training, management and market intelligence, concerned with technical information about building services and construction, and the testing of new products and innovative systems.

BURA: British Urban Regeneration Association (www.bura.org.uk)

Formed in 1990 to provide a forum for the exchange of ideas, experience and information for the emerging regeneration sector. BURA promotes best practice in all aspects of regeneration, and helps to understand and define the mechanisms for success.

CABE: Commission for Architecture and the Built Environment (www.cabe.org.uk)

CABE's principle aim is to lift the profile of design quality in the built environment by encouraging designers, investors and users to raise their expectations about the buildings and communities with which they are involved. They have produced a number of excellent generic design guide publications, and also offer direct help on individual projects.

CAT: Centre for Alternative Technology (www.cat.org.uk)

An organisation concerned with developing practical solutions to sustainability and the use of ecologically sound technologies.

CEBE: Centre for Education in the Built Environment (ctiweb.cf.ac.uk)

Part of the Subject Network of the Higher Education Academy, the Centre provides support to enhance the quality of learning and teaching in the UK for the subject communities of architecture, landscape, urban planning, housing, transport, construction and real estate.

CIAT: Chartered Institute of Architectural Technologists (www.ciat.org.uk)

The professional body for architectural technologists, specialists in the science of architecture, building design and construction that form the link between concept and construction of building projects from conception through to completion.

CIBSE: Chartered Institution of Building Services Engineers (www.cibse.org.uk)

The learned body concerned with all aspects of design, installation, maintenance and manufacturing associated with building engineering services including lighting, heating, ventilation, air conditioning and lifts which, in commercial buildings, may account for as much as 40 per cent of the total construction cost and up to 50 per cent of UK energy use.

CIC: The Construction Industry Council (www.cic.org.uk)

CIC is the representative forum for the professional bodies, research organisations and specialist business associations in the construction industry. It endeavours to promote quality, continuous improvement and sustainability in the built environment.

CIEH: Chartered Institute of Environmental Health (www.cieh.org.uk)

Aims to maintain, enhance and promote improvements in public and environmental health.

CIH: Chartered Institute of Housing (www.cih.org)

Aims to maximise the contribution that housing professionals make to the well-being of communities.

CIOB: Chartered Institute of Building (www.ciob.org.uk)

The leading professional body for managers in construction aiming to establish, promote and maintain standards of excellence in the construction industry for the benefit of customers, clients and contractors.

CIRIA: Construction Industry Research and Information Association (www.ciria.org.uk)

CIRIA works with the construction industry, government and academia to improve the performance of those in the construction and related industries. It runs collaborative projects, workshops, seminars and conferences, and publishes best practice guides.

CITB: The Construction Industry Training Board (www.citb.org.uk)

The CITB is concerned with the image of the present day construction industry, and how best to promote professionalism. It is looking at ways of attracting new people into the industry to enable it to perform effectively and competitively, and deliver sustainable development. It seeks to ensure that recruits acquire essential skills, are suitably qualified and able to work safely.

Civic Trust (www.civictrust.org.uk)

The Civic Trust promotes high standards of planning and architecture in cities, towns and villages, helping to develop dynamic partnerships between communities, government and business to deliver regeneration and local improvement.

Concrete Society (www.concrete.org.uk)

Promotes excellence in design, construction, appearance and performance of concrete and encourages innovation.

Countryside Agency (www.countryside.gov.uk)

The Countryside Agency is the statutory body working to make the quality of life better for people in the countryside and the quality of the countryside better for everyone.

CPRE: Campaign to Protect Rural England (www.cpre.org.uk)

The CPRE, a registered charity founded in 1926, seeks to influence policy and awareness at local and national levels about the quality, character and diversity of the English countryside. Its focus is on issues such as land use planning, transport, natural resources, farming, the rural economy and quality of life.

DEFRA: Department for Environment, Food and Rural Affairs (www.defra.gov.uk)

The government department responsible for economic, social and environmental concerns in the context of promoting sustainable development.

Design Council (www.design-council.org.uk)

The Design Council promotes the role of design in the UK economy.

DTI: Department of Trade and Industry (www.dti.gov.uk)

This government department supports UK business interests and helps people and companies to become more productive by promoting enterprise, innovation and creativity.

English Heritage (www.english-heritage.org.uk)

English Heritage exists to protect and promote England's spectacular historical environment and ensure that its past is researched and understood.

English Partnerships (www.englishpartnerships.co.uk)

English Partnerships is a national regeneration agency whose purpose is to promote new, and sustainable, growth in urban areas suffering from the decline in traditional manufacturing industries. The Government's *Sustainable Communities Plan* outlined their strategic role in achieving a target of 60% of new housing to be built on Brownfield sites.

Environment Agency (www.environment-agency.org.uk)

The Environment Agency is a recently established authority based on the previous Inland Waterways Authority with powers to control development close to existing water courses, and where construction activity and subsequent use may affect water table levels and introduce risks of pollution and contamination.

The Environment Agency provides environmental protection and improvement working with businesses and other organisations to prevent damage to the environment by providing education, guidance, and where necessary, enforcing regulations through prosecutions.

FMB: Federation of Master Builders (www.fmb.org.uk)

Currently the largest trade association in the UK building industry, the FMB protects the interests of small- and medium-sized building firms.

Forum for the Future (www.forumforthefuture.org.uk)

Recognised as the UK's leading sustainable development charity, promoting sustainable development through education. Foundation in 1996 by environmentalists Jonathon Porritt, Sara Parkin and Paul Ekins, they work with companies, local authorities, regional bodies and universities to overcome barriers to more sustainable practice.

Groundwork (www.groundwork.org.uk)

An environmental charity, a partnership with local people, Local Authorities and business to promote economic and social regeneration.

HBF: The House Builders Federation (www.hbf.co.uk)

The principal trade federation for private sector housebuilders and voice of the house building industry in England and Wales.

HM Land Registry (www.landreg.gov.uk)

Concerned with land registration providing confidence when dealing with property.

HSE: Health and Safety Executive (www.hse.gov.uk)

The HSE is responsible for the regulation of almost all the risks to health and safety arising from work activity in Britain.

IBC: Institute of Building Control (www.rics.org/AboutRICS/RICSforums/RICSBuildingControlForum)

The IBC, now known as the building control forum, represents the building control discipline within the RICS, but with a remit to provide a mechanism for all those interested, involved or affected by building control issues.

ICE: Institution of Civil Engineers (www.ice.org.uk)

Seeks to advance the knowledge, practice and business of civil engineering, to promote the breadth and value of the civil engineer's global contribution to sustainable, economic growth, and ethical standards, and to include in membership all those involved in the profession.

ICW: Institute of Clerks of Works (www.icwgb.org)

Provides a central organisation to protect and promote the interests, professional, educational and social welfare of its membership.

IEE: The Institution of Electrical Engineers (www.iee.org.uk)

Promotes the advancement of electrical, electronic and manufacturing science and engineering. It represents the profession, sets standards, issues regulations and offers guidance on best practice.

IHBC: Institution of Historical Building Conservation (www.ihbc.org.uk)

Exists to establish the highest standards of conservation practice to support the effective protection and enhancement of the historical environment.

IStructE: The Institution of Structural Engineers (www.istructe.org.uk)

The world's leading professional body for structural engineering concerned with buildings and other civil engineering structures in the built environment.

Joseph Rowntree Foundation (www.jrf.org.uk)

The Joseph Rowntree Foundation sponsors charitable research into the causes of social problems, and as a social landlord seeks to create new housing based on innovative design ideas and forms of tenure. The organisation developed from the success of the Rowntree cocoa factory in York (1891) and the establishment of the model village at New Earswick in 1904, a balanced community open to any working people.

LI: The Landscape Institute (www.l-i.org.uk)

Promotes the highest standards in the practice of landscape architecture, concerned with design and master planning, site surveys and ecological assessments, plant and environmental management, and other specialist scientific skills.

NALC: National Association of Local Councils (www.nalc.gov.uk)

Represents some 10,000 Parish and Town Councils in England and Community Councils in Wales, working to develop community involvement in local government through effective, democratic leadership.

National Trust (www.nationaltrust.org.uk)

Founded in 1895 by three Victorian philanthropists, Miss Octavia Hill, Sir Robert Hunter and Canon Hardwicke Rawnsley, its interest is to act as a guardian for the nation in the acquisition and protection of threatened coastline (approximately 600 miles), countryside, including forests, woods, fens, farmland, downs, moorland, islands, archaeological remains, nature reserves, whole villages and buildings (presently some 200 buildings and gardens of outstanding interest and importance), all generally held in perpetuity to secure their future protection.

NHBC: National House-Building Council (www.nhbc.co.uk)

The NHBC is the standard setting body and leading warranty and insurance provider for new homes in the UK. Its main customers are homebuyers and registered housebuilders. The NHBC registers builders, regulates standards, promotes sustainability, inspects construction work and offers Building Control services.

ODPM: Office of the Deputy Prime Minister (www.odpm.gov.uk)

The government department responsible for local and regional government, housing, planning, fire safety, regeneration, social exclusion and neighbourhood renewal.

OS: Ordnance Survey (www.ordnancesurvey.co.uk)

Ordnance Survey is the internationally recognised market leader in its field. The OS produce and offer traditional walking maps, large scale locality and site maps and a variety of digital mapping products.

Peabody Trust (www.peabody.org.uk)

A housing association, charity, developer and community regeneration agency aiming to work with local communities and government to build lasting, sustainable communities.

RDAs: Regional Development Agencies (www.englandsrdas.com)

The eight RDAs (plus the London Development Agency) were established under the Regional Development Agencies Act 1998 to co-ordinate regional economic development and regeneration, to help to promote business efficiency, investment, competitiveness, skills and employment possibilities and to contribute to sustainable development in their areas.

RIBA: Royal Institute of British Architects (www.riba.org.uk)

The RIBA exists to advance architecture and promote excellence in the profession. It is concerned with creating better communities, better buildings and a better environment, and sponsors prestigious awards, lectures, exhibitions and events.

RICS: Royal Institute of Chartered Surveyors (www.rics.org.uk)S

Representing property professionals, the RICS is a leading source of information and independent advice on land, property, construction and associated environmental issues, concerned with economics, valuation, finance, investment and management of all the world's physical assets.

RTPI: Royal Town Planning Institute (www.rtpi.org.uk)

Exists to advance the science and art of planning, the conservation and development of the built environment for the benefit of the public, and is concerned with the characteristics of places, sustainability of communities and integration of activity.

SCI: The Steel Construction Institute (www.steel-sci.org.uk)

Develops and promotes the use of steel in construction.

SPAB: Society for the Protection of Ancient Buildings (www.spab.org.uk)

An expert national pressure group fighting to save old buildings from decay, demolition and damage.

TCPA: Town and Country Planning Association (www.tcpa.org.uk)

Campaigns for high-quality, intelligently built homes in well-designed communities empowered to influence the decisions that affect them, and a sustainable future.

Twentieth Century Society (www.c20society.org.uk)

The Twentieth Century Society exists to safeguard the heritage of architecture and design in Britain from 1914 onwards.

Urban Splash (www.urbansplash.co.uk)

Set up in 1993 by Tom Bloxham and Jonathan Falkingham, Urban Splash has developed many under-used historical buildings and Brownfield sites principally in the north of England, creating new mixed-use spaces, stimulating broader regeneration of urban communities.

Urban Task Force (www.odpm.gov.uk)

Acting under the remit of the Office of the Deputy Prime Minister (ODPM), the group comprising politicians, academics and practitioners under the chairmanship of Richard Rogers (Lord Rogers of Bankside), attempted to identify causes of urban decline in England and to recommend ways of supporting government policy to encourage people back into towns and cities. The Task Force presented some 100 recommendations for urban regeneration focused on ideas about quality of design, transport, management, regeneration, skills, planning, investment and social well-being, which informed the Urban White Paper (2000) and the Sustainable Communities Plan (2003).

The Victorian Society (www.victorian-society.org.uk)

Concerned with the study and protection of Victorian and Edwardian architecture and other arts.

Glossary

Accommodation The usable spaces within a building.
Activity and flow The way that people use spaces inside and outside buildings.
Aesthetics The visual appearance of a building or its constituent parts.
Affordability Creation of new buildings, notably starter housing or small work units, which could be purchased or rented by those on relatively low incomes.
Agent Someone acting on behalf of a client.
Allocation Identification of potential land-use options by a local authority on their local plan.
Amenity Essential support facilities close to a development site.
Anthropometrics The study of the human form and performance.
Appointment The formal arrangements regarding employment between client and advisor or contractor.
Approval/s Formal acknowledgement by a statutory authority of the acceptability of design and/or construction proposals.
Approved documents The separate sections of the Current Building Regulations.
Best practice Advice or guidance disseminated to others working in the same area.
Blight Declining value in an area associated with uncertain planning intensions.
Brief Statement of intent or requirement for development produced by clients, developers or authorities.
British Standards Published documents defining the qualities and performance of materials.
Brownfield land A site which has been developed previously in its history.
Bubble diagram A brainstorming technique as a way of identifying elements.
Building designer An architect, architectural technologist or any other specialist involved in the creation of new buildings.
Building Regulations Statutory controls concerning the construction of buildings.
Built environment Developed land containing buildings and associated infrastructure.
Catchment area A notional measure of likely users of a facility.
Circular Government dissemination of advice on interpretation of national policy.
Client Person or organisation commissioning and funding a new building.
Code of Conduct Professional behaviour established and policed by an institute.
Codes of Practice Published documents defining recognised methods of using materials.
Commuted payment Financial contribution to off-site provision for community benefit associated with securing Planning Permission.
Conservation area Protection for a locality with characteristics defined in the local plan as being of architectural or historic merit.

Constraints Factors which limit or determine a particular course of action.
Consultations Contact with and advice from those with an interest in the development of a site.
Contaminated land A site containing buried harmful chemicals.
Curtilage The extent of a site around any building up to its boundary.
Defensible space The understanding or clarity of ownership at the junction between public and private property.
Density The intensity of development of the built environment in terms of population, number of buildings, floor area or height.
Design and build An arrangement where responsibility for the design and construction of a building rests with the same company.
Design guides Publications disseminating good practice, or recommended approach.
Design proposals A suggested design solution for a new (or part) building.
Desire line The shortest route between A and B created through use, ignoring a longer formally constructed route.
Development The design and construction of new buildings on a site, or alterations and extensions to existing buildings.
Duty of care A professional approach to ensuring best possible services are offered.
Ecology The interaction of the inhabitants of the total environment.
Elevations The external walls of a building or a line of buildings.
Embodied energy The total energy consumed in the extraction, manufacture, transport and assembly of building materials and products: a life-cycle energy audit.
Environment A locality or wider area; the buildings and spaces around them at any scale.
Environmental assessment A review of the likely impact of development proposals.
Environmentally friendly A development, process or use of materials which minimises environmental damage.
Feedback Information received about the efficacy of design proposals.
Fenestration The windows and doors on a buildings elevation.
Flexibility Design options to permit or encourage alternative use.
Form The shape and appearance of buildings and their constituent elements.
Full Planning Permission Local authority approval of detailed, fully illustrated development proposals.
Function The way in which buildings or their constituent elements work in practice.
General contractor A builder who co-ordinates all construction works on site.
Good practice guides Published documents disseminating previous experience of successful design solutions.
Green belt A notional boundary on the local plan between the built environment and the open countryside preventing uncontrolled expansion.
Greenfield site A site which has not been developed previously in its history.
Greenhouse effect The increase in carbon dioxide in the atmosphere which is believed to be causing global warming.
Health and Safety Measures taken to prevent injuries during and after construction.
Hectare The metric measure of land area 10,000 m^2 (2.471 acres).
Holistic design An approach to design incorporating every possible consideration.
Inception The beginning of the design process.
Infrastructure Normally refers to roads, sewers and services directly associated with a development site, but may be extended to social, welfare and amenities of a locality.
In-house labour A company's own directly employed staff.

Kerb appeal The immediate visual impact of a building, particularly first impressions of residential development.

Lateral thinking A way of looking at problems/solutions in a different way than might be expected.

Life-cycle costing Consideration of costs in use over and above costs for design and build.

Listed building Protection for specific buildings with characteristics defined by the Local Planning Authority as being of architectural or historic merit.

Local Plan Definition of land-use drawn up by a local authority after extensive consultations.

Local Planning Authority The function of councils controlling development in their own area.

Measurement Qualifying and quantifying all the elements of any building.

Mixed-use development A planned or designed mixture of uses within a complex or single building development.

Needs The requirements of the occupants of a building.

Neighbourhood A locality around a development site.

Occupation Temporary ownership of a site by the contractor.

Orientation The position of significant features of buildings in relation to the site boundary, or to direct sunlight or prevailing winds.

Outline Planning Permission Local authority approval of principles of development only without extensive detail.

Overshadowing The restriction or denial of direct sunlight by development on adjacent sites.

Party A person or organisation who is involved with a project or contract.

Passive solar gain Natural penetration of warmth into buildings.

Pedestrianisation Removal of vehicles from an urban area allowing free access to people on foot.

Performance A measure of, or requirement for how materials, assemblies or whole buildings should behave in use.

Photovoltaics Equipment for converting solar power to electricity.

Planning Brief Local authority guidance on development potential or requirements for a site or area.

Planning conditions Additional requirements in association with Planning Permission.

Planning Gain An obligation on a developer in association with Planning Permission to provide additional off-site benefit to the community.

Planning Policy Guidance (PPG) Government dissemination of advice on interpretation of national policy.

Plant Equipment and machinery used on site in the construction process.

Pre- and post-contract Stages of work before and after an exchange of contracts to start construction on site.

Presentation drawings Non-technical illustrations of design proposals.

Procurement The way in which a client gets their new building.

Programme A plan of the timescale for design and construction works.

Progress The extent of completed design or construction works measured against the original programme.

Quality assurance Independent support for selection of services and materials.

Recycling Constructive reuse of redundant or waste materials.

Regulations Statutory controls limiting or controlling design and construction.

Remediation Treatment of contaminated land prior to redevelopment.

Risk assessment An appreciation of possible problems in adopting certain design decisions.

Section agreements Legal agreements between developers and local authorities regarding adoption of public amenities such as roads and sewers.

Secured by design An initiative promoted by the police authority to minimise crime by the way that buildings and sites are developed.

Services Input into the site and building to create habitable environmental conditions: power, lighting, water and drainage, for example.

Site A piece of land or a building which can be developed.

Site analysis An exploration of the implications of the site for any subsequent design proposals.

Sketch A rough visual presentation of a design idea.

Socio-economics The general condition of an area in terms of how it is being used.

Specifications Written descriptions of materials and elements of construction.

Speculative development New buildings constructed for sale or rent.

Statutory authorities External bodies who have a duty to control development.

Substructure The elements of buildings below ground floor level.

Superstructure The elements of buildings above ground floor level.

Sustainable development Defined by the World Commission on Environment and Development (1987) as 'development that meets the needs of the present without compromising the ability of future generations to meet their own needs'.

Technical drawings Line drawings to an accurate scale, sometimes referred to as 'mechanical drawings'.

Tolerance The relationship of one element of construction to another, particularly if fixed together.

Topography The shape of the landscape or any particular site.

Traffic calming Highway design features which slow traffic speeds in order to promote safety.

U-value A measure of heat movement through materials and elements of construction.

Vernacular Use of materials and construction techniques which are typical of a locality or region.

Walkover First impressions of a site or building to assess limitations.

Working drawings Technical drawings showing how elements of construction fit together.

Professional journals and publications (Sources of reference)

Architects Journal (www.ajplus.co.uk)
Architectural Review (www.arplus.com)
Architectural Technology (www.biat.org.uk/index.jsp)
Architecture Today (www.architecturetoday.co.uk)
Architecture & Urbanism (www.bdpworld.com)
Barbour Index (www.barbour-index.co.uk)
BRE Digests (www.bre.co.uk)
BS Handbook 3 Summaries of BS for Building (www.bsonline.bsi-global.com)
BSRIA Technical Notes (www.bsria.co.uk)
Building (www.building.co.uk)
Building Design (www.bdonline.co.uk)
Building Engineer (www.abe.org.uk/journal.jsp)
Building and Environment (www.elsevier.com)
Concrete (www.concrete.org.uk)
Construction Manager (www.construction-manager.co.uk)
Journal of Architectural Conservation (www.donhead.com)
Journal of Architecture (www.tandf.co.uk/journals)
Laxton's Building Price Book (http://books.elsevier.com)
Media UK (www.mediauk.com/magazines)
RIBA Journal (www.ribajournal.com)
RIBA Product Selector (www.productselector.co.uk)
Spon's Architects' and Builders' Price Book (www.pricebooks.co.uk)
Vernacular Architecture (www.ccurrie.me.uk/vag)
What's New in Building (www.wnibonline.com)
See also www.architecture.com for journals indexd by the British Architectural Library.

Index

PEARLS
of WISDOM

Ph M Re

REV

dition

McGraw-Hill
Medical Publishing Division

New York Chicago San Francisco Lisbon London
Madrid Mexico City Milan New Delhi
San Juan Seoul Singapore
Sydney Toronto

The McGraw·Hill Companies

Physical Medicine and Rehabilitation Review, Second Edition

2 3 4 5 6 7 8 9 0 CUS/CUS 0 9 8 7 6

ISBN 0-07-146446-8

Notice

Medicine is an ever-changing science. As new research and clinical experience broaden our knowledge, changes in treatment and drug therapy are required. The authors and the publisher of this work have checked with sources believed to be reliable in their efforts to provide information that is complete and generally in accord with the standards accepted at the time of publication. However, in view of the possibility of human error or changes in medical sciences, neither the authors nor the publisher nor any other party who has been involved in the preparation or publication of this work warrants that the information contained herein is in every respect accurate or complete, and they disclaim all responsibility for any errors or omissions or for the results obtained from use of the information contained in this work. Readers are encouraged to confirm the information contained herein with other sources. For example and in particular, readers are advised to check the product information sheet included in the package of each drug they plan to administer to be certain that the information contained in this work is accurate and that changes have not been made in the recommended dose or in the contraindications for administration. This recommendation is of particular importance in connection with new or infrequently used drugs.

The editors were Catherine A. Johnson and Marsha Loeb.
The production supervisor was Phil Galea.
The cover designer was Handel Low.
Von Hoffmann Graphics was printer and binder.

This book is printed on acid-free paper.

Library of Congress Cataloging-in-Publication Data

Physical medicine & rehabilitation review / Robert Kaplan [editor-in-chief].—2nd ed.
p. ; cm.—(Pearls of wisdom)
Rev. ed. of: Physical medicine & rehabilitation / Robert J. Kaplan. 1st ed. c2003.
ISBN 0-07-146446-8
1. Medicine, Physical—Examinations, questions, etc. 2. Physical therapy—Examinations, questions, etc. I. Title: Physical medicine and rehabilitation review. II. Kaplan, Robert J. III. Kaplan, Robert J. Physical medicine & rehabilitation. IV. Series.
[DNLM: 1. Physical Medicine—Examination Questions. 2. Rehabilitation—Examination Questions. WB 18.2 P57783 2006]
RM701.6.K36 2006
615.8'2—dc22

2005053351

INTERNATIONAL EDITION ISBN: 0-07-110886-6